Contributions to Management Science

The series *Contributions to Management Science* contains research publications in all fields of business and management science. These publications are primarily monographs and multiple author works containing new research results, and also feature selected conference-based publications are also considered. The focus of the series lies in presenting the development of latest theoretical and empirical research across different viewpoints.

This book series is indexed in Scopus.

More information about this series at http://www.springer.com/series/1505

Paola Demartini • Lucia Marchegiani •
Michela Marchiori • Giovanni Schiuma
Editors

Cultural Initiatives for Sustainable Development

Management, Participation and Entrepreneurship in the Cultural and Creative Sector

Springer

Editors
Paola Demartini
Department of Business Studies
Roma Tre University
Roma, Italy

Lucia Marchegiani
Department of Business Studies
Roma Tre University
Roma, Italy

Michela Marchiori
Department of Business Studies
Roma Tre University
Roma, Italy

Giovanni Schiuma
Department of Mathematics, Computer
Sciences and Economics
University of Basilicata
Potenza, Italy

ISSN 1431-1941 ISSN 2197-716X (electronic)
Contributions to Management Science
ISBN 978-3-030-65686-7 ISBN 978-3-030-65687-4 (eBook)
https://doi.org/10.1007/978-3-030-65687-4

This Springer imprint is published by the registered company Springer Nature Switzerland AG.
The registered company address is: Gewerbestrasse 11, 6330 Cham, Switzerland

Contents

Connecting the Dots: A Proposal to Frame the Debate Around Cultural Initiatives and Sustainable Development

Paola Demartini, Lucia Marchegiani, Michela Marchiori, and Giovanni Schiuma

Abstract Sustainability and sustainable development have received increasing attention in the academic world, across many disciplines and fields. Several conceptual models have been proposed to apply the sustainability perspective to culture and cultural heritage. These models break down the broad concept of sustainability into distinct dimensions, with relation to specific domains where a given impact might be detected. Instead, it is necessary to adopt an integrated perspective and an interdisciplinary approach. Hence, first, we propose a multifaceted view to crystallise the current debate of what sustainable development entails in the cultural heritage forum. Then, we use it to map out the contributions included in this book. In particular, we suggest the following dimensions: (1) the levels of analysis (micro, meso, macro); (2) the four domains in which the sustainability concept unfolds (cultural, social, economic, and environmental) according to the Hangzhou Declaration (2013); and (3) the regenerative processes of cultural initiatives that are consistent with the sustainability perspective (management, participation, entrepreneurship). After a brief overview of each contribution included in the book, we conclude by pinpointing a list of themes that should be further explored to enrich the discussion about cultural initiatives and sustainability.

Keywords Cultural initiatives · Cultural heritage management · Cultural management · Sustainability · Sustainable development - SDGs · Cultural participatory initiatives · Cultural entrepreneurship

P. Demartini (✉) · L. Marchegiani · M. Marchiori
Department of Business Studies, Roma Tre University, Roma, Italy
e-mail: paola.demartini@uniroma3.it; lucia.marchegiani@uniroma3.it; michela.marchiori@uniroma3.it

G. Schiuma
Department of Mathematics, Computer Sciences and Economics, University of Basilicata, Potenza, Italy
e-mail: giovanni.schiuma@unibas.it

P. Demartini et al. (eds.), *Cultural Initiatives for Sustainable Development*, Contributions to Management Science,
https://doi.org/10.1007/978-3-030-65687-4_1

1 The Debate on the Role of Culture in Promoting Sustainable Development

Sustainability and sustainable development have received increasing attention in the academic world, across many disciplines and fields. The concept of sustainability has been discussed for decades, but a unique definition has not been provided. Yet, sustainability, in its broader meaning, refers to the preservation of the environmental status quo for future generations. In fact, the consideration that resources are finite and cannot be wasted or irresponsibly exploited led to the affirmation of sustainability as a long-term view of the consequences of how resources are used. Under the management lens, sustainability entails a broader view of how organisations develop and grow, moving into the future under the perspectives of profit, planet, and people (Elkington 1997). Social and environmental concerns can be seen from a wider standpoint, as social and global goals and not just developed from a business perspective. Hence, the term sustainable has been increasingly associated with development, defined by the UN Brundtland Commission (Brundtland 1985) as: "development that meets the needs of the present without compromising the ability of future generations to meet their own needs" (WCED 1987, p. 40).

In this vein, cultural resources are also subject to a perspective based on sustainability. Undoubtedly, cultural heritage belongs to humanity, and it should be preserved for future generations. Just like natural resources and the environment are common goods, cultural artefacts, such as urban landscapes, monuments and museums, are also "goods" that benefit specific communities. Moreover, they can be key elements for local development, helping to improve the community's quality of life. Yet, the relationship between sustainability, sustainable development, and culture is highly debated and discussed. In fact, a universal understanding of culture in promoting sustainable development has not been reached yet. On the one hand, culture is sensed as being bound to the past, holding conservative power. On the other hand, it is believed to have the strength to promote truly sustainable development, which involves not only progress but also the promotion of human and social capital. As witnessed by the European Parliament Resolution of September 8, 2015, "Towards an integrated approach to cultural heritage for Europe", sustainable cultural heritage management can lead to the maximisation of the intrinsic, economic and societal value of cultural heritage.

The so-called cultural turn in international aid and development (Labadi 2019) gained momentum over the disillusionment of models based only on economic growth, in favour of a more central role of culture, people and human development.

At a European level, cultural heritage is widely recognised as an important driver for sustainable development (Helly and Galeazzi 2016). In fact, the EU has progressively incorporated culture into its development policies, considering culture as a basic element that drives the success of other development goals (human development, social cohesion, green environment and gender-balanced opportunities). Indeed, promoting culture has been, since 2007, one of the three main objectives of the European Agenda for Culture in which the cultural sector is regarded as an increasing source of job creation, contributing to growth throughout Europe.

Numerous lobbying attempts have advocated for a central part of culture in the UN debate on sustainable development, such as the #culture2015goal, backed by ICOMOS and Culture Action Europe, among others.

Nevertheless, neither the Millennium Development Goals (MDG 2000–2015) nor the more recent Sustainable Development Goals (SDGs 2015–2030) defined by the United Nations directly pay tribute to a central role of culture and cultural heritage as pillars of sustainable development.

This matter has led to an interesting debate and juxtaposed opinions, briefly summarised below.

Some claim that the framework of the 17 SDGs only marginally and indirectly recognised culture as a driver of sustainable development (Labadi 2019; Torggler et al. 2015). The closest reference to culture is made in Goal 11, and specifically with Target 11.4, which aims to "strengthen the effort to protect and safeguard the world's cultural and natural heritage". And yet, this formulation appears weak as it does not link culture to any of the pillars of sustainable development, such as poverty reduction or gender equality, for example. Moreover, it betrays the outdated idea that protecting and safeguarding cultural heritage automatically leads to development.

An additional reference to culture in the SDGs is found in Goal 4 on inclusive and equitable quality education, with Target 4.7, as it aims for the appreciation of cultural diversity and culture's contribution to sustainable development. However, it does not recognise the importance of local cultural contexts and the need to include culture and the arts in education policies, as it had been indeed advocated by UNESCO (Labadi 2019).

Moreover, the SDGs framework includes targets that indicate sustainable tourism to create jobs for the local population. Given the relevance of cultural tourism, these targets are certainly indirectly impacting culture and cultural heritage. Nevertheless, sole focus on tourism does not recognise the ambivalence of its negative effects on local culture.

Some others have rather an opposite opinion (Wiktor-Mach 2020), stating that the SDGs "represent a significant step forward with regard to the acknowledgment of the role of culture in development processes"[1] (Culture 2030 Goal Campaign 2019, p. 26). Audrey Azoulay, Director-General of UNESCO, optimistically opens up the Culture for the 2030 Agenda publication, stating that "[. . .] with the adoption of the 2030 Agenda for Sustainable Development in 2015, the international community also recognised the role of culture as a driver of sustainable development" (UNESCO 2018).

Nevertheless, the same sources recognise that, "References to culture in the SDGs are scarce and do not sufficiently acknowledge the many ways in which cultural

[1]Culture2030Goal Campaign (2019), "Culture in the Implementation of the 2030 Agenda". Published in Barcelona, Paris, Harare, Sydney, Montreal, The Hague and Brussels, in the frame of the first UN SDG Summit that took place on September 24 and 25, 2019.

aspects influence and contribute to sustainable development" (Culture 2030 Goal Campaign 2019, p. 5).

All considered, it is interesting to inquire and further debate the role of cultural initiatives in the face of the most modern perspectives on sustainable development. From this perspective, this book aims to unfold the entrepreneurial processes and socio-economic dynamics related to managerial practices of cultural organisations and cultural interventions. This objective also implies tackling new sources, dimensions and characteristics of knowledge that support the cultural sectors.

The translation of managerial paradigms and technologies into the cultural and creative sector is subject to critical reflection. Each chapter contributes to highlighting the critical perspectives in the related area of interest. This book constitutes the second collection of studies that has been submitted in response to a call launched by an interdisciplinary and international research team based at Roma Tre University. The research team holds the vision of providing readers with a platform to better understand, reflect and discuss the sometimes clashing and mutually fertilising areas of arts, culture, business, management and innovation. The previous book, edited by Piber and titled "Management, Participation and Entrepreneurship in the Cultural and Creative Sector" (Piber 2020), put in evidence the societal relevance of the arts, culture, cultural entities and heritage as collective memories and reservoirs of experience. Adding up to those findings, the contributions collected in this book share the underpinnings that cultural heritage should be conceived and preserved in line with the principles of sustainable development. Each one of them provides an original view of the interplay between management, participation, and entrepreneurship in the sustainable development debate. The latter (management, participation, and entrepreneurship) have been highlighted by management scholars as the main regenerative process of cultural initiatives consistent with a sustainability view (e.g. Ebewo and Sirayi 2009). In the past, cultural initiatives have been extensively studied through specific disciplinary lenses and approaches, to scrutinise specific themes associated with various disciplines, such as cultural economics, management, architecture, engineering and sociology. At odds, with the aim of investigating cultural initiatives against the sustainability paradigm, it is necessary to adopt an integrated perspective and an interdisciplinary approach. Hence, this book gathers all these viewpoints and offers an integrated approach to sustainability by providing the publication outlet to authors representing diverse backgrounds and fields of specialisation.

The importance of cultural and arts management is not a new issue (e.g. Schiuma 2011). However, in light of the sustainability debate, the management of cultural initiatives and organisations is enlightened with a new perspective on how cultural organisations adopt and adapt established management practices to embrace a multi-stakeholder perspective and consider all the stakeholders' needs (e.g. ForHeritage project).

Moreover, participation and the participatory approach in culture have become increasingly important, as participation fosters socio-economic development, urban regeneration and capabilities development through art and culture (Biondi et al. 2020). A participatory approach allows us to involve different actors and stakeholders, who could contribute commitment, passion, expertise and/or their knowledge to various processes in the cultural and creative sectors.

Cultural entrepreneurship has the potential to create new ways of connecting the past, represented by tangible and intangible cultural heritage, with the renewed meanings of the present and projecting them to the future. In this sense, cultural entrepreneurship fosters sustainable development through innovation and growth in the context of the creative industries.

2 A Framework for Discussing the Debate on Cultural Initiatives and Sustainable Development

From a sustainability perspective of culture and cultural heritage, several conceptual models have been proposed that break down the broad concept of sustainability into distinct dimensions, with relation to specific domains where a given impact might be detected. Back in 2013, the Hangzhou Declaration "Placing Culture at the Heart of Sustainable Development Policies" maintained the critical role of culture in achieving sustainable development. Moreover, the Declaration addressed culture's contribution to the social, environmental and economic development pillars of sustainable development. Later on, the "Cultural Heritage Counts for Europe" manifesto discussed the distinct and yet interconnected domains defined as cultural, social, economic and environmental (CHCfE 2015). Based on a literature review recently released by the SoPHIA European project,[2] the four domains can be described as follows (SoPHIA Consortium 2020):

- The cultural domain embraces the proper cultural sphere and its relations with the citizens' well-being, sense of belonging, identity and cultural memory. Studies that fall under the cultural domain unfold the understanding of the relation between people and heritage.
- The social domain is related to the vast array of interaction between culture and people, being communities, organisations or individuals. This interaction spurs a wide range of values that needs to be integrated into cultural policies and practice.

---

[2]The SoPHIA Social Platform is a project funded by the Program of the European Union for Research and Innovation Horizon 2020 (GA n°870,954). SoPHIA aims to set up a network of stakeholders, both researchers and practitioners active in the field of heritage; to stimulate a structured interaction and participation among the stakeholders involved; to provide sustainable and useful tools to raise knowledge on the formulation and implementation of relevant policies and practices; and to promote the project and its outputs at the European and international levels, towards the general public, target communities and policy makers. The social platform proposes a participative model for research and policy-making related to heritage impact assessment and quality of interventions, and, as a collaborative platform, it addresses the complex challenge of setting up a structured network across a broad range of players such as international and umbrella organisations, national authorities and public policy makers, researchers, NGOs as well as public and private organisations.

Social values and interactions have a primary role in both the conservation and valorisation of cultural heritage.
- The economic domain is connected with the valorisation of the cultural heritage, with innovative and sustainable managerial practices of cultural initiatives and organisations and the economic impact of cultural interventions.
- The environmental domain offers room for discussion about how to overcome the repercussions of worsening phenomena such as climate change, overtourism and growing urbanisation. Aspects related to sustainable urban planning are also included in the environmental domain.

From another point of view, following the renowned ChCFE report, three levels of analysis are now widely accepted: macro, meso and micro. The macro level comprises a broader global perspective, like the European coverage or elsewhere in the world. The meso level entails local, regional and national levels. Finally, the micro level includes case studies at a local scale.

3 Applying the Framework: Mapping the Contributions Included in This Book

We propose to use these multiple views to crystallise the current debate of what sustainable development entails in the cultural heritage forum and map out the contributions of our book. In particular, we propose to classify each contribution according to the level of analysis (micro, meso, macro) and the regenerative process (management, participation, entrepreneurship) addressed by the author(s). As a third dimension of the mapping, we considered the four domains in which the sustainability concept unfolds (cultural, social, economic and environmental), highlighting the main themes faced in each contribution (see Fig. 1).

Subsequently, as shown in Tables 1, 2 and 3, we will present a brief summary of the main highlights of the discourse on culture initiatives for sustainable development that have emerged from the mapping of this book's contributions. The latter is divided into three sections.

3.1 Section 1: The Micro Level of Analysis

The first section includes contributions that scrutinise cultural organisations (museums, performing art organisations) or actors (cultural entrepreneurs and workers) (see Table 1).

Chapters "Aligning Market Strategies, Digital Technologies, and Skills: Evidence from Italian Museums" and "Exploring the Financial Strategies of Private Museums: The Case of the *Opera di Santa Maria del Fiore* in Florence" investigate the management of museum organisations with particular reference to various

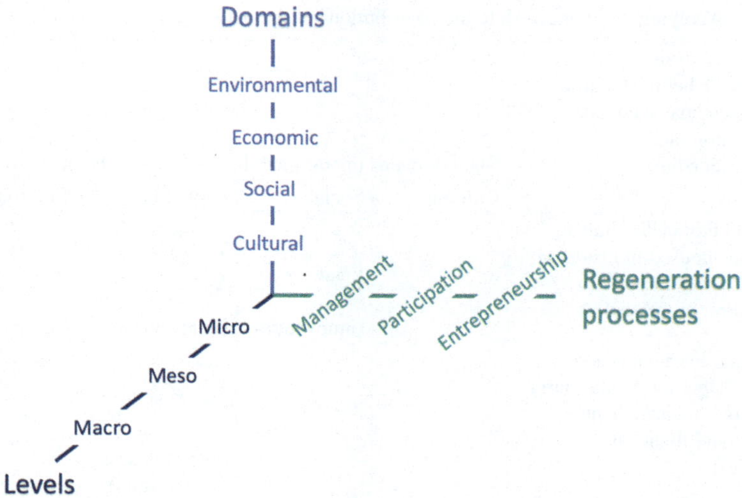

Fig. 1 Conceptual framework to map out the contributions included in this book

processes such as digitisation, fundraising and citizen participation, which are all relevant for museums to not only preserve cultural heritage but also to regenerate it and to keep culture alive over time in order to be able to hand it over to new generations.

Cori and Fraticelli (chapter "Aligning Market Strategies, Digital Technologies, and Skills: Evidence from Italian Museums") aim to shed light on the digitisation processes in Italian museums housed in historical sites. Evidence reveals that mimetic isomorphism represents the main driver of the digitisation process in Italian museums. Yet, the development of digital skills also plays a meaningful role, as the latter favour widespread innovative capacity, which is able to guide the subsequent steps of the digitisation process. Hence, it is the theme of innovation, seen in its economic and social implications that emerge as a crucial aspect for the enhancement of museum management from a perspective of sustainable development.

Romolini, Fissi, Gori and Contri (chapter "Exploring the Financial Strategies of Private Museums: The Case of the *Opera di Santa Maria del Fiore* in Florence") aim to thoroughly analyse the financial structure of a private museum by exploring the case of the "Opera di Santa Maria del Fiore" in Florence. In a context in which public financial resources are increasingly scarce, it is vital for managers to acquire managerial skills to attract private investors and donors or to identify new forms and alternative sources of funding for the sustainability of museum organisations. Hence, this study offers interesting insights for museum managers, who play a pivotal role in facilitating private investments and donations for cultural institutions.

The work by Zbuchea, Romanelli and Bira (chapter "Through the Public's Lens: Are Museums Active Members of Society? An Investigation During the COVID-19 Pandemic") proposes an investigation of the way museums have reacted and tried to support their communities and society in the times of COVID-19. While a

Table 1 Applying the framework to the contributions of the book—the micro level

Level of analysis Authors, "label for the title" (Processes: management, participation, and entrepreneurship)	Main domains of sustainability (addressed themes)			
Micro level	Cultural	Social	Economic	Environmental
Cori and Fraticelli, "Italian museums housed in historical sites" (management: digitisation processes)		✓ (innovation)	✓ (innovation)	
Romolini, Fissi, Gori and Contri, "Opera di Santa Maria del Fiore" (management: funding and financial strategies)			✓ (attractiveness for investors)	
Zbuchea, Romanelli and Bira, "Museums and the COVID-19 pandemic" (participation)		✓ (activism)		
Scapolan and Giannecchini, "The leadership dance" (management: leadership)		✓ (relational resources)	✓ (attractiveness)	
Trevisan, "The case of the performing arts" (management: innovation processes)	✓ (innovation)		✓ (innovation)	
Montanari, Mizzau and Razzoli, 'Start me up' (cultural entrepreneurship)			✓ (cultural work)	
Bizjak, Pareschi and Ripetta, "The puntOorg experience" (academic entrepreneurship)	✓ (education)			

participatory approach to engage the audience in developing and delivering the museum offer is widely investigated in the literature, museum activism and social involvement with society and communities are scarcely investigated. Hence this chapter offers some clues to reflect on the public image of museums, drawing the attention of managers, curators and museum professionals to the expectations of the contemporary audience and the developments of present-day society. As a result,

Table 2 Applying the framework to the contributions of the book—the meso level

Meso level	Cultural	Social	Economic	Environmental
Barreca and Meneguzzo, "Puglia Music Industry" (management: networking, collaborative processes)		✓ (social capital)	✓ (regional attractiveness)	
Palmi, Esposito and Prete, The "Francigena Route" (management: digitisation processes)			✓ (regional attractiveness/ cultural tourism)	
Catalfo, Giustra and Cardillo "The Taormina Film Fest" (management: impact assessment)			✓ (regional attractiveness)	
Pareschi and Ponzoni, "Festivaletteratura" (management and participation: audience development)	✓ (identity)	✓ (participation)		
Bertoni, Dubini, Monti, "Participatory events in the urban context" (participation)		✓ (participation)		
Del Baldo and Demartini, "Urbino per Bene–Urbino for Good" (participation)	✓ (identity, sense of place; education)	✓ (social capital, participation)		
Borin, Jolivet, "Rural cultural centres" (cultural entrepreneurship)	✓ (identity)	✓ (social capital)	✓ (regional attractiveness/ prosperity)	

this contribution highlights new facets, such as social involvement and activism, as emerging key elements for the management of museum organisations.

Chapters "The Leadership Dance in a Performing Arts Organization" and "Interpretive Innovation in the Performing Arts: The Role of Organizations" investigate

Table 3 Applying the framework to the contributions of the book—the macro level

Macro level	Cultural	Social	Economic	Environmental
Cicerchia, "Culture Indicators" (management: standards for monitoring culture contribution to SDGs)	✓ (knowledge and skills)	✓ (inclusion and participation)	✓ (prosperity and livelihoods)	✓ (environment and resilience)
Macrì and Cristofaro, "Europeana" (management: digitisation processes)		✓ (inclusion)		✓ (heritage at risk)
Lazzaro and Noonan, "The contribution of crowdfunding regulation" (cultural entrepreneurship)			✓ (attractiveness for investors)	

the performing arts organisations, analysing dynamics that can lead to both the development of the organisation and a regeneration of cultural heritage. Namely, the following works address two main questions: how do leadership forms change over time, and how do organisations support innovation in the performing arts?

Scapolan and Giannecchini (chapter "The Leadership Dance in a Performing Arts Organization") point out that leadership in cultural organisations is strategic since leaders, besides being responsible for the internal management, represent the image of the organisation, which in turn affects the capability of the organisation to attract vital resources to guarantee the artistic quality of the production and maintain the survival of the organisation over time. Based on a case study of "Fondazione Nazionale della Danza *Aterballetto*", this chapter shows that various factors and dynamics at multiple levels (individual, organisational and environmental) may shape the leadership structure and how it changes over time.

Trevisan (chapter "Interpretive Innovation in the Performing Arts: The Role of Organizations") deems that understanding how cultural and art institutions innovate is a way to understand how our cultural heritage is renewed and how new cultural assets are created. Performing arts, in particular, have the peculiarity of matching tradition with novelty; therefore, performing arts organisations have such innovative endeavours as a central part of their mission. Namely, this chapter detects the role of performing arts organisations in setting up complex evaluative processes that lead to a regeneration of cultural heritage and sustain its transmission. Hence, in this contribution, innovation emerges as a central theme that affects several sustainability dimensions.

This first section concludes with two contributions (chapters "'Start Me Up': The Challenge of Sustainable Cultural Entrepreneurship for Young Cultural Workers" and "Organizing Academic Entrepreneurship Drawing on Cultural Knowledge: The puntOorg Experience") that address cultural entrepreneurship as a regenerative process through which cultural workers create artistic or cultural as well as economic and social value.

Based on 120 interviews with cultural workers at an early stage of their entrepreneurial activity, chapter "'Start Me Up': The Challenge of Sustainable Cultural Entrepreneurship for Young Cultural Workers", by Montanari, Mizzau and Razzoli, reveals that sustainability is perceived as a multifaceted concept. First, cultural workers interpret sustainability in creative terms, as the capability of developing new creative ideas continuously. Then, they interpret sustainability in economic terms by reckoning how to sustain their activities economically. Finally, cultural workers interpret sustainability in social terms as they derive a sense of purpose from the social impact of their work. In addition, findings reveal that young cultural entrepreneurs put into practice a varied set of actions aimed at figuring out a personal path to sustainability. Indeed, cultural workers face several challenges in such an uncertain and fast-changing environment and struggle to find their way for innovation, growth and sustainable development.

Bizjak, Pareschi and Ripetta (chapter "Organizing Academic Entrepreneurship Drawing on Cultural Knowledge: The puntOorg Experience") present the case of *puntOorg* to reflect on the interplay between cultural entrepreneurship and academic entrepreneurship, especially in the light of university reforms that have reframed the role of professors and researchers working in academia. With the rise of an entrepreneurial dimension characterising the cultural work, the same dimension can be applied to cultural management scholars. Hence, as a first outcome, this chapter offers an original point of view of being academic entrepreneurs. In addition, as the authors claim: "In the perspective of the 'anthropology of organisations', *puntOorg* is an artefact [...], that makes tangible the meeting between University and business, so to generate retention of value, valid in the long run". Hence, this contribution offers food for thought to reflect on the role that institutions of higher education and scholars can have in promoting sustainable development.

3.2 *Section 2: The Meso Level of Analysis*

This section of the book includes contributions that inquire into cultural initiatives (such as cultural events, festivals, networks) that involve many actors and have an impact on a regional scale, thus implying a meso level of analysis (see Table 2). Indeed, consistent with the studies on the influence of the cultural sector on sustainable development, more attention has been put on the interpretation of culture, not as a *silo* but instead connected to its broader environment, territory and its stakeholders. Along with our framework, we mapped out the following contributions according to the main regenerative processes of management, participation and entrepreneurship.

Chapters "Promoting Collaboration Through Creative Networks: The Puglia Music Industry" and "Change in Perspectives in Cultural Tourism: A Sustainable

Managerial Model for Cultural Thematic Routes Creating Territorial Value" develop two case studies located in the same region of Southern Italy. Authors address different managerial issues, namely, networking and digitalisation, both aimed at fuelling regional attractiveness and pursuing a sustainable development approach.

Barreca and Meneguzzo (chapter "Promoting Collaboration Through Creative Networks. The Puglia Music Industry") draw on the idea that public networks are considered a significant solution to deal with the many wicked and complex problems characterising our society and an effective tool to foster strategies for regional attractiveness and sustainable development. Their contribution aims to investigate collaborative processes through the development of networks in the creative context and depicts the early development of a cultural and creative network in the music industry in the Apulia Region.

Palmi, Esposito and Prete (chapter "Change in Perspectives in Cultural Tourism: A Sustainable Managerial Model for Cultural Thematic Routes Creating Territorial Value") propose a sustainable managerial model to avoid the negative externalities due to mass tourism and its unprofitable concentration in some artistic city centres or in some periods of the year. This model is based on an appropriate diversification of the cultural heritage offer and the use of new technologies. It is applied to a specific case study: the "Francigena Route" in the Apulia Region. Practical implications are strategic guidelines to both cultural and touristic operators and policy makers to increase regional attractiveness.

Chapter "Detecting the Social and Economic Impact of Cultural Initiatives: a Case Study of the Taormina Film Fest", by Catalfo, Giustra and Cardillo, presents a further managerial perspective regarding the theme of impact assessment applied to interventions on intangible cultural heritage. This research builds on a case study and focuses on the evaluation of the social, cultural and economic values generated by the *Taormina Film Fest*, a rich cultural event in the northeaster Sicilian territory. Evidence shows that a holistic impact assessment appears to be a relevant tool for monitoring the changes that interventions in the cultural sector can bring to society.

Subsequently, chapters "A Struggle of Capitals Over the Identity and the Cultural Offering of Festivaletteratura: the Organizational Impact of Audience Development", "Participatory Event Platforms in the Urban Context: The Importance of Stakeholders' Meaning of "Participation"", and "Cultural Heritage Through The "Youth Eyes": Towards Participatory Governance And Management Of UNESCO Sites" analyse, from different angles, the process of involvement of citizens and stakeholders in the design/implementation of cultural initiatives and the process of engagement of users/audiences in the production and consumption of cultural goods and services. According to our framework, all these contributions address themes that refer to the cultural and social dimensions of sustainability, which are strictly related. For example, we can refer to the feeling of belonging to a present community (social dimension) or to a community of the past (sense of identity) that culture may nurture and enhance.

Pareschi and Panzoni (chapter "A Struggle of Capitals Over the Identity and the Cultural Offering of Festivaletteratura: the Organizational Impact of Audience Development") explore the case of *Festivaletteratura*, the most important Italian literary festival, which has experienced an internal struggle over its cultural offering and cultural identity in the last few years. Indeed, a new generation of managers within *Festivaletteratura*'s organisational bodies has attempted to modify the cultural offering of the festival and, consequently, the festival's identity, hopefully aimed at meeting the preferences of new audiences. Conversely, some other agents, not willing to change the cultural offering of the festival, relied on symbolic value acquired over the years. Findings open up to interesting reflections on how a past success formula has crystallised a power structure that determined *Festivaletteratura*'s success. At the same time, faithfulness to the original spirit has prevented the festival from attracting younger audiences, thus endangering its survival and the natural evolution of the festival.

Bertoni, Dubini and Monti (chapter "Participatory Event Platforms in the Urban Context: The Importance of Stakeholders' Meaning of "Participation"") examine how participation is interpreted and enacted in iterative urban events. Based on the longitudinal analysis of an iterative participative cultural initiative that has taken place in Milan since 2012, their investigation contributes to the understanding of how participatory cultural initiatives are structured and unfold. Main considerations arise around the specificity and the intrinsic fragility of such initiatives, "where participation becomes a resource over time—if successfully managed—or liability in itself". However, according to a sustainable development perspective, it is confirmed that the participatory model has great potential to create values for citizens and different stakeholders.

In their contribution, Del Baldo and Demartini (chapter "Cultural Heritage Through The "Youth Eyes": Towards Participatory Governance And Management Of UNESCO Sites") deem that reinterpreting cultural heritage in the light of perceived cultural values of local communities is an obligation and an opportunity for politicians and city managers that aim to foster urban cultural-led regeneration projects. Based on the *Urbino per Bene—Urbino for Good* project, launched in 2017 by the municipality of Urbino (Italy), this chapter discusses how youths experience a World Heritage Site and how the Public Administration has intervened to reconcile multiple and often conflicting interests. What emerges from the study is that without a real understanding of the values and needs of the local communities, it is difficult to engage citizens and young people in protecting and enhancing their cultural heritage, in turn, making it difficult to achieve the impacts expected from interventions in culture that should support the regeneration of the city's fabric.

Finally, in chapter "Entrepreneurial Cultural Ecosystems in Rural Contexts: Some Insights From Rural Cultural Centers in France", Borin and Jolivet address the debate on cultural entrepreneurship in reference to sustainable development. To this purpose, the concept of entrepreneurial ecosystems in the cultural and creative sector is gaining momentum, and the authors deeply discuss it. A cultural ecosystem, indeed, develops not merely based on the need for physical proximity with customers and competitors, or for policy or financial incentives, but mainly with the purpose to focus on producing collective cultural and social values. Moreover, the authors apply this concept to rural territories, a topic rather underexplored, and carried out a qualitative empirical investigation based on two significant case studies located in the region of Auvergne-Rhône-Alpes in France. Hence, they provide insights on a promising phenomenon and contribute to the debate on the need for a more balanced development between urban and rural regions.

3.3 Section 3: The Macro Level Analysis

This last section of the book entails contributions that address issues regarding a broader global scale (see Table 3).

As far as the concept of this book is concerned, Cicerchia (chapter "Culture Indicators for Sustainable Development") tackles a core issue as the author inquiries into *The Culture 2030 Indicators* report, released by UNESCO in 2019. The latter is a framework of thematic measures, structured in four domains (environment and resilience, prosperity and livelihoods, knowledge and skills, inclusion and participation) that aim to assess and monitor culture's contribution to the national and local implementation of the SDGs. Interestingly, the author critiques both the design of the new framework, with a detailed analysis of its domains and indicators, and its practical application to the institutions, organisations and enterprises of the cultural sector.

The main conclusions, and input for further reflection, are that the challenge of providing an appropriate set of indicators is difficult and twofold: "it involves both working on consistent conceptual definitions and political choices—the latter seldom universally shared—on the one hand, and the availability or at least feasibility of pertinent, timely and reliable data at a global scale, on the other".

Subsequently, in chapter "The Digitalisation of Cultural Heritage for Sustainable Development: The Impact of Europeana", Macrì and Cristofaro analyse the digitalisation process for cultural heritage as a way to contrast its risk of damage and loss. In addition, and drawing on the case of the *Europeana* portal, the authors deem that the digital platform guarantees the widest possible use of its contents and, at the same time, re-use in different areas, thus ensuring the effective and economical use of resources. For all these reasons, *Europeana* can be considered as an example of added value for the cultural policies of the EU and its Member States. Hence, these processes act as multipliers of creativity in which the innovation resulting from digitalisation promotes shared knowledge and intercultural dialogue and supports sustainable development.

Finally, in chapter "The Contribution of Crowdfunding Regulation to Cultural Entrepreneurship in a Supportive Ecosystem", Lazzaro and Noonan explore how an effective crowdfunding regulation can support cultural entrepreneurship. Indeed, the rapid emergence of donation and reward crowdfunding for cultural and creative projects and ventures points to many benefits and barriers. Appreciating these benefits and barriers is central to regulating crowdfunding operations and growth. Insofar, public policy specific to crowdfunding for cultural and creative projects is relatively under-developed, and the major areas of improvements include taxation and matching subsidies, consumer and investor protection and information provision. In addition, authors discuss the critical design features of the two main markets for crowdfunding, namely, the USA and EU, and identify how efficiency and equity might be affected by policy. Hence, authors recommend key policy priorities towards a harmonised, balanced and supportive regulation and offer food for thought for future research in this field, crucial for the sustainable development of the cultural sector.

4 Concluding Remarks and Further Research

To give a definitive statement about the debate on cultural initiatives and sustainability is far beyond the scope of this introduction. Nonetheless, it is interesting to draw some remarks from the discussion presented above. As we stated, this book investigates the relationship between cultural initiatives and sustainability by adopting the multidimensional framework that combines the dimensions of sustainability (cultural, social, economic, environmental); the three levels of analysis (micro, meso and macro); and the regenerative managerial processes (management, participation, entrepreneurship).

From the micro level perspective, many contributions included in the book insist on the theme of innovation, which takes on several nuances across all the domains. First, the digitalisation processes in museums raise managerial issues about digital skills. Furthermore, additional economic and social implications emerge as innovation serves as a mean to renew cultural heritage, create new assets and enrich the performing arts while matching tradition with novelty.

The majority of the contributions in this book refer to the meso level and emphasise the idea of culture as a common good. Moreover, the recurring idea is that the sustainability of cultural initiatives can only be guaranteed if actors, public and private, for-profit and not-for-profit, work in synergy.

The research on cultural initiatives holding a managerial perspective requires the investigation of the value aspects related to the cultural domain. Although managerial scholars rarely possess these aspects in their background, the contributions collected in this book highlight the relevance of the sense of place and identity linked to cultural interventions. This result underlines the importance of the interdisciplinary approach in addressing these issues.

In light of our considerations, we believe that the discussion about cultural initiatives and sustainability is still in its infancy and deserves a long and healthy life. Indeed, we have detected the following themes that should be further explored:

- Impact on economic variables, shifting from a view on competitiveness to a view on prosperity: bringing attention to the quality of income and employment calls for further research in investigating the impact of cultural initiatives on qualified economic indicators.
- Culture and welfare: sustainability calls for a long-term view of citizens' welfare. How are cultural policies designed and enforced to meet this need? Furthermore,

additional research is needed to clarify the role of culture in promoting health and well-being, especially in the contemporary age of hyper-connection and fluidity in society.

- Quality of cultural interventions: adopting a sustainability approach to culture promotes the development of quality in cultural interventions. Evidence-based research could help deepen the understanding of how quality is taken into account in cultural interventions.
- Monitoring: cultural initiatives should be constantly monitored, as monitoring is a fundamental tool at different stages of the initiatives' maturity: (a) ex ante, when monitoring tools are useful in planning the initiatives from a sustainable perspective; (b) ongoing, to constantly be updated on the status of the initiative; and (c) ex post, when it is necessary to understand the several impacts generated by the initiative and the associated legacy for the next generations to come.

We believe that it is a promising stream of research, and we humbly hope that our book contributes to bringing about the understanding of a complex yet very fascinating topic. We are very much grateful to all the authors who have enthusiastically responded to our call for chapters, have carefully followed the instructions given through the blind review process and have patiently followed the Editors' recommendations. We are also in debt to the anonymous reviewers for their dedication and valuable suggestions. Finally, we would like to express our gratitude to the publisher, Springer. We began a fruitful and hopefully prosperous collaboration with them some years ago, and they have accompanied our research journey on cultural initiatives ever since.

References

Barreca, M., & Meneguzzo, M. (2020). Promoting collaboration through creative networks. The Puglia music industry. In P. Demartini, L. Marchegiani, M. Marchiori, & G. Schiuma (Eds.), *Cultural initiatives for sustainable development*. Heidelberg: Springer.

Bertoni, A., Dubini, P., & Monti, A. (2020). Participatory event platforms in the urban context: The importance of stakeholders' meaning of "participation". In P. Demartini, L. Marchegiani, M. Marchiori, & G. Schiuma (Eds.), *Cultural initiatives for sustainable Development*. Heidelberg: Springer.

Biondi, L., Demartini, P., Marchegiani, L., Marchiori, M., & Piber, M. (2020). Understanding orchestrated participatory cultural initiatives: Mapping the dynamics of governance and participation. *Cities, 96*, 102459.

Bizjak, D., Pareschi, L., & Ripetta, S. (2020). Organising academic entrepreneurship drawing on cultural knowledge. The puntOorg experience. In P. Demartini, L. Marchegiani, M. Marchiori, & G. Schiuma (Eds.), *Cultural initiatives for sustainable development*. Heidelberg: Springer.

Borin, E., & Cassandre Jolivet, C. (2020). Entrepreneurial cultural ecosystems in rural contexts: Some insights from rural cultural centers in France. In P. Demartini, L. Marchegiani, M. Marchiori, & G. Schiuma (Eds.), *Cultural initiatives for sustainable development*. Heidelberg: Springer.

Brundtland, G. H. (1985). World commission on environment and development. *Environmental Policy and Law, 14*(1), 26–30.

Catalfo, P., Giustra, M., & Cardillo, A. (2020). Detecting the social and economic impact of cultural initiatives: A case study of the Taormina Film Fest. In P. Demartini, L. Marchegiani, M. Marchiori, & G. Schiuma (Eds.), *Cultural initiatives for sustainable Development*. Heidelberg: Springer.

CHCfE Consortium. (2015). *Cultural heritage counts for Europe*. Krakow: CHCfE Consortium.

Cicerchia, A. (2020). Culture indicators for sustainable development. In P. Demartini, L. Marchegiani, M. Marchiori, & G. Schiuma (Eds.), *Cultural initiatives for sustainable development*. Heidelberg: Springer.

Cori, E., & Fraticelli, F. (2020). Aligning market strategies, digital technologies, and skills: Evidence from Italian museums. In P. Demartini, L. Marchegiani, M. Marchiori, & G. Schiuma (Eds.), *Cultural initiatives for sustainable development*. Heidelberg: Springer.

Culture 2030 Goal Campaign. (2019). *Culture in the implementation of the 2030 Agenda*. Published in Barcelona, Paris, Harare, Sydney, Montreal, The Hague and Brussels, in the frame of the first UN SDG Summit taking place on 24–25 September 2019.

Del Baldo, M., & Demartini, P. (2020). Cultural heritage through the "youth eyes": Towards participatory governance and management of UNESCO sites. In P. Demartini, L. Marchegiani, M. Marchiori, & G. Schiuma (Eds.), *Cultural initiatives for sustainable development*. Heidelberg: Springer.

Ebewo, P., & Sirayi, M. (2009). The concept of arts/cultural management: A critical reflection. *The Journal of Arts Management, Law, and Society, 38*(4), 281–295.

Elkington, J. (1997). *Cannibals with forks—Triple bottom line of 21st century business*. Stoney Creek, CT: New Society.

European Parliament, Resolution of 8 September 2015 Towards an Integrated Approach to Cultural Heritage for Europe, OJ C 316, 22.09.2015, p. 88, available at https://www.europarl.europa.eu/doceo/document/TA-8-2015-0293_EN.html, last access December 2020.

ForHeritage Project. *Excellence in heritage management in central Europe*. https://www.interreg-central.eu/Content.Node/ForHeritage.html. Last retrieved October 23, 2020.

Helly, D., & Galeazzi, G. (2016). *Culture in EU development policies and external action: Reframing the discussion*. (ECDPM Briefing Note 92). Maastricht: ECDPM.

Labadi, S. (2019). *UNESCO, culture, aid, and development in the new millennium. The cultural turn in international aid: Impacts and challenges for heritage and the creative industries*. London: Routledge.

Lazzaro, E., & Noonan, D. (2020). The contribution of crowdfunding regulation to cultural entrepreneurship in a supportive ecosystem. In P. Demartini, L. Marchegiani, M. Marchiori, & G. Schiuma (Eds.), *Cultural initiatives for sustainable development*. Heidelberg: Springer.

Macrì, E., & Cristofaro, C. L. (2020). The digitalisation of cultural heritage for sustainable development: The impact of European. In P. Demartini, L. Marchegiani, M. Marchiori, & G. Schiuma (Eds.), *Cultural initiatives for sustainable development*. Heidelberg: Springer.

Montanari, F., Mizzau, L., & Razzoli, D. (2020). 'Start me up'. The challenge of sustainable cultural entrepreneurship for young cultural workers. In P. Demartini, L. Marchegiani, M. Marchiori, & G. Schiuma (Eds.), *Cultural initiatives for sustainable development*. Heidelberg: Springer.

Palmi, P., Esposito, M., & Prete, M. I. (2020). Change in perspectives in cultural tourism: A sustainable managerial model for cultural thematic routes creating territorial value. In P. Demartini, L. Marchegiani, M. Marchiori, & G. Schiuma (Eds.), *Cultural initiatives for sustainable development*. Heidelberg: Springer.

Pareschi, L., & Ponzoni, N. (2020). A struggle of capitals over the identity and the cultural offering of *Festivaletteratura*: The organizational impact of audience development. In P. Demartini, L. Marchegiani, M. Marchiori, & G. Schiuma (Eds.), *Cultural initiatives for sustainable development*. Heidelberg: Springer.

Piber, M. (2020). *Management, participation and entrepreneurship in the cultural and creative sector*. Heidelberg: Springer.

Romolini, A., Silvia Fissi, S., Gori, E., & Contri, M. (2020). Exploring the financial strategies of private museums. The case of the *Opera Di Santa Maria Del Fiore* in Florence. In P. Demartini, L. Marchegiani, M. Marchiori, & G. Schiuma (Eds.), *Cultural initiatives for sustainable development*. Heidelberg: Springer.
Scapolan, A. C., & Gianecchini, M. (2020). The leadership dance in a performing arts organization. In P. Demartini, L. Marchegiani, M. Marchiori, & G. Schiuma (Eds.), *Cultural initiatives for sustainable development*. Heidelberg: Springer.
Schiuma, G. (2011). *The value of arts for business*. Cambridge: Cambridge University Press.
SoPHIA Consortium. (2020). *Review of research literature, policy programmes and (good and bad) practices*. Released on September 8, 2020. Retrieved from https://sophiaplatform.eu/uploads/sophiaplatform-eu/2020/10/21/a4309565be807bb53b11b7ad4045f370.pdf
Torggler, B., Murphy, R., France, C., & Balta Portoles, J. (2015). *UNESCO's work on culture and sustainable development: Evaluation of a policy theme*. IOS/EVS/PI/145/Rev.2. Paris: UNESCO.
Trevisan, P. (2020). Interpretive innovation in the performing arts: The role of organizations. In P. Demartini, L. Marchegiani, M. Marchiori, & G. Schiuma (Eds.), *Cultural initiatives for sustainable development*. Heidelberg: Springer.
UNESCO. (2018). *Culture for the 2030 Agenda*. Paris: UNESCO.
Wiktor-Mach, D. (2020). What role for culture in the age of sustainable development? UNESCO's advocacy in the 2030 Agenda negotiations. *International Journal of Cultural Policy, 26*(3), 312–327.
WCED. (1987). *Our common future*. New York: Oxford University Press.
World Commission on Environment and Development. (1987). *Our common future* (p. 27). Oxford: Oxford University Press. ISBN 019282080X.
Zbuchea, A., Romanelli, M., & Bira, M. (2020). Through the public's Lens: Are museums active members of society? An investigation during the COVID-19 pandemic. In P. Demartini, L. Marchegiani, M. Marchiori, & G. Schiuma (Eds.), *Cultural initiatives for sustainable development*. Heidelberg: Springer.

Paola Demartini is Full Professor of Managerial and Financial Accounting at Roma Tre University, Department of Business Studies. She is the Head of the Corporate Governance Lab, which includes a special section on the Governance of Cultural Heritage and Contemporary Culture. Since 2000, she is member of the supervisory committee of PhD courses on Financial Accounting and Governance disciplines, where she taught, among others, Performance Management and Business Evaluations. Since 2016 she is involved in the project "Cultural Heritage and Contemporary Culture: Transferring knowledge, Managing Outcomes, and Enabling Sustainability", a collaboration between Corporate Governance Lab (Rome Tre University) and Institut für Organisation und Lernen (Innsbruck University). Currently she is member of the leading research team engaged in a H2020 project titled SOPHIA—Social Platform for Holistic Heritage Impact Assessment.

Lucia Marchegiani is Associate Professor of Business Organization and Human Resources Management at Roma Tre University, where she teaches Human Resources Management and Knowledge Management in the Digital Age. Her research interests cover topics such as Creative and Cultural Industries, Innovations and Organizations, Experimental Organizational Behavior, Knowledge Management and Social Media, Management and Sustainability, and Participatory Approaches in Culture. She has been visiting scholar at the Copenhagen Business School (DK), IESEG Business School (F), University of Copenhagen (DK), and ISC Business School at Paris (F). She has been Chair of the "International Conference on Sustainable Cultural Heritage Management", and Scientific Coordinator of the NetMUSE international research proect. Currently she is member of the leading research team engaged in a H2020 project titled SOPHIA—Social Platform for Holistic Heritage Impact Assessment.

Michela Marchiori is Full Professor of Business Organization and Change Management at the Department of Business Studies in Roma Tre University. Since 2016 she has been member of the project "Cultural Heritage and Contemporary Culture: Transferring knowledge, Managing Outcomes, and Enabling Sustainability" founded in collaboration between Corporate Governance Lab (Rome Tre University) and Institut für Organisation und Lernen (Innsbruck University). She is the current Director of the post-graduate, lifelong learning "Master in Management, Promotion, Technological Innovations in Cultural Heritage" and Director of the 2-year-long post-graduate, lifelong learning "Master in Economics and Management of Culture Heritage" both founded by the Department of Business Studies. Currently she is leader of a research team engaged in a H2020 project titled SOPHIA—Social Platform For Holistic Heritage Impact Assessment.

Giovanni Schiuma is Chairman of the Arts for Business Institute and Director of the CLabUnibas (Centre for Entrepreneurship and Innovation Development) of the University of Basilicata. He is also the Director of the Master in Business Administration and Professor in Innovation Management. He has authored over 100 scholarly publications, including "The Value of Arts for Business", published by Cambridge University Press. He is widely recognised as one of the world's leading experts in the strategic knowledge management for company value creation dynamics, and arts-based management for organisational development and innovation, and chairs the International Forum of Knowledge Assets Dynamics, an international network for those interested in the role of knowledge and innovation for organisations value creation. He is Chief Editor of two international journals: Knowledge Management Research and Practice, published by Taylor & Francis, and Measuring Business Excellence, published by Emerald. Currently he holds the position of Visiting Professor at University of the Arts London, where he directed the Innovation Insights Hub and held the first international professorship chair in Arts Based Management. He is also Visiting Professor at Graduate School of Management of the St. Petersburg University, Adjunct Professor at Tampere University of Technology, and he is Fellow of the Royal Society of Arts. Previously he served as Vice Mayor for Strategy Planning and Innovation of Matera city from July 2015 to September 2016 and has held appointments at Cranfield School of Management, University of Cambridge, Kozminski International Business School of the University of Kozminski and Liverpool Business School.

Part I
The Micro Level of Analysis

Aligning Market Strategies, Digital Technologies, and Skills: Evidence from Italian Museums

Enrico Cori and Fabio Fraticelli

Abstract This chapter aims at shedding light on the degree of alignment between market strategies and digitalization processes in Italian museums housed in historical sites. The on-the-field analysis, namely, highlights choices concerning the adoption of digital technologies and the development of digital skills.

The research is qualitative in nature and is based on a multi-case study. The sample has been built according to the following criteria: (a) combination of historical and artistic importance of the site and relevance of the collections housed in it and (b) variety of the sample, in terms of ownership, year of foundation, annual number of visitors, and characteristics of the collections hosted.

The variety of the analyzed sample allowed us to identify different approaches to digitalization processes. Moreover, our research provided new insights in order to understand how held or accessible digital skills drive decision-making process regarding the deployment of technologies.

Research findings provide indications to the museums' managers that are currently engaged in the challenge of digitalization, in order to foster a better alignment between market strategies and digital technologies.

Keywords Museums · Digitalization processes · Digital skills · Strategy-technology alignment · Organization choices

1 Introduction

The fruition of the cultural heritage has always represented, for Italy, one of the most relevant items for the purpose of tourism development in individual cities or territories. Digital technologies adoption processes can play a key role in developing museums' ability to attract audiences and, consequently, boosting tourist flows,

---

E. Cori (✉) · F. Fraticelli
Department of Management, Polytechnic University of Marche, Ancona, Italy
e-mail: e.cori@univpm.it

P. Demartini et al. (eds.), *Cultural Initiatives for Sustainable Development*, Contributions to Management Science,
https://doi.org/10.1007/978-3-030-65687-4_2

especially where the single museum or the local museum system represent a reference point for a tourism destination.

While digitization (Gobble 2018) is the shift from analogic to digital (like the mere conversion of documents from paper to an electronic format), in this chapter, we refer to digitalization as a process that, by adopting a broader change mindset, gets to transform the business and organizational model (Følstad and Kvale 2018; Ghobakhloo 2018; Komulainen and Saraniemi 2019; Lin et al. 2018). It includes embracing new ways of doing business, using new capabilities to serve customers, and designing organizational processes to become more agile (Dörner and Edelman 2015).

Digitization and digitalization are therefore necessarily correlated processes. The first one refers to the adoption of digital technologies to digitize the artifacts and the artworks in the museum's collection. At a deeper level, digitalization is a proper organizational change that involves digital skills development/acquisition. In this paper, we are going therefore to refer mainly to digitalization.

There is a widespread awareness that cultural organizations can benefit in many ways from the adoption of digital technologies (Cori and Fraticelli 2018; Clini et al. 2020). Collection, conservation, and access to the cultural heritage in novel, accessible, and attractive ways demand for digitizing museums, as well as for designing methodologies to represent, manage, and exploit cultural heritage data at different levels. In Italy, as well as in many other countries, heritage organizations are currently engaged in choices regarding the adoption of digital technologies. However, either the literature or the empirical evidence suggest that not always these choices are the result of a clear perception and of a careful reflection about opportunities and limitations linked with the adoption of digital artifacts and devices.

This chapter aims to shed light on the link between digitalization processes and market strategies in Italian museums housed in historical sites. Consistently with the exploratory nature of research, we adopted a qualitative approach; this has been carried out through the multi-case method. We wonder how digital skills, held or accessible by the cultural organizations, do affect the organization's capability to align technological innovation and strategy in Italian museums.

The chapter is structured as follows. The next section proposes a literature review concerning the meanings that the terms "digitalization" and "digital skills" assume in museums; then, two studies that help focusing the relationship between market strategies and the adoption of digital technologies in museums are illustrated. In the third paragraph, the results of the analysis conducted at eight Italian museums are presented, and some propositions are formulated. Finally, the fourth section aims at identifying four different approaches to the digitalization process, characterized by a variable degree of alignment between corporate strategy, digital technologies, and digital skills.

2 Museums and Digital Technologies: An Intriguing Topic

The general framework of reference is constituted by the relationships that exist between the firm's strategic choices, the digitalization processes, and the development of digital skills. The literature review has been circumscribed to the analysis of these relationships in museums, as well as to the identification of some possible classifications and taxonomies, related to competitive strategies, technologies, as well as the so-called digital skills within cultural organizations.

Research on the processes of digitalization of cultural heritage, and in particular of the museums' collections, has initially privileged the analysis of technologies for the preservation of works of art, and for their remote fruition (through the Internet), and then moved on the role and the impact of digital technologies for on-site fruition. To date, two themes still seem to us insufficiently investigated in the literature: that of the relationship between choices relating to technology and the market orientation of museum institutions, as well as the question of the adequacy of digital skills and the connected degree of technology acceptance.

2.1 Digital (Virtual) Museum or Museum's Digitalization?

Even a hasty analysis of the varied literature on museums and digital technologies cannot fail to highlight the absence of uniquely interpretable terms and expressions, used to define the field of investigation. What is meant by "virtual museum"? And for "digital museum"? Are they synonymous expressions or do they indicate different realities?

Semantic uncertainty can be attributed to the relative youth of the phenomenon under study, but also to the fact that the attention of scholars of different disciplines converges on it and that within these fields, the same meaning is not always attributed to the same term (Pescarin 2014).

This applies in particular to the term "virtual," which with reference to museums first identified something opposing to the museum in a physical sense (online collection), then including the virtual dimension of a tangible collection, housed in a well-defined physical-architectural space (Negri 2012). In this second sense, the virtual dimension is initially limited to forms of digital reproduction, which can be used offline or online, via the Internet, or even not usable by the public, but created for the sole purpose of conservation and digital reproduction. Subsequently, the virtual dimension tends to contemplate new forms, which allows not only the remote use (*online fruition*) of the contents of the "physical" museum, but an enriched fruition during the museum going-experience (*on-site fruition*), as in the case of the VR/AR devices.

This debate is still blooming: just recently, Evrad and Krebs (2017) proposed a conceptual analysis of "real" (visiting the museum) and "virtual" (visiting its website) experiences of museums and emphasized how the two types of visit can be considered the complement of each other.

For the purposes of our study, the second meaning, that contemplates a virtual representation of physical contents of the museum, can be accepted without any doubt. Indeed, the choice to refer the virtual/digital attributes to a tangible collection is in our case reinforced by the decision to combine the artistic value of the structure used as a museum and the collections housed in it. Our field of study is in fact defined by the "container-content" set, both physically well identifiable.

However, despite the fact that our interest is toward the application of digital technologies in the context of traditional museums, we prefer not to use the expression "virtual" (or digital) museum, for the risk that it is associated with a predefined state, reached or achievable, with a preconceived idea that perhaps implies the total digitization of the works housed in the museum and/or the online communication of all its contents. Far from wishing to identify a hypothetical point of arrival or to outline its characteristics or, even less, to imagine it as a desirable state, we aim at analyzing: the paths taken by museums; their choices regarding the audience segments to be intercepted or retained, that is, the strategies of audience development (Kirchberg and Kuchar 2014; Cerquetti 2016); the adoption of digital technologies; finally, the development of digital skills. For this reason, we will refer, from now on, to the "digitalization processes," regardless of the fact that these processes have just started by the museum or are already at an advanced stage.

The literature is quite rich in references to the opportunities offered by the digital (or virtual) museum, both in terms of dissemination of digital content and in terms of enriching the visiting experience (Li et al. 2012: 646); in some cases, it also focuses on the possible limits and the risk of "hegemony of the digital as a representational frame" (Geismar 2013: 256). However, past research highlights an almost complete absence of a reflection on digitization processes, on what drives museums toward the choices of adoption of digital technologies, on *whether and how* the choices regarding the targeted audience intertwine and integrate with those who inherent the introduction of technologies and tools, and on *whether and how* organizational choices are considered relevant and contiguous to the technology adoption decisions.

For this reason, in the following two paragraphs, we will deepen, through the analysis of the available literature, two aspects that from our point of view are decisive for the effectiveness of the digitalization processes: the first concerns the development and the degree of diffusion of digital skills; the second issue concerns the relationship between market strategies pursued by the museum and choices for the adoption of digital technologies and tools.

2.2 In Search of Digital Skills

While several contributions have focused on the output of digitalization processes occurring in museums (Vom Lehn and Heath 2005; Bertacchini and Morando

2013),[1] the literature on the skills required to activate, manage, and keep alive a digitalization process seems less developed.

In a knowledge perspective (Hatlevik et al. 2015), digital skills are intended to cover a wide area of technology-related concepts, such as information management, collaboration, communication and sharing, creation of content and knowledge, ethics and responsibility, evaluation and problem solving, and technical operations (Ferrari 2012).

In order to understand the role played by digital skills in the museums' digitalization process, we will analyze two main issues:

- Which digital *skills* are typically involved into a digitalization process
- Which *actors* provide such skills at different stages of the digitalization process

An initial attempt to address the first issue consists in distinguishing between *technical* digital skills (technological skills on solutions, platforms, programming languages, which typically characterize ICT roles) and *applied* digital skills (ability to use software and devices by the decision makers and the employees involved in routine tasks), following what proposed by AICA and Assintel (2017).

Recently, further attempts to classify the wide plethora of skills that are involved in the digitization process have been proposed by the e-Cult Skills project[2] in 2015 and the Mu.SA (Museum Sector Alliance) project in 2017 (Homem et al. 2017).

According to the taxonomy proposed Mu.SA, digital skills can be grouped into four clusters, embedded in the following job profiles: Digital Strategy Manager, Digital Collections Curator, Digital Interactive Experience Developer, and Online Community Manager (Table 1).

The taxonomy from Mu.SA Project is helpful to understand which digital skills are required along a digitalization process, but it seems to underestimate the question of how the overall cognitive work required in the process itself is shared among different actors, inside and outside the museum. Indeed, this seems to be critical to assess the ability of the museum to sustain over time its own choices concerning the digitalization process. In our opinion, such an ability depends on the extent to which the digital skills involved in the process are either available internally or accessible through network relationships.

How a digitalization process unfolds is still under debate. While Preuss (2016) proposes a six-step process (from content specification to digital preservation), Herbert (2017) highlights the following five phases: bridge, uncover, iterate, leverage, and disseminate. According to this contribution, a digitalization process starts by recognizing the (digital) "gaps" between the organization, current and potential customers, and the occurring environmental changes. In order to understand the size

[1]In a nutshell, when we talk about digitalization output, we refer to benefits for museums connected, on the one hand, to the possibility of reaching a wider audience and generating, through online access to museum collections, a greater number of visitors and, on the other hand, to the possibility of enriching and making the visit experience more enjoyable, thanks to on-site digital technologies, such as virtual/augmented reality devices, touchscreens, and so on.

[2]http://groupspaces.com/eCult/

Table 1 Digital Skills for museums' digitalization processes (Mu.SA Project 2017)

Mu.SA job profile	Main tasks	Digital skills
Digital Strategy Manager	– Definition of digital strategy – Financial planning of digital resources – Management of relations with stakeholders – Benchmarking – Definition of solutions *user-centered* – Performance analysis	– Management and implementation of digital projects – Ability to use digital platforms and tools – Implementation of *audience development* strategy
Digital Collections Curator	– Metadata production according to international standards – Implementation of cataloging and archiving standards – Management, accessibility, and security control of digital data	– Ability to use software – Digitization of collections – Ability to catalogue and preserve digital collections – Understanding and application of the *user experience* processes
Digital Interactive Experience Developer	– Design and development of solutions for digital interactive experiences, in relation to the various audience targets – Development of tools for the accessibility to cultural heritage – Mediation between the museum and *high-tech* companies	– Ability to use digital technologies – Ability to analyze users' needs – Ability to design interactive virtual exhibitions – Ability to use interactive software
Online Community Manager	– Design and implementation of the online audience development plan, in line with the museum's communication strategy – Analysis of users' needs – Production of online content and experiences – Evaluation of the effectiveness and efficiency of online activities	– Ability to analyze web data – Ability to use digital tools for online activities (CMS, blog, social network, etc.) – Ability to create graphic representations

of the gap and to identify a proper strategy to fill it, an organization needs to access to resources that are often located outside its boundaries (therefore, those resources can either come from external consultants, partners, or suppliers). Herbert considers the digitalization as a process through which the organization progressively takes ownership of the key digital skills that are required to ensure the evolution of the process itself. In this perspective, as the digitization process proceeds, an organization should increasingly rely more on the (applied and technical) digital skills held internally.

Moreover, in order to be adequately supported, a digitalization process should progressively move from a centralized to a decentralized governance model (Barnes et al. 2018). At the very beginning of this transition, one or few roles own most of the digital skills required, while progressively those skills are shared with a number of actors inside the museum, until digital skills become a widespread asset among the museum's employees. When an organization moves from a centralized to a decentralized governance of the digitalization process, it usually becomes increasingly aware about how to make choices concerning the use of technology effective. This scenario implies that digital skills are diffusely and internally available among the museum employees.

2.3 Market Strategy and Digital Technologies: A Missing Link?

In museum organizations, the term strategy has been used from time to time to indicate choices and orientations referable to different areas.

For example, the belief has spread that museums should develop an "edutainment" strategy (a neologism resulting from the fusion of the words education and entertainment), aimed at identifying forms of fruition of the collections that can integrate both entertainment and education purposes (Balloffet et al. 2014; Addis 2005). When this occurs, it seems more appropriate to speak of institutional purpose, rather than strategy, since the focus is on the social role played by cultural institutions.

In recent years, the term strategy has been repeatedly associated with the choices regarding the adoption of digital technologies by the museum (digital strategy) (Cigola et al. 2014; Kane et al. 2015; Cavriani 2016; Giannini and Bowen 2018); this led to the identification of a variety of approaches to the introduction of digital technologies (Gombault et al. 2016; Dewdney 2018). Here one cannot fail to notice how the use of the term "strategy" mainly refers to technology choices; actually, these are narrowly related with strategy, but should not be confused with it.

In our opinion, with reference to museums too, the use of the term strategy must be correctly referred to the choices relating to which services to provide and which segment(s) of audience to contact. In this perspective, the contributions of Kotler and Kotler (2000), Kotler et al. (2008), and that of Bakhshi and Throsby (2012) still represent valid points of reference.

The first proposes three museum strategies, aimed at "building audience, support and income" (Kotler and Kotler 2000: 273). They are, respectively, the museum going-experience improvement strategy, the community service strategy, and the market repositioning toward entertainment strategy.

The first strategy focuses on the improvement of the museum going-experience, through "richer exhibits and programs, better services and design elements and more accessible and comfortable facilities" (2000: 275–6). Generating a deeper going-experience implies the redesign of the very idea of fruition of the cultural heritage, through the development of "activities in which visitors can directly participate" (2000: 276), such as "media and interactive elements to expand the visitor's sense of immediacy and participation." According to the authors, this strategy "has the advantage of appealing to several different audience segments" (2000: 277); definitely, pursuing a going-experience strategy means acting on the levers of "range of visitor experiences, which includes visual, sensory, and aesthetic experience, recreational, sociable, and learning experiences, and the experiences of celebration and enchantment (and) levels of intensity (. . .), from relatively passive viewing of collections to active immersion experience and applied learning" (2000: 279).

Through the second strategy, the museum is primarily oriented at providing a community service, so raising its image and the local impact. Initially focused on educational purposes, in recent years, this strategy moved "to a focus on community needs in a broader sense" (2000: 282). In terms of audience, this strategy "gives the most emphasis to identifying, segmenting, and targeting the public, the consumers and the members of a given community and region (. . .) museum offerings and programs have to be tailored to meet the needs of different segments" (2000: 282).

Finally, the third strategy aims at repositioning museums in the competitive arena, toward the industry of entertainment, "in order to attract an entirely new audience to make itself competitive with other leisure activity organizations, or to become a place which is popular and entertaining with a broad and diverse audience" (2000: 282). This strategy, in the authors' view, implies "a substantial move away from a museum's traditional audience and thus the need to build entirely new constituencies" (2000: 282).

The study by Bakhshi and Throsby (2012) is oriented to analyze the role of new technologies as an engine of innovation in artistic and cultural institutions. They assume that "digital technologies have the potential to allow arts and cultural organizations to achieve a step increase in the audiences" (2012: 206). The starting point of the analysis is the identification of four categories of innovation, one of which concerns the so-called innovation in audience reach, which is based on the one side in the "generation of new audiences through use of digital technologies" (2012: 208), on the other in a deepening of the engagement with the public.

The authors, taking up a previous elaboration (McCarthy and Jinnett 2001), distinguish between audience broadening, diversifying, and deepening, where the first refers to the "capturing a larger share of the population already known to be audiences," the second to the attraction of "new groups of consumers that do not currently attend," and the third to the "increasing and/or intensifying the engagement of audiences" (Bakhshi and Throsby 2012: 209).

According to them, especially the third type, strategies can benefit from technologies that facilitate the interaction between art organizations, including museums, and their respective audiences, in order to enable the latter to access cultural experience without any space-time constraint.

If Kotler and Kotler indirectly address the issue of the link between market strategies and the role of digital technologies, a link that still appears in an embryonic phase, in Bakhshi and Throsby the question of the possible impact of new technologies on public acquisition strategies occupies a central place in the analysis, probably also by virtue of the fact that their study is conducted more than a decade after the previous one. In fact, in the lapse of time between the two studies, the digital applications in support of the fruition of the museum's collections have multiplied, thus revealing their enormous potential more and more clearly.

Differences aside, the two researches not only provide a consolidated basis for analyzing the current and prospective positioning of museums, but to date represent two significant exceptions in the panorama of studies on the management and organization of the museums themselves. Indeed, looking at most of the studies, it seems that the introduction of digital technologies in museums is considered a sort of independent variable, with potential consequences on strategy and organization. A possible explanation can be looked for in the relevance of institutional pressures, of a mimetic or normative type, which could push museums to adopt digital technologies, regardless of an assessment relating to the alignment with strategies and characters of the organization.

In this panorama, in which studies on digitalization and those on market strategies of the cultural organizations seem to proceed on parallel tracks, both Kotler and Kotler and Bakhshi and Throsby seem to have the right awareness that strategy choices and technological choices are not only closely interrelated, but that the latter are somehow instrumental to the former.

In our opinion, the almost total absence of a debate about the relationship between the role of digital technologies and market strategies can be regarded as the main gap in the literature. Filling this gap really seems desirable, in order to support museum organizations to pursue adequate levels of efficiency and effectiveness in the digitalization processes underway.

The investigation carried out at some Italian museums was just aimed at investigating what pushed museums to adopt digital technologies for the fruition and whether a corporate strategy is clearly identifiable at the origin of the choices made or in progress. In other words, we have tried to understand whether market strategies and choices related to digitalization are sufficiently aligned. At the same time, we have tried to understand whether the decision to introduce digital platforms/tools is supported by a corresponding commitment in the development of digital skills, in order to adequately oversee the ongoing digitalization process.

3 A Survey on Italian Museums

3.1 Methodology, Sample Selection, and Data Collection

The first aim of our research was to assess the coherence between digital technologies in use and the competitive strategy pursued by museums. Besides, we wanted to understand how digital skills, held by the museum, do affect the organization's capability to match technological innovation and strategy.

We adopted a qualitative research design that we carried out through the multi-case study approach. Indeed, the multi-case method makes it possible to highlight a variety of situations, even diverging from one another, with the ultimate aim of contributing to the elaboration of an explanatory theory (Eisenhardt 1989; Eisenhardt and Graebner 2007; Yin 2009).

The sample has been built according to the following criteria: (a) combination of historical and artistic importance of the site and relevance of the collections housed in it and (b) variety of the sample, according to the logic of multi-case method. The choice to select a sample consisting of art collections hosted in ancient palaces, castles, or villas is linked to the belief that this combination extolls the opportunities inherent in the digital technologies for on-site fruition (e.g., virtual/augmented reality devices). The construction of the sample took place according to the "convenience sampling" approach (Mayan 2016: 62), based on the availability to be interviewed and to comply with the research protocol (recording of the interviews, authorization to publish excerpts) by the directors of museums which met criterion (a). A significant variety of the sample was obtained by considering the following characteristics: ownership, year of foundation, annual number of visitors, and characteristics of the collections hosted. The main data relating to each museum are summed up in Table 2.

Data was collected through semi-structured interviews addressed to the managers of the museums, with the aim of drawing up deeper insight into the features of digitizing processes. Interviews were performed in the period November 2018–April 2019, either at the office of the manager or by telephone. They have lasted from 30 min to an hour and followed a precompiled track. However, we left managers free to investigate further points. Three main questions were posed about the following issues: type of market strategy pursued by the museum; type of digital technologies/devices in use and reasons that has pushed museums to their adoption; and range of digital skills held by managers or employees in the museum or externally accessible (at universities, museum networks, consultants, etc.). The data collected through the interviews were integrated with those obtained from the analysis of the museums' website. This with the aim of analyzing the main features of the museum offer, the types of services offered, and the methods of communication and interaction with the relevant public.

As regards market strategies, we referred to the distinction between audience broadening, diversifying, and deepening (Bakhshi and Throsby 2012, built on McCarthy and Jinnett 2001), while as of digital technologies, the taxonomy proposed in Styliani et al. (2009) and the distinction by Bonacini (2011) between *remote* fruition technologies (e.g., website, social networks, multimedia, database) and *on-site* fruition technologies (e.g., info-points, touchscreens, VR/AR devices) were used. Lastly, regarding digital skills, we adopted the distinction between *technical* and *applied* skills, proposed by AICA and Assintel (2017).

3.2 Introducing Digital Technologies: Between Institutional Pressures and Market Strategy

The analysis of the sample of museums has led to a significant wealth of information, related on the one hand to the market strategies pursued, on the other to the

Table 2 The sample

	Museum	Place	Opening	Ownership	Visitors (2017)	Type of heritage
1	National Gallery of the Marche at Palazzo Ducale	Urbino	1912	Public (government)	164,000	Renaissance painting
2	Museums at Palazzo Ducale	Mantua	1881	Public (government)	323,000	Thematic exhibitions
3	Museum at La Venaria Reale	Turin	2008	Public-private network	1,049,000	Paintings and sculptures from the fifteenth to the seventeenth century Thematic exhibitions
4	Museums at Castello Sforzesco	Milan	Late 1800s	Public (municipality)	450,000	Museums of archaeological finds, wooden sculptures, ancient art, decorative arts, musical instruments. Renaissance sculptures and paintings
5	San Domenico and Palazzo Romagnoli Museums	Forlì	2005	Public (municipality)	20,000	Frescoes and paintings from the Middle Ages Archaeological museum (forthcoming opening)
6	Pontificial Museum Santa Casa	Loreto (AN)	Late 1800s	Public (government)	20,000	Paintings and tapestries from the fifteenth to the nineteenth century
7	Museums at Castello Scaligero	Malcesine (VR)	1960	Public (municipality)	150,000	Natural history museum and museum of the Venetian galleys
8	Balì Museum	Colli al Metauro (PU)	2004	Private	50,000	Science museum

digitalization processes, and to the methods in which digital skills are developed or accessed.

Evidence gathered allow us to identify some common traits: among these a sufficient awareness of the potential of digital technologies in the development of the attraction capacity of the museums themselves, and the common recognition of the scarcity of financial means to support the digitalization processes. Moreover, the fear that technologies allowing remote access to collections are at the expense of the decision to visit the museum is resized from the analysis.

Below, we illustrate the results of the analysis following the outline of the interviews conducted. We will first analyze the strategic behavior of museums, then the choices of introducing digital technologies, and, finally, the substance of the digital skills that support the processes in place.

The market strategies of the museums being analyzed are mainly oriented toward increasing the number of visitors (audience broadening). This trend is present in the different types of museums, but especially in the state museums, evidently pressured by the need to reach targets concerning the number of visitors. The strategy of expanding the public is absent in the museums that have already a high number of visitors and who believe that the time has come to deepen the relation with the existing public or to approach segments of the population that are traditionally not inclined to visit museums.

The strategy is rarely clear and unambiguous. In most cases, the will to pursue multiple strategic goals is declared. The search for audience broadening is combined with other strategies, aimed now at audience deepening, now at audience diversifying. Finally, in a couple of cases, attempts of audience deepening and diversifying are combined.

Especially the strategies aimed at audience deepening or diversifying may offer more precise indications about the digitalization path to go. For example, if audience deepening can be fostered through enhanced interactivity and technological convergence, as suggested by Bakhshy and Throsby (2012: 209) or by expanding the visitor's sense of immediacy and participation, as proposed by Kotler and Kotler (2000: 278), museums can pursue an audience diversifying strategy, which implies identifying, segmenting, and targeting the public, through the use of different digital technologies according to the different audiences. Thus, tools and devices for "gamification" experiences could be introduced for the "school students" segment, while other applications or devices, for example, AR/VR, could favor an entertainment experience for the "tourists" segment; finally, technologies closer to those in use to supporting professional work could favor a more in-depth use by the "passionates and scholars" segment. Notwithstanding this, the awareness of the need to pursuing a careful alignment between strategies and technologies rarely emerges from the interviews, but in the case of La Venaria Reale:

> Digital technology must be a tool to support an idea, a goal, it must not be an end in itself (...); the leap in quality is made by the contents, not the technology itself (...); when I have a strong content, I have a nice idea that goes well with the general experience that the public is doing in my museum and that is visible through an innovative digital technology, then the combination is perfect. (Interview with Matteo Fagiano, Social Media Manager at La Venaria Reale)

The analyzed museums highlight different combinations of remote and on-site fruition technologies; however, if we also consider the digital devices to be installed in the short term, apart from a couple of extreme situations, differences appear to be fairly limited, in relation to the high variety of the sample. At the moment, we can say that none of the analyzed museums have reached the "digital transformation" stage; however, as at least in one case (Museums at Castello Sforzesco), this goal seems to be really at hand.

As regards remote access technologies, the use of websites and social networks is almost generalized, even if their contents vary considerably; in particular, the choice of making a virtual tour possible is an exception (Museums at Castello Sforzesco) and is openly opposed in other cases. The construction of digital archives is also quite widespread, but it follows different approaches and logics from case to case.

A particularly meaningful experience is that from the Museums at Castello Sforzesco, in Milan. Here a digital archive allows on-site and online visitors to "discover" works not on exhibition. This represents, among evidence we collected, an opportunity that would not be possible without digital technology:

> An important project is linked to the digitization of works that are usually not exhibited (...), such as graphic and photographic works, prints and drawings. Two portals have been created (...) that allow you to search and consult all the heritage that is preserved in the archives. (Interview with Fiorella Mattio, Museum Curator at Castello Sforzesco)

As regards on-site fruition technologies, touchscreen applications are the most widespread, while there is a certain hesitancy with the adoption of VR/AR instruments that is presumably due to the challenging management of these tools. On-site fruition technologies is the category in which a more pronounced dynamic is expected in the coming years; however, this dynamic appears to be conditioned by the future availability of resources.

As for the factors that have influenced the decision to adopt digital technologies, it is quite evident that the example of other museums has played an important role, together with the proposals formulated by IT consultants. This seems true especially for on-site fruition technologies. These two modes often occur, regardless of the proprietary nature and the size of the museums that have been studied. However, the conditioning of choices made by other museums and of proposals coming from external consultants seems greater where an intentional and conscious approach to the market of potential users is missing or weak or when market strategies carried out by the museums do not demand for a specific set of digital technologies and devices.

In other words, the evidence from our investigation suggests that the combination of not always clear and unequivocal strategies and of institutional pressures, the relevance of which has already been highlighted in the literature (Rasmussen 2019), is at the origin of a framework in which mimetic isomorphism represents the main driver for the digitalization process in Italian museums. In our opinion, in the medium-long term this risks unduly reducing management's spaces of action, especially when a strategy of differentiation of the cultural offer is pursued, in line with the tourism development choices of the territory.

3.3 *Developing Digital Skills: When Organization Choices Make the Difference*

With reference to digital skills, the survey has allowed us to gather information regarding both the competences currently held and those that are expected to be developed in the coming years. At the moment, none of the sample museums can be said to be completely self-sufficient in terms of digital skills. However, the differences from one museum to another are considerable, with regard not only to the stock of skills currently held but also the intention to develop them in the coming years. Indeed, this may prefigure different paths of development by the museums.

At one extreme, we find museums that, while having at their disposal only some applied skills concentrated in a few roles, do not feel the need to rebalance a situation of marked dependence on the outside (universities, software companies, digital consultancy, etc.). At the other extreme, there are museums that, despite being already in possession of some technical skills, other than the applied ones, show a clear orientation to further develop them and to make them widespread among all the roles. This in order to reduce their dependence on the outside, which does not give sufficient guarantees regarding the development of the systems and their maintenance. Research evidence shows how this approach is followed in particular by the civic museums in Forlì and by the Museo del Balì:

> We try to develop digital skills internally; (. . .) the limit of relations with specialized external companies is linked to the loss of digital skills when the contractual relationship comes to end. (Interview to Deanna Conficconi, Manager at San Domenico & Palazzo Romagnoli Museums)

> The objective is to develop resources internally (digital skills); it is preferable to involve employees in a training course rather than relying on an external consultant, because the resources thus developed remain available to the museum. (Interview to Francesca Cavallotti, Scientific director, Balì Museum)

In the cases just mentioned, against the common goal of building a stock of digital skills in the long term, the internal development of those skills does not provide for any formalized plan nor the identification of an ad hoc role that guides and address such process. These choices are primarily motivated by the small size of the museum and the limited amount of financial resources. However, the idea of internal development is declined differently in the two cases. For the San Domenico Museum, it primarily means to rely on belonging to the network of civic museums of Forlì and, recently, also to a wider regional circuit (Coordination of the Houses-Museums of Romagna), in a logic of access to skills that are developed at network level. In the Balì Museum, however, developing resources internally means first of all leveraging the personal attitudes and inclinations of the employees, so encouraging and orienting an updating process of digital skills. In addition to specific training courses that are provided in correspondence with the adoption of particular systems or devices, a continuous learning process can be observed, carried out in self-study mode (in particular with the help of webinars) but in some way connected to the digitalization path of the museum.

Even the simple availability of widespread applied skills seems to constitute a good result, because it is believed that an adequate level of digital literacy can favor the development of ideas and proposals to make the digital equipment of the museum more in line with the current and potential audience, as especially evidenced in the case of La Venaria Reale, in Turin. Evidence from this case suggests that, even where a role specifically dedicated to digitalization process is lacking in the organization (e.g., digital manager), a further driver of digitalization may come from the employees themselves:

> Overall, the employees have gained the necessary skills not only to make the best use of the resources currently in use, but also to suggest some possible development hypotheses to be presented to those who will then make the decisions. (...) Proposals (for the introduction of new technologies) mature at various levels internally, by those directly involved in the process of providing the services; everyone can have a proposal linked to digital innovation; (...) in a widespread way it is possible to identify elements of innovation. (Interview to Matteo Fagiano, Social Media Manager at La Venaria Reale)

As mentioned above, findings of our research suggest that the choices for the introduction of digital technologies in use in Italian museums are often the result of mimetic isomorphism processes. However, institutional pressures cannot be regarded as the only driver of the digitalization processes. Indeed, the case of La Venaria Reale demonstrates how the choices concerning work organization and operators' skills not only improve the quality of fruition, thanks to a careful running of current digital devices, but also favor the development of a widespread innovative capacity, able to guide the subsequent steps of the digitalization process. Looking ahead, even the museums that are clearly oriented to develop both technical and applied skills within them, and in a widespread way (primarily the Civic Museums of Forlì and the Balì Museum), could replicate such virtuous processes, thanks to an approach that enhance diffuse abilities of evaluating and choosing digital infrastructures and devices.

The analysis suggests that the pursue of a widespread possession of applied skills, and sometimes also technical ones, among the employees, is intentional, rather than casual. This may be read as an attempt to foster an emergent digitalization strategy that is built on the capacity of employees to assess the preferences of the audience, so supporting managers in pursuing a better alignment between technologies in use on the one hand and audience needs and expectations on the other. It seems that the proprietary nature of the museum can be a relevant factor for the possibility of developing a wide range of digital skills internally. From this point of view, private museums and civic museums that are part of an integrated circuit seem to be favored: in fact, for the latter, there is the possibility of developing at central level (e.g., the municipality) some technical skills, without the need to develop them internally in each museum.

Definitely, it seems strongly advisable to focus not only, or not so much, on the "maximization" of the stock of skills, but on a wide diffusion of these within the organization. In this way, the digitalization process of museums could take advantage of more and more frequent inputs and benefit from a better tuning on audience expectations. Table 3 presents the results of the on-the-field analysis in a synoptic form.

Table 3 Market strategy, technologies in use, and digital skills

	Museum	Market strategy	Digital technologies		Digital skills	
			Off-site	On-site	Current	To be developed
1	Palazzo Ducale (Urbino)	Audience broadening Audience deepening	Website Social networks Digital archives (*under construction*)	Virtual reality, augmented reality, eye tracking technology (*to be installed*)	Internal, concentrated (applied) External (technical)	No change
2	Palazzo Ducale (Mantua)	Audience broadening	Website Social networks	Touchscreen station (*out of order*) Digital info-point	Internal, concentrated (applied) External (technical)	Development of internal skills (technical)
3	La Venaria Reale (Turin)	Audience deepening	Website Social networks Digital archives Chatbots technology (*to be installed*)	Multimedia device	Internal, widespread (applied) External (technical)	Hiring of a digital manager
4	Castello Sforzesco (Milan)	Audience deepening Audience diversifying	Website Newsletter Social networks Digital archives Virtual tour	QR code Touchscreen stations Multimedia device Virtual reality	Internal, widespread (applied) Internal, concentrated, and external (technical)	Less use of external skills
5	San Domenico and Palazzo Romagnoli (Forlì)	Audience deepening Audience broadening	Website Newsletter Social networks PatER portal	Touchscreen stations Smartphone e tablet apps QR code (*to be installed*)	Internal, widespread (applied) Internal, concentrated (technical)	Deepening of internal skills (technical)
6	Pontificial Museum Santa Casa (Loreto, AN)	Audience broadening	Website Social networks (*to be installed*)	Touchscreen stations (*to be installed*)	External (applied and technical)	Deepening of internal skills (applied)

7	Castello Scaligero (Malcesine, VR)	Audience broadening Audience diversifying	Website (municipality) Digital archives	Touchscreen stations Multimedia device	Internal, widespread (applied) External (technical)	Internal development (applied)
8	Balì Museum (Colli al Meaturo, PU)	Audience deepening Audience diversifying	Website Newsletter Social networks Online use of astronomical observatory	QR code Multimedia device (*to be installed*)	Internal, widespread (applied) Internal, concentrated, and external (technical)	Deepening of internal skills (technical)

4 Emerging Approaches to Digitalization Within Museums

The findings of this study allows us to answer, albeit partially and provisionally, to the research question on which this chapter rests, namely, the role of digital skills, held or accessible by the cultural organizations, in fostering the organization's capability to align technological innovation and market strategy.

The analysis allows us to hypothesize the existence of four different approaches toward the development of digital skills in the Italian museums. The identification of these approaches takes into account not only the current consistency of digital skills but also and above all two other aspects: the degree of concentration/diffusion of skills within the museum and short-term development intentions. So, we can read the approaches described below primarily as managers' propensity to develop digital skills.

A first approach sees the museum completely dependent on the outside as regards digital skills and oriented toward acquiring only basic applied skills. The second type of approach characterizes museums that hold applied skills, covered by one or a few roles, and which are oriented toward the progressive diffusion of these skills. In the third approach, the museum is characterized by the widespread possession of applied skills and by an orientation to integrate these skills with the acquisition of a technical knowledge base. Finally, the fourth and more advanced approach characterizes museums which, in addition to having applied skills spread throughout the organization, have a basis of technical skills and are oriented toward deepening them, with the aim of minimizing dependence on the outside.

Each approach to the development of digital skills can be considered adequate in relation to the size of the museum, its current ability to attract the public, the available resources, and, last but not least, the stage of progress of the digitalization process. However, what we are interested in is to understand whether the identified approaches, from the basic to the more advanced one, allow the alignment between market strategies and digital technologies. In fact, we can think that each of them corresponds to an increasing level of awareness regarding the potential of the fruition's digital technologies. This in turn can favor the pursuit of consistency between the choice to focus on one or more market segments (types of audience) and the decisions to introduce some digital devices.

In this perspective, the basic approach to the development of digital skills can be justified if the museum adopts digital tools (website, social networks, etc.) with the main purpose of "telling" about the activities carried out in the museum itself, according to a "storytelling" approach, and where the strategy pursues a generic expansion of the number of visitors.

The second approach may prove suitable for those museums that integrate traditional methods of remote fruition (website, newsletter, etc.) with tools to facilitate the orientation of visitors within the museum (touchscreens, info-points, etc.), in a logic which can be defined as "navigating," in association with a strategy aimed at audience broadening.

The third approach seems consistent with a greater level of digitalization, characterized by the introduction of a mix of tools, although not particularly advanced, for both remote access and on-site use (digital archives, virtual tours, multimedia devices, etc.) in a "wikipedian" logic, as it is oriented toward a fruition that goes beyond the mere narration of the heritage. This approach also seems consistent with the museum's willingness to focus on deepening the relationship with the current audience ("audience deepening"), but also at diversifying it.

Finally, the more advanced approach is appropriate if the museum chooses to adopt, in addition to a wide range of remote access modes, some of the more sophisticated tools to support the on-site use of the collections (virtual reality, augmented reality, eye tracking, etc.) which allow visitors to improve the live experience and to access specific contents that integrate physical accessibility, with a prevailing emphasis, although not exclusive, on "entertainment." This approach seems capable of supporting concurrent strategies, e.g., oriented both at deepening the relationship with the current audience and at diversifying the audience itself.

The perspective adopted here assigns a key role to the current "endowment," as well as to the propensity to develop (internal) digital skills by museums. Especially in the absence of a close correlation between strategic and technological choices, we believe that the possession of adequate skills related to the evaluation and the management of innovative fruition technologies can consciously orient the digitalization processes in museums.

This perspective suggests how some organizational choices play a key role in promoting or inhibiting the achievement of an adequate alignment between market strategy and digital technologies in museums. Both the recruiting criteria and the choices about the organization of work can in fact direct the museum toward a situation in which the digital skills, few or many that are, are the prerogative of one or more specialist roles or, conversely, are widespread in the organization. Moreover, they can also stimulate or inhibit employees' ability to evaluate any misalignment between audience expectations and level of technology acceptance on the one hand and digital devices in use on the other.

5 Concluding Remarks

This chapter aimed at shedding light on the degree of alignment between market strategies and digitalization processes in the Italian museums housed in historical sites. The on-the-field analysis, namely, highlighted choices concerning the adoption of digital technologies and the development of digital skills. The variety of the analyzed sample allowed us to identify four different approaches to digitalization processes. Moreover, our research provided new insight in order to understand how held or accessible digital skills drive decision-making process regarding the deployment of technologies. A framework emerges in which mimetic isomorphism represents the main driver for the digitalization process, especially as regards the decision of introducing particular tools and devices for the on-site fruition. Yet, also some choices concerning work organization play a meaningful role, insofar as they favor the development of a widespread innovative capacity within the organization; in its turn, this capacity seems able to influence the subsequent steps of the digitalization process.

Research findings have some managerial and theoretical implications. As for the first ones, our research provides indications to the museums' managers that are currently engaged in the challenge of digitalization, in order to foster a better alignment between market strategies and digital technologies. Moreover, it helps managers to develop awareness of how some organizational choices, through the development of spread digital skills within the museum, can improve the oversight of the digitalization processes, so helping to mitigate their external dependence. As far as theoretical implications are concerned, this research contributes to strengthening a holistic approach to the study of digitalization processes in museums. In particular, it highlights that these processes can benefit from the analysis of organizational choices and of their impact on the development of skills that are able of supporting managers when implementing the processes themselves.

The fact that we have simply obtained a "snapshot" of the progress of digitalization processes, whereas the ongoing process should benefit for a prolonged observation, represents a limitation of on-the-field research. A number of observations over time could offer larger and more significant insights to the understanding of the digitalization processes in museums.

On the one hand, this leads us to consider the opportunity to continue the research according to a longitudinal approach, deepening the analysis of the same sample. On the other hand, a second line of research development can lead to testing the proposed approaches on a large-sized sample, using quantitative research tools. This, in order to understand if significant relationships can be identified between the choice of a specific approach to the digitalization process and some characteristics of museums.

References

Addis, M. (2005). New technologies and cultural consumption—Edutainment is born! *European Journal of Marketing, 39*(7–8), 729–736.

AICA & Assintel. (2017). *Osservatorio delle competenze digitali*. Retrieved August 18, 2020, from https://www.agid.gov.it/sites/default/files/repository_files/osservatorio_competenze_digitali_2017.pdf

Bakhshi, H., & Throsby, D. (2012). New technologies in cultural institutions: Theory, evidence and policy implications. *International Journal of Cultural Policy, 18*(2), 205–222.

Balloffet, P., Courvoisier, F. H., & Lagier, J. (2014). From museum to amusement park: The opportunities and risks of edutainment. *International Journal of Arts Management, 16*(2), 4–18.

Barnes, S. A., Kispeter, E., Eikhof, D. R., & Parry, R. (2018). *Mapping the museum digital skills ecosystem phase one report*. Leicester: University of Leicester.

Bertacchini, E., & Morando, F. (2013). The future of museums in the digital age: New models of access and use of digital collections. *International Journal of Arts Management, 15*(2), 60–72.

Bonacini, E. (2011). *Nuove tecnologie per la fruizione e valorizzazione del patrimonio culturale*. Roma: Aracne.

Cavriani, E. (2016). Everyone's collections at art museums: Groundbreaking digital business strategy as cornerstone for synergies. *Sinergie Italian Journal of Management, 99*. https://doi.org/10.7433/s99.2016.07.

Cerquetti, M. (2016). More is better! Current issues and challenges for museum audience development: A literature review. *Journal of Cultural Management and Policy, 6*(1), 30–43.

Cigola, M., Gallozzi, A., Ceccarelli, M., Carbone, G., De Stefano, C., & Di Freca, A. S. (2014). Strategie robotiche ed informatiche per la fruizione museale. *SCIRES-IT-SCIentific RESearch and Information Technology, 4*(1), 59–68.

Clini, P., Quattrini, R., Bonvini, P., Nespeca, R., Angeloni, R., Mammoli, R., Dragoni, A. F., Morbidoni, C., Sernani, P., Mengoni, M., Leopardi, A., Silvestrini, M., Gambelli, D., Cori, E., Gallegati, M., Tamberi, M., Fraticelli, F., Acciarri, M. C., & Mandolesi, S. (2020). Digit (al)isation in museums: Civitas project—AR, VR, multisensorial and multiuser experiences at the Urbino's Ducal Palace. In G. Guazzaroni & A. S. Pillai (Eds.), *Virtual and augmented reality in education, art, and museums* (pp. 194–228). Hershey: IGI Global.

Cori, E., & Fraticelli, F. (2018). *Digitizing cultural heritage: Evidence from Italian museums*. Paper presented at Recent Advances in Information Technology, Tourism, Economics, Management and Agriculture (ITEMA), Graz.

Dewdney, A. (2018). *What Is the current fascination with VR on the part of museums and art galleries?* Paper presented at Contemporary Art society Annual Conference. The Virtual in Museums: Hot Medium? London.

Dörner, K., & Edelman, D. (2015). *What 'digital' really means*. Retrieved August 5, 2020, from http://www.mckinsey.com/industries/technology-media-and-telecommunications/our-insights/what-digital-really-means

Eisenhardt, K. M. (1989). Building theories from case study research. *Academy of Management Review, 14*(4), 532–550.

Eisenhardt, K. M., & Graebner, M. E. (2007). Theory building from cases: Opportunities and challenges. *Academy of Management Journal, 50*(1), 25–32.

Evrard, Y., & Krebs, A. (2017). The authenticity of the museum experience in the digital age: The case of the Louvre. *Journal of Cultural Economics, 42*(3), 353–363.

Ferrari, A. (2012). *Digital competence in practice: An analysis of frameworks*. JRC technical reports. Luxembourg: European Commission.

Følstad, A., & Kvale, K. (2018). Customer journeys: A systematic literature review. *Journal of Service Theory and Practice, 28*(2), 196–227.

Geismar, H. (2013). Defining the digital. *Museum Anthropology Review, 7*(1–2), 254–263.

Ghobakhloo, M. (2018). The future of manufacturing industry: A strategic roadmap toward industry 4.0. *Journal of Manufacturing Technology Management, 29*(6), 910–936.

Giannini, T., & Bowen, J. (2018). *Of museums and digital culture: A landscape view*. Paper presented at EVA, London.

Gobble, M. M. (2018). Digital strategy and digital transformation. *Research Technology Management, 61*(5), 66–71.

Gombault, A., Allal-Chérif, O., & Décamps, A. (2016). ICT adoption in heritage organizations: Crossing the chasm. *Journal of Business Research, 69*(11), 5135–5140.

Hatlevik, O. E., Ottestad, G., & Throndsen, I. (2015). Predictors of digital competence in 7th grade: A multilevel analysis. *Journal of Computer Assisted Learning, 31*(3), 220.

Herbert, L. (2017). *Digital transformation: Build your organization's future for the innovation age*. London: Bloomsbury.

Homem, P. M., Remelgado, P., & Medina, S. (2017). *Museum professionals in the digital era: Agents of change and innovation*. Retrieved August 5, 2020, from http://www.project-musa.eu/wp-content/uploads/2017/03/MuSA-Museum-professionals-in-the-digital-era-full-version.pdf

Kane, G. C., Palmer, D., Phillips, A. N., Kiron, D., & Buckley, N. (2015). Strategy, not technology, drives digital transformation. *MIT Sloan Management Review and Deloitte University Press, 14*, 1–25.
Kirchberg, V., & Kuchar, R. (2014). States of comparability: A meta-study of representative population surveys and studies on cultural consumption. *Poetics, 43*, 172–191.
Komulainen, H., & Saraniemi, S. (2019). Customer centricity in mobile banking: A customer experience perspective. *International Journal of Bank Marketing, 37*(5), 1082–1102.
Kotler, N., & Kotler, P. (2000). Can museums be all things to all people? Missions, goals, and marketing's role. *Museum Management and Curatorship, 18*(3), 271–287.
Kotler, N. G., Kotler, P., & Kotler, W. I. (2008). *Museum marketing and strategy: Designing missions, building audiences, generating revenue and resources*. Hoboken, NJ: Wiley.
Li, Y. C., Liew, A. W. C., & Su, W. P. (2012). The digital museum: Challenges and solution. In R. Sebrle & S. W. H. Changmei (Eds.), *ICIOT 2012 8th International Conference on Information Science and Digital Content Technology* (pp. 646–649). Sydney: Springer.
Lin, D., Lee, C., Lau, H., & Yang, Y. (2018). Strategic response to Industry 4.0: An empirical investigation on the Chinese automotive industry. *Industrial Management and Data Systems, 118*(3), 589–605.
Mayan, M. J. (2016). *Essentials of qualitative inquiry*. London: Routledge.
McCarthy, K., & Jinnett, K. (2001). *A new framework for building participation in the arts*. Santa Monica, CA: Rand Corporation.
Negri, M. (2012). The virtual museum, a shift in meaning. In A. Nicholls, M. Pereira, & M. Sani (Eds.), *The learning museum network project—The Kenneth Hudson seminar 5. Report 1*. Bologna: Istituto dei Beni Culturali.
Pescarin, S. (2014). Museums and virtual museums in Europe: Reaching expectations. *SCIRES-IT-SCIentific RESearch and Information Technology, 4*(1), 131–140.
Preuss, U. (2016). Sustainable digitalization of cultural heritage—Report on initiatives and projects in Brandenburg, Germany. *Sustainability, 8*(9), 891.
Rasmussen, C. H. (2019). Is digitalization the only driver of convergence? Theorizing relations between libraries, archives, and museums. *Journal of Documentation, 75*(6), 1258–1273.
Styliani, S., Fotis, L., Kostas, K., & Petros, P. (2009). Virtual museums, a survey and some issues for consideration. *Journal of Cultural Heritage, 10*(4), 520–528.
Vom Lehn, D., & Heath, C. (2005). Accounting for new technology in museum exhibitions. *International Journal of Arts Management, 7*(6), 11–21.
Yin, R. K. (2009). *Case study research. Design and methods* (2nd ed.). Thousand Oaks, CA: Sage.

Enrico Cori, PhD in Business Administration at the University of Pisa, is Full Professor of Business Organization and Organization in Public and Non-profit Sectors at the Department of Management of the Polytechnic University of Marche.

His research interests mainly concern the organizational control processes, the organization of family SMEs, and the organizational dynamics in the cultural heritage.

Since 2017 he has coordinated the Business Administration Curriculum of the PhD in Management and Law of the Polytechnic University of Marche.

Since 2018 he has been a member of the CIVITAS interdepartmental project (ChaIn for excellence of reflectiVe societies to exploit dIgital culTural heritAge and museums).

Fabio Fraticelli, PhD in Business Administration at the University of Pisa, is Chief Operating Officer at TechSoup Italy and Adjunct Professor of Organization in Public and Non-profit Sectors at the Department of Management of the Polytechnic University of Marche.

His research interests mainly concern digital transformation in non-profit organizations, as well as agile project management and startups.

Since 2018 he has been a member of the CIVITAS interdepartmental project (ChaIn for excellence of reflectiVe societies to exploit dIgital culTural heritAge and museums).

Exploring the Financial Strategies of Private Museums: The Case of the *Opera di Santa Maria del Fiore* in Florence

Alberto Romolini, Silvia Fissi, Elena Gori, and Marco Contri

Abstract Many studies pointed out that museums are usually unable to support their activities independently as self-generated revenue is not enough to fulfil their institutional tasks. Literature has so far devoted limited attention to financing ways of private museums. However, private museums represent an interesting context of analysis as they could be more capable to self-finance their activities as well as to diversify income sources.

This chapter aims to analyse in-depth the financial structure of a private museum, by exploring the case of the Opera di Santa Maria del Fiore in Florence with a semi-structured interview and secondary sources.

The results show that culture is not always a sector that "lives" only thanks to public funding. Indeed, the Opera shows a general increase in the revenue from tickets and the pursuit of financial equilibrium. The main challenge of this institution seems to be the identification of new "mindful" users and new resources for the building maintenance and restoration.

At the academic level, this study addresses the call for more in-depth research on financing ways of private museums, thereby filling the literature gap. Furthermore, as concerns practical implications, this study offers a "food for thought" for museum managers, which plays a pivotal role in facilitating private investments and donations to cultural institutions.

A. Romolini (✉)
Facoltà di Economia, Università Telematica Internazionale Uninettuno, Rome, Italy
e-mail: a.romolini@uninettunouniversity.net

S. Fissi · E. Gori
Dipartimento di Scienze per l'Economia e l'Impresa, Università degli Studi di Firenze, Florence, Italy
e-mail: silvia.fissi@unifi.it; elena.gori@unifi.it

M. Contri
Dipartimento di Economia e Management, Università degli Studi di Pisa, Pisa, Italy
e-mail: marco.contri@phd.unipi.it

P. Demartini et al. (eds.), *Cultural Initiatives for Sustainable Development*, Contributions to Management Science,
https://doi.org/10.1007/978-3-030-65687-4_3

Keywords Museum funding · Financial strategies · Financial sustainability · Private museums

1 Introduction

Museums are "non-profit, permanent institution[s] in the service of society and its development" (ICOM 2017, p. 3), committed to creating long-term value by adopting a multi-stakeholder approach (Burton and Scott 2003). Indeed, so that museums achieve their mission and ensure their survival, they have to handle relations with several key actors such as visitors, museum staff, competitors, donors and government (Bertacchini et al. 2018; Camarero and Garrido 2012); in this way, museums create socio-cultural and economic benefits for local communities (Stylianou-Lambert et al. 2014).

To accomplish their institutional tasks, museums need to have permanent income sources, including *self-generated revenue*, *public grants* and *private contributions* (namely, sponsorships and donations) (Fedeli and Santoni 2006). All of these resources define the income side of a museum's financial structure; such a structure depends on the governance of museum itself, that is, the systems and structures that define policies and long-term strategies, provide leadership and management and coordinate resources and procedures (Camarero et al. 2019). Indeed, the governance influences the legal form (public or private) of a museum, and, in turn, the legal form affects both sources of funding—linked to a greater or lesser extent to public subsidies—and the organisation of the museum itself (Vicente et al. 2012). In this regard, Camarero et al. (2019) identified three modes of museum governance:

1. *Public museums directly run by central or local government*
2. *Autonomous public museums*
3. *Private museums*

The first category refers to museums directly managed and almost entirely funded by a central or local public authority. These museums are integrated as a part of the public entity itself; accordingly, they have no independent legal status, and all decisions other than cultural ones are taken at a central or local level. Public subsidies cover their budget, and museums have no budgetary independence: this means that any additional revenue (from admission fees, donations, etc.) is part of government revenue and thus not freely usable for the benefit of the museum. As a result, directors have low incentives to engage in a managerial style and are unlikely to seek further income from sponsors and donors.

Autonomous public museums remain under the ownership and control of the public authorities but have managerial and financial autonomy. In recent years, in the vein of the reforms undertaken by public administrations over the last decades (Hood 1995; Osborne 2006), many public museums (especially in continental Europe) have moved from being governmental departments to being more independent public bodies under different institutional forms, such as foundation and consortium

(Vicente et al. 2012). This kind of museum still receives public support (usually in the form of annual grants) that, however, no longer covers all institution costs. Moreover, these institutions can retain the additional resources they independently earn. Consequently, museum managers might be more prone to developing more consumer-oriented programmes and seeking financial supporters.

Finally, private museums are privately owned institutions independent from public control. Public funding is typically lacking, and hence the financial sustainability depends on self-generated revenue and private contributions.[1] Accordingly, private museums are more likely to adopt a more business-like approach to attract a greater number of visitors as well as develop loyal relationships with sponsors and donors (Camarero et al. 2011).

Over the last few decades, museums have been facing substantial changes in their political, cultural and social environment (Proteau 2018). In particular, the financial crisis has led to a reduction of self-generated revenue and a substantial downturn in public and private contributions for cultural institutions (Bonet and Donato 2011). All this has meant that museums, whether public or public, have been challenged to redefine their finances (Lindqvist 2012; Sargeant and Jay 2014).

Even though the financial dimension represents a crucial issue for museums (Kotler et al. 2008), only a few studies have analysed in-depth how these institutions finance their activities (Lindqvist 2012). Furthermore, among these studies, most papers focus on public museums (e.g. Romolini et al. 2020), while private realities remain still underexplored (e.g. Yermack 2017). Yet, investigating financial strategies adopted by private institutions might represent an interesting topic; indeed, as noted earlier, they are expected to be more engaged in a market-orientated management style (Bertacchini et al. 2018; Cole 2008), and therefore they could be more prone to and capable of planning and implementing managerial (Camarero and Garrido 2012) and financial (Lindqvist 2012) strategies.

In this vein, the purpose of this chapter is to fill the literature gap by investigating the financial strategies of an Italian private museum, specifically the Opera di Santa Maria del Fiore (hereinafter, also simply "Opera") in Florence, which can be considered as a best practice for its financial self-sufficiency.

The remainder of the chapter is structured as follows. Section 2 reviews the relevant literature on museum finances, while Sect. 3 describes the research methodology adopted in this study. Then, Sect. 4 shows and discusses the results of the case analysis, and, finally, Sect. 4.3 explains the main conclusions, outlining the limitations and providing suggestions for future research.

[1]Nevertheless, the government can still support the museum's activity either directly, through occasional subsidies, or indirectly, through tax incentives for sponsorships and donations.

2 Literature Review

As stated earlier, a museum's financial structure refers to the mix of income—which includes *self-generated revenue*, *public grants* and *private contributions*—through which the organisation supports its activities and accomplishes its institutional tasks (Fedeli and Santoni 2006). Such a structure plays a pivotal role in the definition and implementation of a museum's mission and strategies (Kotler et al. 2008), in this way representing a crucial issue for this kind of institution (Proteau 2018).

Self-generated revenue comprises income earned by admission fees or generated by commercial activities, such as offering products and services for which a price is charged (museum shop and café sales, space rentals for special events, etc.). Indeed, museums, despite being non-profit organisations wherein social goals (conservation, custody, education, etc.) prevail, may also pursue commercial activities, in the sense that they may offer visitors an alternative leisure activity aimed at increasing visitor numbers and hence earning additional revenue (Camarero and Garrido 2012). Obviously, this kind of income is closely related to the museum's capacity to offer attractive services (Vicente et al. 2012). From a broader perspective, Camarero et al. (2019) highlighted the need for cultural entities to develop strategies both to retain current visitors, creating a sense of belongingness and identification with the institution, and to attract new visitors, offering new and experiential activities. However, many studies pointed out that museums are usually unable to support themselves independently (Bertacchini et al. 2018) as self-generated revenue is not enough to satisfy visitors' cultural and social desires as well as the economic needs expressed by staff (Camarero et al. 2019). The structural deficit of museums stems from the fact that these institutions have substantial fixed costs (in particular, related to conservation, restoration and staff) (Frey and Meier 2006) and usually cannot charge them directly to visitors (Lindqvist 2012).

Public contributions comprise subsidies and grants allocated by a public authority. The particular nature of museums makes it usually essential for them to resort to public support, which can be justified by several arguments (Baumol and Bowen 1966; Frey and Meier 2006). Among others, museums generate positive externalities in their production and consumption (Fernández-Blanco and Prieto-Rodríguez 2020); this means that they create value or benefits to society as a whole (i.e. also to individuals and firms not involved in the production processes and to whose not paying for receiving such benefits), for which they are not compensated in monetary terms (Camarero et al. 2019). Furthermore, heritage goods may be seen as "merit goods" (Musgrave 1959), in the sense that they are assessed—by definition—as being intrinsically valuable and hence worthy of public support (Frey 2020). Finally, government support may be justified as a means of ensuring free and equal access for everybody (Camarero et al. 2011).

As for private contributions, the recent financial crisis has emphasised that building and maintaining long-term relationships with sponsors and donors is essential to the survival of many non-profit organisations like museums (Camarero et al. 2019). Therefore, museums should implement fundraising strategies (Romolini et al. 2020; Sargeant and Jay 2014), which can be regarded as those initiatives aimed at raising additional money from companies, philanthropic foundations and individual donors for specific projects or for general funds (Woodward 2012). In this sense, the increasing relevance of adopting marketing strategies devoted to making the institution more people-oriented (Cole 2008) and thus more "appealing" to potential sponsors and donors (Siano et al. 2010) is widely acknowledged. In this regard, Blasco López et al. (2019) found that over the last years, many museums have progressively introduced marketing strategies to improve visitor engagement.

As already pointed out in the introduction, only a few studies so far have deeply investigated financial strategies adopted by museums to support their activities. Research on this topic focused mostly on public museums and confirmed that, despite a considerable decrease in recent years (Bonet and Donato 2011), public funding remains high for this kind of institution (Romolini et al. 2020) that hence has fewer incentives to introduce real financial strategies (Vicente et al. 2012). Conversely, private museums, relying more on self-generated revenue and private contributions, have greater incentives to plan effective financial strategies, preferably in a long-term perspective as suggested by Lindqvist (2012); in this way, these institutions seek to broaden their income base as well as raise additional contributions by sponsors, donors and friends' associations, thereby ensuring the financial sustainability of the organisation itself (Bertacchini et al. 2018). For this reason, it is important to understand more about the financial strategy of private museums.

Based on this, the present paper aims to contribute to filling the literature gap by exploring the financial strategies of the Opera di Santa Maria del Fiore, a prestigious private museum located in Florence.

3 Methodology

Considering the novelty of the topic, this study has an exploratory nature. Accordingly, we adopted a qualitative approach that is particularly suitable when little is known about a certain phenomenon (Lune and Berg 2017). Specifically, we used the case study method, which enables researchers to obtain in-depth and comprehensive information about the phenomenon in its real context (Yin 2018).

In particular, this research puts forth the case of Opera di Santa Maria del Fiore in Florence (hereinafter, also simply "Opera"), which may be considered an interesting case for exploring the financing system of a private museum. The rationale for our choice is twofold. First, the Opera is one of the Italian Fabbricerie, which are ancient institutions funded in Italy in the second half of the fifth century and responsible for the administration of the asset of some particularly important churches (especially, cathedrals) (Caron 1967; Greco 2005, pp. 4–5). As the Fabbricerie are institutions born in Italy, this is without a doubt the best context to study this phenomenon. Second, the cathedral of Santa Maria del Fiore, which is included in the cultural complex of Opera, is one of the most famous and visited Italian cathedrals, located in an international touristic destination like the city of Florence.

To collect data, we opted for a mixed methodology that is built on interviews and analysis of secondary sources (i.e. the official website and annual reports of the Opera). Indeed, the mixed method is widely used in the field of accounting studies, as it allows to obtain a more comprehensive picture of the phenomenon under study and strengthen the findings' validity (Creswell and Clark 2017).

As concerns the interview design process, we followed the suggestions provided by Qu and Dumay (2011) and then we carefully considered (1) who to interview, (2) what type of interview to conduct and (3) how to analyse the interview data. Thus, we decided to perform two online face-to-face semi-structured interviews with the key person of the Opera, that is, the President. The interview protocol was developed and discussed among the authors; consistently with the exploratory nature of this research, most of the interview questions were "how" and "why" questions and were structured accordingly from broad to specific issues. Such interviews were conducted by two of the authors in the Italian language between March and April 2020 and lasted an average of 45 min. President's answers were recorded on a digital device, then transcribed onto paper and finally submitted to a content analysis in order to explore the key elements of the Opera financing model, with a particular focus on (1) the traditional and innovative financial strategies implemented in recent years and (2) those planned for the next years, as well as (3) the Opera's organisational structure.

Additionally, secondary sources (especially, balance sheets and income statements) were used to supplement the information obtained through the interviews and hence to increase the validity of the results through the triangulation of data collection (Lune and Berg 2017).

4 Results and Discussion

The "Fabbricerie" were founded in Italy in the period 468–496 when the ecclesiastical patrimony administered by the bishop was split into four parts by juridical division during the Pope Simplicus and Pope Gelasius pontificates. One of these parts, named "Fabbriceria", had to manage the church maintenance (Caron 1967; Greco 2005, pp. 4–5).

Afterwards, "Fabbricerie" sprang up in different areas of Italy and assumed different names: "Fabbricerie" in the north, "Opere" in Tuscany and Umbria and "Cappelle" in the Neapolitan and "Maramme" in Sicily (Greco 2005, pp. 6–11; Rivella 2005, p. 5). Bellucci et al. (2020) also specify that the Fabbricerie are not only a single Italian phenomenon, but they are also common at the European level.

According to law, in Italy the Fabbricerie are private-law entities regulated by the L. 848/1929 (Concordat Laws between the Italian State and the Catholic Church, Lateran Treaty), and in 1984 the law was revised. Moreover, according to the most recent indications of the Council of State, they are able to acquire the qualification of non-profit organisation (ONLUS).

From an accounting point of view, according to the Lateran Treaty in 1929 and Italian laws (D.P.R. 33/1987 and 227/1999), they must draw up a budget for the next year by 30 November and a financial statement by 31 March. The Board of Directors approves both documents. For the Fabbricerie that have also the qualification of a non-profit organisation, in addition to the use of the acronym ONLUS, there are further legal dispositions: the adoption of accrual accounting and the prohibition of distributing profits.

Currently, in Italy there are 24 Fabbricerie, and 50% of them are located in Tuscany (Table 1).

Table 1 The Fabbricerie in Italy

Name	Province
Fabbriceria "Opere Riunite del Duomo e della Chiesa Monumentale di S. Maria Nuova"	Arezzo
Fabbriceria "Opera di Santa Croce"	Florence
Fabbriceria della Chiesa Monumentale di S. Maria all'Impruneta	Florence
Fabbriceria di Santa Maria del Fiore—Opera del Duomo di Firenze	Florence
Fabbriceria "Opera di S. Maria delle Grazie di Pietracupa"	Florence
Opera Mediceo Laurenziana	Florence
Fabbriceria della Parrocchia di Santa Maria Assunta di Carignano	Genoa
Veneranda Fabbrica del Duomo di Milano	Milan
Cappella del Tesoro di San Gennaro	Naples
Fabbriceria "Maramma di Monreale"	Palermo
Fabbriceria "Maramma di Palermo"	Palermo
Veneranda Arca di Sant'Antonio in Padova	Padua
Fabbriceria Sagrestia della Concattedrale di Todi	Perugia
Opera della Primaziale Pisana	Pisa
Fabbriceria "Opera del Duomo di Prato della Chiesa Cattedrale Monumentale di Santo Stefano"	Prato
Fabbriceria della Basilica Cattedrale di Parma	Parma
Fabbriceria della Chiesa Cattedrale Monumentale di Santo Stefano Martire	Pavia
Fabbriceria Opera Laicale della Cattedrale di Chiusi	Siena
Fabbriceria della Chiesa Cattedrale di Pienza	Siena
Fabbriceria del Duomo di Siena denominata Opera della Metropolitana	Siena
Fabbriceria "Opere Ecclesiastiche di Montepulciano"	Siena
Fabbriceria "Opera di S. Maria della Stella" ossia "Opera del Duomo di Orvieto"	Terni
Fabbriceria del Duomo di Venzone	Udine
Procuratoria di San Marco	Venice

Source: Adapted from Bellucci et al. (2020)

4.1 The Opera di Santa Maria del Fiore in Florence

The Opera di Santa Maria del Fiore is a lay institution founded on 8 September 1296 by the Florentine Republic to supervise the construction of the new cathedral and its bell tower. After the completion of the Dome of Brunelleschi and the consecration of the Cathedral of Santa Maria del Fiore by Pope Eugene IV on 25 March 1436, the main task of the Opera became maintaining and embellishing the monumental complex. The Baptistery of San Giovanni (in 1777) and the Museum of the Opera di Santa Maria del Fiore (in 1891) joined the monumental complex. The museum was founded to show the works of art that, over the centuries, had been removed from the Cathedral and the Baptistery.

Nowadays, the Opera follows, as a "Fabbriceria", the Concordat Laws of 1929 and 1984, and a Board of Directors, composed of seven members, manages it. The President is the legal representative, and he is elected among the Board's members.

Since 1998, the Opera has been a non-profit organisation (ONLUS), regulated by its own statute. Its institutional aims, as well as any Fabbriceria, are the following (Opera di Santa Maria del Fiore 2018, p. 4):

1. Maintenance and restoration of the church
2. Asset management
3. Support for the expenses of furnishings, liturgical objects, ornaments and installations necessary for the church and the sacristy

The Administrative Secretary runs the general direction of the Opera.

From an organisational point of view, the Opera has undergone in the last 10 years important transformations that resulted in a dimensional growth consequent to the increase of visitors. This growth is part of the increase in tourist flows in the city of Florence (Irpet 2019, pp. 8–9) and of the policy of pedestrianisation of the old town which started at the end of 2009. The increase in tourists lead to a growth from about 8000 visitors a year to the current 1.2 million for the museum with a significant impact on the Opera's cash liability:

> Actually, there was no museum as it counted few visitors a year. The management was very simple and familiar, and it led to accumulating a lot of liquidity that we decided to invest in improving the museum and in making us better know. (The President)

In particular, the available resources allowed the management to invest in the expansion and in the renovation of the museum. These works ended in 2015, and they were carried out by a pool of Florentine architects, including Magni, Guerrini and Natalini.

The organisational change also engaged the staff that increased significantly in number with new recruiting staff and with the creation of new organisational units. In particular, the marketing area includes statistical and market analysis activities and the drafting of the marketing plan. The Board of Directors also launched research activities, for example, the two projects in outsourcing handled by an external specialised company at the end of 2016, in order to better understand the point of view of visitors (Opera di Santa Maria del Fiore 2016, p. 29).

Table 2 The number of tickets according to the type

Year	Full	Reduced	Florence card	School	Tour	Free of charge	Total	Total variation (%)
2016	1,132,285	53,419	93,380	31,523	17,759	45,574	1,373,940	–
2017	1,110,865	51,652	98,515	61,420	20,339	45,446	1,388,237	1.04
2018	1,022,016	51,239	107,365	17,234	35,103	34,833	1,267,790	−8.68

The commercial area was set up for advertising and selling the entrance fees. The general management implemented new products, in addition to the cumulative ticket (including a visit to Museo dell'Opera di Santa Maria del Fiore—Cathedral Museum, Brunelleschi's Cupola—Dome, Giotto's Bell Tower, Baptistery of San Giovanni and Archaeological site of Santa Reparata, while the access to the Cathedral is free and not bookable), introduced in 2013, to engage direct sales relationships with the customer. The five standard product types are cumulative and reduced ticket, Firenze Card, school tickets, tour tickets and free tickets (Opera di Santa Maria del Fiore 2017, p. 53). Table 2 indicates the number of tickets sold for each type for the 3-year period 2016–2018.

The Opera established stable commercial relationships with the city's tour operators and with Italian and foreign universities and schools. Furthermore, it signed with agreement with the Florentine tourist guides who were engaged to explain the Opera museum.

When in 2017 the new Board of Directors took charge, they appointed a General Manager supported by an external advisor in order to simplify and to improve the efficiency of the organisational structure. Nowadays, there are six area chiefs and only one manager for the technical area, responsible for maintenance and restoration:

> The current organisational structure includes the General Manager and the Manager of the "Technical" area, and two Director's staff offices: "Legal Affairs" and "Information Technology Systems". Moreover, there are the areas of "Personnel", "Security", "Events and Communication" (from an institutional and commercial profile linked to the sale of tickets) and "Commercial" (sale and organisation of free activities). (The President)

4.2 *The Fundraising Strategy*

The Opera's fundraising strategy is based on self-generated revenue which, for some years, has allowed the Fabbriceria to be in balance but, often, to generate profits as shown in Table 3. The revenue from the core activities (institutional management) increases each year with the big contribution coming from tickets. Considering the level of revenue, the management has increased the expenses in order to develop the organisational structure of the Opera. However, the net income remains in balance during all the period analysed. The Opera does not receive funding from public entities at all. It focuses on its ability to produce revenue through ticketing, and it uses these funds in preserving, restoring and maintaining the monumental complex.

Table 3 The value generated in the Opera (in euros)

	2018	2017	2016
Revenue from institutional management	21,817,848	20,113,716	20,004,974
Tickets	20,467,470	18,834,397	18,756,035
Others	1,350,378	1,279,319	1,248,939
Non-operating revenue	858,943	759,277	692,402
Financial and asset management	48,444	84,182	167,774
Total revenue	22,725,235	20,957,175	20,865,150
Expenses from institutional management	19,753,674	18,635,506	14,980,320
Non-operating expenses	347,191	422,329	314,104
Financial and asset management	317,847	50,475	325,096
General expenses	2,209,770	1,733,092	1,479,092
Total expenses	22,628,482	20,841,402	17,098,612
Net income from institutional management	2,064,174	1,478,210	5,024,654
Net income from non-operating management	511,752	336,948	378,298
Net income from financial and asset management	−269,403	33,707	−157,322
Net income	96,753	115,773	3,766,538

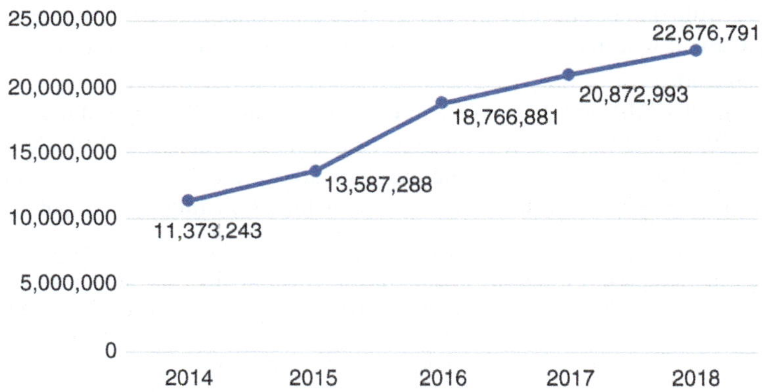

Fig. 1 The revenue trend for the 2014–2018 (in euros)

Most of the revenue arises from ticket sales with an annual visitor average in the last 3 years of 1343 million euros. In the same period, the revenue is constantly increasing, as shown by Fig. 1.

The increasing revenue from ticket sales rose significantly, at the end of 2015, with the opening of the renovated museum. Other factors that have influenced the increase in revenue were the pedestrianisation of the historic centre and the growth of tourists that have allowed to increase the visitors, first, of the Dome and then, through the cumulative ticket, of the Museum. During the high season, the Opera sells around 4500 cumulative tickets per day; however, for security and protection

Fig. 2 Revenue vs. visitors in the Opera di Santa Maria del Fiore di Firenze

reasons, the number of Dome's visitors is fixed, and not all buyers can enter the Dome.

Given the above situation, the Opera does not need to find new commercial strategies for increasing the number of visitors, but it has to differentiate the visitors in order to find a new "category" of users:

> Starting from the point that the monumental complex has too many visitors compared to how many it is able to welcome and also to make the Church usable and considering that the Fabbriceria is already profitable, the challenge is modifying the visitors' typologies. The peak of 1.4 million users that we reached in 2017 was no longer sustainable. (The President)

Therefore, the strategy was to increase the cumulative ticket price to select a new type of "mindful" visitor interested in visiting not only the Dome but also the historical and artistic aspects of the monumental complex. This visitor's typology will be willing to pay a higher cumulative ticket and to join a guided tour. Despite the increase of the price of the cumulative ticket, the Opera's revenue did not decrease, testifying to the validity of the strategy as evidenced by the 3-year comparison between revenue and visitors (Fig. 2).

As concerns the sales channels, traditionally, the majority of tickets are sold by the ticket office located in Piazza San Giovanni, and unfortunately this generates queues and long waiting times. However, this situation has been changing since 2019, as the Opera has been trying to persuade users to buy the ticket on-line, not necessarily before, considering that they are able to buy tickets on the day they decide to visit the complex by using the Opera's free Wi-Fi. In particular, according to the 2018 annual report, the sales from the ticket office in the square are 49.88% of the total ticketing revenue, while on-line sales stop at 34.62%. The strategy for the future is to overturn this relationship.

Additional revenue arises from the bookshop and from public restrooms located inside the "Art and Culture Center" next to the Opera. Other revenue is related to real estate management, mainly commercial and residential buildings, which allow to obtain about 1 million euros of rentals. In the future, the Opera is also going to set up a wardrobe service, free for those who have tickets and paying for those who do not.

The Opera's development strategies are based on several pillars:

> Our medium-term strategy involves: higher attention to the monuments and the identification of the more engaged visitor compared to the simple tourist to accompany during the visit; the use of available resources to focus on restoration and conservation; looking for patrons who are willing to link their name to the restorations of the Opera. About this latter pillar, we can say that the Opera already has the resources to operate autonomously, but if we find those who are willing to finance us, we can do more and better by improving and renovating our image. (The President)

With regard to innovative financing ways, the Opera intends to take particular action in the area of sponsorship by looking for patrons who are willing to support the restoration of the monumental complex:

> The idea is to find external sponsors for two reasons: to link our monuments with the names of the rich and famous (we hosted the Apple CEO for a dinner) in order to get closer to the younger audience and, at the same time, to be closer to Florence and to avoid the feeling that the Opera is an institution far from people. Hence the organisation of events, such as concerts in the cathedral and in the museum and the exhibition of the Opera's choir by singing in the churches from the outskirts to the heart of the city and the free access to monumental complex for Florentine inhabitants.

4.3 Discussion

Considering the financial analysis, the case study shows that in the last few years the institution has realised a fast growth in the number of visitors and tickets. The general result is an increase in the revenue and the achievement of a stable economic and financial equilibrium. This is an interesting aspect, as literature point out the decrease of ticket sales and, consequently, of revenue because of the economic crisis which started in 2008 (Lindqvist 2012). Additionally, contrary to prior research (Camarero and Garrido 2008), the Opera shows a balance between income and corresponding cost, and the latter are covered by revenue for tickets sales. Previous studies (e.g. Camarero et al. 2019) highlight the importance of additional contributions from sponsors and donors for the survival of museums; however, the Opera has not sponsorships and donors from external entities. Indeed, the revenue comes essentially from tickets sale.

Moreover, considering that the Opera has too many visitors compared to how many it is able to welcome, the institution has not urgency to develop new marketing strategies in order to increase the number of users. In the light of this consideration, the management has decided to change the visitors' typology, moving towards a "mindful" user interested in understanding the historical and artistic history of all the monumental buildings. In other words, the Opera has been trying to select a new visitor willing to pay a higher price ticket and experience a guided tour. The main problem of this institution is not to find additional financial resources in the short term but to identify new users and new resources for the building maintenance and restoration.

Finally, according to Bertacchini et al. (2018), the focus of the management is not only based on the efficiency and the maximisation of the revenue from visitors but also on the value generation for stakeholder according the Fabbriceria mission. In this perspective, the Opera has also been trying to attract innovative forms of financing through crowdfunding, international sponsorship and events.

5 Conclusions

Studies about financial strategies in cultural institutions are mainly concentrated on public museums. However, we can observe a gap in the literature concerning how private museums are managed and financed without receiving public funds. The aim of this paper is to fill the gap in order to provide a better understanding regarding the financial strategies of a private museum like the Opera of Santa Maria del Fiore in Florence.

The balance sheet data and the Opera's strategies show that culture is not always a sector that "lives" only thanks to public funding. The Opera di Santa Maria del Fiore in Florence, in fact, does not only carry out the core museum activities of protection, display and acquisition of heritage but also has a perishable nature of one's "heritage" superior to that of any other traditional museum. The conclusion is that it is possible to protect, manage and restore the artistic and monumental heritage and, at the same time, to carry out research and training activities using own funds in conditions of economic and balance equilibrium.

This research has numerous limitations. First, the Opera di Santa Maria del Fiore is located in one of the best-known art cities in the world. Hence, the centre of the city of Florence is visited every year by millions of people, and this situation is a strong point for the capacity of visitors' attraction of the Opera. Probably, the conclusions regarding the financial strategies could be different if we consider an institution located in a different touristic destination. For this reason, the results obtained up to now must be compared by expanding the number of cases analysed to verify the possibility of disseminating and generalising results.

Second, the results obtained could change according to other types of private museums analysed and future trends. Furthermore, future research can be carried out with different approaches, including quantitative methods.

References

Baumol, W. J., & Bowen, W. G. (1966). *Performing arts. The economic dilemma*. New York: The Twentieth Century Fund.

Bellucci, M., Manetti, G., & Sibilio, B. (2020). Dalla rendicontazione alla comunicazione dialogica: il caso delle fabbricerie italiane. In B. Sibilio & A. Matacena (Eds.), *I musei ecclesiastici. Proposte di valorizzazione* (pp. 329–354). Milan: FrancoAngeli.

Bertacchini, E. E., Dalle Nogare, C., & Scuderi, R. (2018). Ownership, organization structure and public service provision: The case of museums. *Journal of Cultural Economics, 42*(4), 619–643. https://doi.org/10.1007/s10824-018-9321-9.

Blasco López, M. F., Recuero Virto, N., & San-Martín, S. (2019). The cornerstones of museum performance. A cross-national analysis. *Museum Management and Curatorship, 34*(2), 211–233. https://doi.org/10.1080/09647775.2018.1516562.

Bonet, L., & Donato, F. (2011). The financial crisis and its impact on the current models of governance and management of the cultural sector in Europe. *Journal of Cultural Management and Policy, 1*, 4–11.

Burton, C., & Scott, C. (2003). Museums: Challenges for the 21st century. *International Journal of Arts Management, 5*(2), 56–68.

Camarero, C., & Garrido, M. J. (2008). Improving museums' performance through custodial, sales, and customer orientations. *Nonprofit and Voluntary Sector Quarterly, 38*(5), 846–868. https://doi.org/10.1177/0899764008319230.

Camarero, C., & Garrido, M. J. (2012). Fostering innovation in cultural contexts: Market orientation, service orientation, and innovations in museums. *Journal of Service Research, 15*(1), 39–58. https://doi.org/10.1177/1094670511419648.

Camarero, C., Garrido, M. J., & Vicente, E. (2011). How cultural organizations' size and funding influence innovation and performance: The case of museums. *Journal of Cultural Economics, 35*(4), 247–266. https://doi.org/10.1007/s10824-011-9144-4.

Camarero, C., Garrido, M. J., Vicente, E., & Redondo, M. (2019). Relationship marketing in museums: Influence of managers and mode of governance. *Public Management Review, 21*(10), 1369–1396. https://doi.org/10.1080/14719037.2018.1550106.

Caron, P. G. (1967). Fabbricerie (ad vocem). In *Enciclopedia del Diritto* (Vol. XVI, pp. 196–207). Milan: Giuffrè.

Cole, D. (2008). Museum marketing as a tool for survival and creativity: The mining museum perspective. *Museum Management and Curatorship, 23*(2), 177–192. https://doi.org/10.1080/09647770701865576.

Creswell, J. W., & Clark, V. L. P. (2017). *Designing and conducting mixed methods research*. Thousand Oaks: Sage.

Fedeli, S., & Santoni, M. (2006). The government's choice of bureaucratic organization: An application to Italian state museums. *Journal of Cultural Economics, 30*(1), 41–72. https://doi.org/10.1007/s10824-006-9005-8.

Fernández-Blanco, V., & Prieto-Rodríguez, J. (2020). Museums. In R. Towse & T. Navarrete Hernández (Eds.), *A handbook of cultural economics* (pp. 349–357). Cheltenham: Edward Elgar.

Frey, B. S. (2020). Public support. In R. Towse & T. Navarrete Hernández (Eds.), *A handbook of cultural economics* (pp. 449–456). Cheltenham: Edward Elgar.

Frey, B. S., & Meier, S. (2006). The economics of museums. In V. Ginsburgh & D. Throsby (Eds.), *Handbook of the economics of arts and culture* (pp. 1017–1047). Amsterdam: Elsevier.

Greco, G. (2005). Un "luogo" di frontiera: l'Opera del Duomo nella storia della Chiesa locale. Premessa storica sulle Fabbricerie. In *Opera Primaziale Pisana (a cura di), La natura giuridica delle fabbricerie, giornata di studio, Pisa 4 maggio 2004* (pp. 3–19). Pontedera: Staperia Bandecchi & Vivaldi.

Hood, C. (1995). The "new public management" in the 1980's: Variations on a theme. *Accounting, Organizations and Society, 20*(2), 93–109. https://doi.org/10.1016/0361-3682(93)E0001-W.

ICOM—International Council of Museums. (2017). *Statutes*. Paris: ICOM.

Irpet—Istituto Regionale Programmazione Economica della Toscana (Regional Institute for Economic Planning of Tuscany). (2019). *Il rapporto sul turismo in Toscana. La congiuntura 2018*. Florence: Irpet.

Kotler, N. G., Kotler, W., & Kotler, P. (2008). *Museum strategy and marketing. Designing missions, building audiences, generating revenue and resources*. San Francisco: Jossey Bass Wiley.

Lindqvist, K. (2012). Museum finances: Challenges beyond economic crises. *Museum Management and Curatorship, 27*(1), 1–15. https://doi.org/10.1080/09647775.2012.644693.
Lune, H., & Berg, B. L. (2017). *Qualitative research methods for the social sciences*. Boston: Pearson.
Musgrave, R. A. (1959). *The theory of public finance*. New York: McGraw-Hill.
Opera di Santa Maria del Fiore. (2016). *Annual report*. Retrieved March 25, 2020, from https://duomo.firenze.it/getFile.php?id=172
Opera di Santa Maria del Fiore. (2017). *Annual report*. Retrieved March 25, 2020, from https://duomo.firenze.it/getFile.php?id=171
Opera di Santa Maria del Fiore. (2018). *Annual report*. Retrieved March 25, 2020, from https://duomo.firenze.it/getFile.php?id=170
Osborne, S. P. (2006). The new public governance? *Public Management Review, 8*(3), 377–387. https://doi.org/10.1080/14719030600853022.
Proteau, J. (2018). Reducing risky relationships: Criteria for forming positive museum-corporate sponsorships. *Museum Management and Curatorship, 33*(3), 235–242. https://doi.org/10.1080/09647775.2018.1467274.
Qu, S. Q., & Dumay, J. (2011). The qualitative research interview. *Qualitative Research in Accounting & Management, 8*(3), 238–264. https://doi.org/10.1108/11766091111162070.
Rivella, M. (2005). Le fabbricerie nella legislazione concordataria. In *Osservatorio delle libertà ed istituzioni religiose, Relazione alla giornata di studi su "Le fabbricerie, diritto, cultura, religione"* (Ravenna 10 dicembre 2005). Retrieved March 25, 2020, from https://archivio.olir.it/areetematiche/86/documents/Rivella_%20Fabbricerie.pdf
Romolini, A., Fissi, S., Gori, E., & Contri, M. (2020). Financing museums: Towards alternative solutions? Evidence from Italy. In M. Piber (Ed.), *Management, participation and entrepreneurship in the cultural and creative sector* (pp. 11–32). Cham: Springer.
Sargeant, A., & Jay, E. (2014). *Fundraising management. Analysis, planning and practice*. London: Routledge.
Siano, A., Eagle, L., Confetto, M. G., & Siglioccolo, M. (2010). Destination competitiveness and museum marketing strategies: An emerging issue in the Italian context. *Museum Management and Curatorship, 25*(3), 259–276. https://doi.org/10.1080/09647775.2010.498984.
Stylianou-Lambert, T., Boukas, N., & Christodoulou-Yerali, M. (2014). Museums and cultural sustainability: Stakeholders, forces, and cultural policies. *International Journal of Cultural Policy, 20*(5), 566–587. https://doi.org/10.1080/10286632.2013.874420.
Vicente, E., Camarero, C., & Garrido, M. J. (2012). Insights into innovation in European museums. The impact of cultural policy and museum characteristics. *Public Management Review, 14*(5), 649–679. https://doi.org/10.1080/14719037.2011.642566.
Woodward, S. (2012). Funding museum agendas: Challenges and opportunities. *Management Leisure, 17*(1), 14–28. https://doi.org/10.1080/13606719.2011.638202.
Yermack, D. (2017). Donor governance and financial management in prominent US art museums. *Journal of Cultural Economics, 41*(3), 215–235. https://doi.org/10.1007/s10824-017-9290-4.
Yin, R. K. (2018). *Case study research and applications: Design and methods*. Los Angeles: Sage.

Alberto Romolini is currently Associate Professor in Business Administration at the Università Telematica Internazionale Uninettuno, Rome, Italy. He is also Vice-Dean of the Faculty of Economics in the same university. He holds a Ph.D. in Public Management. His research interests are in the field of corporate social responsibility and nonfinancial reporting, public management in local authorities and health-care entities, museums, history of accounting and tourism management.

Silvia Fissi is Assistant Professor at the Department of Business and Economics of Florence University. She holds a Ph.D. in Planning and Control from the Università degli Studi di Firenze and a Laurea (Master degree) in Economics from the same university. Her research interests include public management, local authorities, tourism management, museums, corporate social responsibility and accounting history.

Elena Gori is Associate Professor of Financial Accounting. She holds a Ph.D. in Planning and Control from the Università degli Studi di Firenze. Her research interests include public management, local authorities, tourism management, museums, corporate social responsibility and accounting history. She is director of the Centro Interuniversitario di Studi sul Turismo (Interuniversity Centre of Tourism Studies) of the Università degli Studi di Firenze.

Marco Contri is currently a Ph.D. Student in Business Administration and Management at the University of Pisa (Italy). He previously obtained a BSc. in Business Administration and MSc. in Accounting from the University of Florence (Italy). His main research interests include public management, museums and sustainability.

Through the Public's Lens: Are Museums Active Members of Society? An Investigation During the COVID-19 Pandemic

Alexandra Zbuchea, Mauro Romanelli, and Monica Bira

Abstract Traditionally, museums have cultural and educational roles. Increasingly, however, museums are also assuming a social role in engaging with visitors and communities and strengthening social involvement coherently given the changing expectations of the public. Museums have become participatory organisations that contribute to the well-being of society. Considering the global health challenge of the COVID-19 pandemic, this chapter investigates the way museums have responded to support their communities, as well as the public's perception of museums' civic profile. An online survey was applied in Italy, Romania, and the Republic of Moldova to identify the dimensions of the activity of museums regarding how the support offered by museums is visible and appreciated. The study shows that museums reacted to the COVID-19 pandemic crisis, developing online and digital sources to adapt and deliver their offers, to have a voice within society. Generally speaking, it seems that the public does not pay much attention to museums concerned in contemporary societal debates. Understanding the public image of museums helps museums redesign their organisation and activities to meet the expectations of contemporary audiences coherently, given the recent societal developments. This investigation is among the first to highlight the way museums have assumed an activist role, providing support to the wider public, by reacting to the COVID-19 pandemic.

Keywords COVID-19 pandemic · Museums' public perception · Museums' activist and social role · Museums as communities

A. Zbuchea · M. Bira
National University of Political Studies and Public Administration, Bucharest, Romania
e-mail: alexandra.zbuchea@facultateademanagement.ro; monica.bira@comunicare.ro

M. Romanelli (✉)
Department of Business and Economics, Parthenope University of Naples, Napoli, Italy
e-mail: mauro.romanelli@uniparthenope.it

P. Demartini et al. (eds.), *Cultural Initiatives for Sustainable Development*, Contributions to Management Science,
https://doi.org/10.1007/978-3-030-65687-4_4

1 Introduction

Contemporary museums have a social role (Knell 2019) within communities (Watson 2007) and society overall (Janes and Sandell 2019), promoting cultural heritage by involving wider audiences in cultural production and awareness of social justice and inclusion, thus strengthening participation and activism, which can enhance the wealth of communities (Black 2005; Janes and Sandell 2019). As an active part of the community, museums stress participation dealing both with the cultural environment and social themes that concern the evolution of communities within society (Simon 2010; Watson 2007; Janes and Sandell 2019).

Museums build relationships and collaborative processes within the community, thus preserving and providing information and knowledge regarding cultural heritage that refers to the history, values, and traditions of a community (Karp 1992; Watson 2007).

Contemporary museums involve the communities by managing and studying heritage together with the public, connecting with the social and political environment in which they are active. They are powerful voices and take a stand against injustice and oppression, contributing to community values becoming activists to support the communities to which they cater, rather than only offering culture.

Activist museums, concerned both with heritage and the existence of a better society, contribute to creating dialogue, promoting ideas, and creating networks to support continuous social and civic involvement. Museums evolve, playing a social role in society, and focusing on culture as well as education and social activism (Emery 2001; Sandell 2003; Black 2005), thus strengthening the relationships between themselves and the community (Simon 2010), attracting visitors to museum offers and enhancing said visitors' experience (Packer and Ballantyne 2002). Museums develop better relationships with and within communities, enabling personal and individual contributions from users of museum collections and heritage because "museums want to have a greater role in our changing society, they must do more than store and present our collective works, they must find a way to make personal connections" (Kopke 2011, p. 411).

The changing practices of museums have led to both new and more significant experiences and expectations, but even to challenges for the museums' audiences. The public, especially those more interested in and connected to the cultural and social environments, requires museums to be both participative and active in society. Museums have begun to be evaluated not only in terms of collections but also concerning their involvement in society. Bearing this in mind, the present chapter, in accordance with the main objective of the research, tries to chart how museums are seen by the larger public as active members in the society. The main objective of the research is to map how the public perceives museums as active members of society. This is achieved by taking part in discussions on the main concern of contemporary society. Understanding the public's perceptions would help museum managers both better design their activity and communicate with their communities.

The advent of the COVID-19 pandemic, due to a new *coronavirus*, led to global health emergency, museums and cultural institutions being forced to stop onsite visits and adopt online and digital strategies (Agostino et al. 2020) to strengthen relationships with audiences to create community engagement (Tully 2020) and bring communities together (Potts 2020) while rediscovering a way to debate questions about the values of society (Christiansen 2020).

The COVID-19 pandemic has had a negative overall impact on society, organisations, and individuals. The influences are not only connected to health but also economic and social frameworks. Museums, which depend on their visitors, are among the organisations which have been very hard hit. The managerial efforts to adapt to the new situation are significant. Additionally, as per the new framework, pictured above, the museums are expected to be voices supporting the communities as they go through these troubled times. Therefore, we also aim to investigate how the general audience perceives the extent to which museums have been helping society during the COVID-19 pandemic.

To reach our goals, we developed an investigation of the way museums in Romania, the Republic of Moldova, and Italy are perceived by the wider public, in terms of activism and contribution during the COVID-19 pandemic. This survey reveals the degree to which museums are considered participatory. Understanding the public image of museums would help managers, curators, educators, and other museum professionals to better design their activity, and adapt to the expectations of the contemporary audience as well as the developments of present-day society.

The present chapter is structured in the following way. After the introduction, museums are presented as people-centred and community-oriented organisations that are embedded within society and which move from participation to activism. In the third section, the theme of museums dealing with engaging visitors' experience and perception is elucidated. In the fourth section, a documental analysis using indirect data sources presents how the museums are coping with the pandemic and, still engaging with their public. The fifth section investigates the public's perception of the museum as a social and proactive agent within society, both generally and specifically, in the times of COVID-19. Finally, conclusions and discussions are outlined.

2 Towards People-Centred and Community-Oriented Museums Embedded Within Society: From Participation to Activism

Production-centred/audience-driven museums support two-way communication, participatory and dialogic interaction; they also contribute to collective expertise using technology to construct a *shared* authority on cultural heritage contents through social media for networking and collaboration between the museum and the public users (Bonacini 2012; Rentschler 2007; Kelly 2010), thus driving the audience, as an active agent, to become the community (Karp 1992). As dynamic educational, cultural, and social-engagement locations and cultural malls (Zbuchea 2015), museums as organisations (Bagdadli 1997) are embedded within the *milieu* of their communities (Janes and Sandell 2019) and develop a community service strategy to support participation involving new audiences (Kotler and Kotler 2007), thus providing a context and space which can bring the past into the present for discussion, critique, and contestation (Lord 2006).

As people-oriented and visitor-centred organisations serving a changing society (Bagdadli 1997; Simon 2010; Falk 2016; Rentschler 2007), museums empower the community, thus contributing to its development and engaging the public in value identification and creation to build the widest possible constituency for ongoing museum support (Scott 2010). Museums enable visitors to contribute their ideas and discuss what they see and perceive during their visit; these visitors can also share knowledge, experience and conversations, thus developing effective relationships with their stakeholders within communities, which build the museums into a community that engages with audiences (Watson 2007).

As civic spaces and agents of social change museums provide an informed debate (Casey 2007) and contribute to social value creation by engaging the community in preserving their knowledge, identity, memory, values, and traditions (Burton and Scott 2007). As sites for critical reflection on the past and the preservation of truth and rights, (Hooper-Greenhill 1995), museums contribute to strengthening the participation and activism of people, groups, and communities, thus giving voice to their audience, and dealing with global challenges that concern society in terms of ethical behaviours, social injustice, and inequality (Sandell 2007).

As drivers of social change, inclusion, regeneration, and justice (Sandell 1998), museums promote activism, mobilising people to co-generate value by relying on a shared authority regarding cultural heritage (Legget 2018) and using collections to support and enable communities to feel a sense of place and control over their cultural heritage (Mears and Modest 2013) through "collaborative participatory ways of working that build relationships and strengthen networks well beyond the museum" (Janes and Sandell 2019, p. 9).

In the twenty-first century, museums involve the community by challenging social inequality, marginalisation, and injustice, thus making people sensitive to environmental questions and opposed to pollution practices. The role of an activist museum is to "act as a catalyst for social regeneration and as a vehicle for empowerment with specific *communities* and also contribute towards the creation of more equitable *societies*" (Sandell 2002, p. 4).

Museums develop effective relationships with communities within society by enhancing those communities (Watson 2007) to develop the potential offered by the collections and exhibitions to "construct more inclusive, equitable and respectful societies" (Sandell 2002, p. 4) increasing social justice (Mears and Modest 2013) and creating inclusive environments which are opposed to social exclusion (Taylor 2017).

As agents of social inclusion, museums promote social responsibility and combat social inequality and exclusion highlighting the challenges of global problems and matters the society is dealing with (Sandell 1998). Museums "must consider their impact on society and seek to shape that impact through practice that is based on contemporary values and a commitment to social equality" (Sandell 2007, p. 110), fostering democratic engagement through participation in museums mobilising local cultural and social activism (Lynch 2011).

3 Museums Dealing with Visitors' Experience and Perception

Traditionally, a museum is defined as "a non-profit, permanent institution in the service of society and of its development, and open to the public, which acquires, conserves, researches, communicates and exhibits, for purposes of education, study and enjoyment" as stated by the 22nd General Assembly of the International Council of Museums in Vienna, Austria, on August 24, 2007. Nowadays, museums play a social role and contribute to local development and well-being of communities (Brown and Mairesse 2018), strengthening the community building in an interactive and experiential framework and providing "a community space where citizens can share their objects and their stories" (Kopke 2011, p. 399).

As already discussed, museums are changing, becoming more audiences-oriented and people-based institutions where people can enjoy and learn from collections that are placed in a trust for society (Rentschler 2007). Thereby, visitors perceive museums as user-friendly and welcoming institutions that democratise cultural heritage contents, promoting *edutainment* as a blend of education and entertainment (Addis 2005) for an improved and co-produced museum experience, where the individual interprets the message in the interaction. Indeed, all these have a positive impact on visitors' experience, as well as the perceived authenticity of the experience (Komarac et al. 2020). Museums can improve visitors' experience and provide community service to address the needs of the audiences, while also cultivating new visitors (Kotler and Kotler 2007). Museums create exhibitions to satisfy a diverse range of leisure-related goals following a visitor-centred perspective (Falk 2016) because "people visit museums in order to satisfy a suite of self-related reasons, reasons associated with the relationship between their own roles and needs and the affordances they perceived the museum possessed" (Falk 2016, pp. 359–360). Museums must also pay attention to attracting new audiences because the ways visitors and non-visitors perceive a museum influences their decision to visit, or not visit, said museum (Stylianou-Lambert 2009). Museums aiming to strengthen the authenticity of the visitor experience have to understand visitors' preparedness to participate in their visitor experience and enable visitors to co-create their experiences in a meaningful way (Hede et al. 2014). Museums must take into consideration both the perceived values of museum experiences and the visitors' mindfulness (Kim Lian Chan 2009). Museums provide benefits and restoration for visitors in terms of well-being as they leave the museum and return to everyday life

(Packer 2008). This is because visitors are expected to live museum experiences so as to improve learning, and thus they perceive the museum as a place where important information is presented in an interesting way (Packer and Ballantyne 2002).

Museums need to offer a context to meet the needs of a broader audience and allow people to live a global experience (leisure, culture, education, and social interaction) (Brida et al. 2016b) supporting increased loyalty and involvement of the visitors in the future (Antón et al. 2018), while also promoting inclusion in the museum's community (Zbuchea 2015) and enabling value to be co-created and personalised experiences to be had, thus increasing the general appeal of museums. Indeed, this could be critical in the case of young people (Nowacki and Kruczek 2020; Manna and Palumbo 2018) and probably of those generally not so interested in museums and culture.

As forums, museums create tailored experiences for visitors who appreciate how the museum becomes a platform and a place where they can talk and engage in discussions. These visitors also expect a societal relevance and agency from museums so that they can transform their views, opinions, and concerns into higher conversations (Bandelli and Konijn 2015).

4 Museums Engaging Visitors and Involving the Community in Times of the COVID-19 Pandemic Crisis

With a rising global health emergency due to the advent of the COVID-19 pandemic, museums as cultural institutions ceased onsite visits, facing a dramatic long-term decrease of income. The museums then increased and relaunched their online presence and digital and audience-oriented strategy to ascertain what people were looking for from cultural institutions. Indeed, the museums achieved this by providing accessible information and materials, strengthening participatory initiatives and conversations through focusing on the community (Potts 2020), and learning to become a space for reflection and meditation (Verdon 2020).

The COVID-19 pandemic health emergency crisis has accelerated digital transformation and stressed the need for people-centred museums that support interaction and participation engaging visitors in active co-production and sharing of cultural contents. The museums have created a new business model by strengthening social value and the potential for innovation with communities in post-pandemic society. The COVID-19 pandemic has driven museums to strengthen their relationships with the public, creating richer digital presentations of collections and exhibitions by reaching broader audiences and reinforcing the social aspect of the museum experience by transforming lives through culture (Potts 2020).

Since museums were forced to briefly stop providing onsite services, they have started to use, with increased frequency, digital technologies and social media for public service delivery online to retain user engagement and ensure that said users have a voice and that the quality of the online experience is high (Agostino et al. 2020). During the COVID-19 crisis, digital tools have been helping to support community-driven museums that have increasingly developed public engagement to maintain their relationship with users who are expected to consolidate online experiences through in-person visits, which will help museums to counteract financial threats "proving their social value to each other, as well as their communities, beyond anything seen before" (Tully 2020, p. 235).

Museums across Europe closed for visitation in March 2020 for approximately 2 months due to the COVID-19 pandemic. All of the museums in Italy were closed by March 8, 2020, while those in Romania closed on March 11, 2020. This inflicted not only a significant negative financial impact but also affected the public missions of museums. To continue to educate and contribute to the cultural development of their communities, and to facilitate access to collections, many museums, all around the world, including in Italy and Romania, enhanced their online communication and developed their online activity, especially on social media. We observe that being active in the times of the pandemic is related to the survival of museums since certain local authorities which finance museums believe that, without visitors, museums will lose their significance and utility; there are several countries where all museum employees have been temporarily fired to save funds which will instead be used to fight the pandemic. Arguments related to the need to preserve and protect heritage, as well as to continue to research collections seem to be not so relevant for some local authorities in countries such as Romania and possibly others.

We will briefly present the approaches used by museums in Romania to adapt to the new situation. A similar evolution has also been registered in other countries. Adapting the online facility has not been simple, since no specific funding was previously considered and the process had to be designed and implemented in a few weeks. Therefore, the approach chosen by museums, at least in the first month following the "shutdown", was to promote and adapt materials they already had (virtual tours, working materials used for educational programs with children were presented online to be downloaded and used at home, etc.). Nevertheless, some museums developed completely new educational programs. For instance, the Bucharest Municipality Museum started 3 weekly online educational programs for children at the beginning of April: a weekly workshop on cartoons developed around the museum and its collections, a weekly interdisciplinary program on "stories and experiments" starting from collections and books, as well as a more traditional colouring program which also highlights the appeal of the museum's art collection. Another example of adaptation is the transfer of the "Reading corner" from the National Museum of Maps and Old Book in Bucharest to YouTube, while also offering additional resources for families. The National Museum of the Romanian Literature is encouraging children (but not only) to read poems and post the records online. Many Romanian museums have transferred their children-focused education to the internet, all of which are free of charge. There are also a few cases that target an adult audience. Ipotesti Museum, a memorial house dedicated to Mihai Eminescu, the most praised Romanian poet, and located in a remote village, launched the

Fig. 1 Easter FB posters of the Antipa Museum in Bucharest, part of a #stayhome campaign

project #Istayathomeandtellstories. Also, an innovative example would be the National Museum of the Romanian Peasant which collects from the wider public personal stories and experiences related to old and dear objects, developing a "museum at home" (https://arhiva.muzeultaranuluiroman.ro/jurnal/particular).

The list of Romanian museums that took their educational and cultural activities online is rather long. Offering online education in this period is helping teachers but especially parents to educate their children at home. Nevertheless, there exist only a few museums that are directly addressing the negative impact of the pandemic and trying to help their audience cope with these times in a more significant way. The National Art Museum of Romania launched an original online guide regarding how to protect against the threat of the virus, illustrated with paintings from the museum's collections. Moreover, the Grigore Antipa National Museum of Natural History launched a #stayhome campaign, necessary considering that part of Romanian society feels that the restriction of movement is not needed and there are debates on how Easter should be celebrated in a period of social distancing (see Fig. 1). Further, the Museum of Visual Art from Galati launched a form of public therapy through an art challenge—inspired by the Italian movement *Andrà tutto bene!*. A more consistent campaign to stimulate proper attitude and behaviour in times of pandemic was proposed by the National Museum of Romanian History (Fig. 2).

In Italy, a group of children's museums (Children's Museum Verona, MUBA—the Museum of Children from Milano, Explora—the Museum of Children from Roma, La Città dei Bambini e dei Ragazzi in Genoa) with the support of several public bodies, developed a "Galactic Guide for Coronavirus for Curious Children" (Fig. 3). In Italy, as in other countries, museums went online as fast as possible and adapted their discourse to the context while also preparing for a long-term "cultural resistance". For instance, their messages contained words such as virus, contagion, etc. (see Fig. 4).

At the beginning of April 2020, the Network of European Museum Organizations (NEMO) (https://www.ne-mo.org/) published the first systematic analysis of how museums are adapting to the pandemic. The sample consisted of 650 museums in

Fig. 2 A selection of the visuals of the National Museum of Romanian History in Bucharest, part of a #stayhome campaign based on images from the museum's archives, developed on Facebook and Instagram

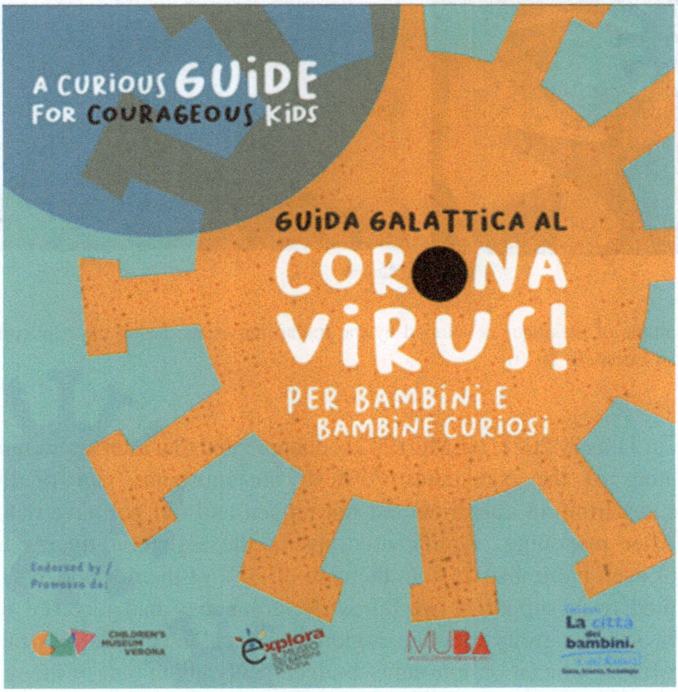

Fig. 3 The cover of a museum-developed Coronavirus guide for children

41 countries. The closure of the museums led to a decrease of around 75% in the funding for museums, while the lending and borrowing activities stopped, most important future exhibitions were postponed, as well as the long-term investments in infrastructure development. Almost two-thirds of the museums ceased their volunteer programmes, and approximately half terminated their contracts with freelancers. Most of the museums had increased their online presence, and almost 14% of them allocated additional funds for online activities, while more than 30% changed staff's responsibilities, making them more online-focused. In a follow up report on the

Fig. 4 A sample of messages developed by the Italian museums during the pandemic in March 2020 (Zardini Lacedelli 2020)

pandemic, in January 2021, NEMO (2021) documented that after seven months from the beginning of the pandemics 70% of the museums part of the survey (600 museums from 48 countries) did not reopen and did not have definite plans for this. After reopening, significant drops in the visitation figures have been reported, of even more than 75% in the case of 20% of the respondents. Museums also benefited from support - two-thirds declaring they had received emergency government support from national, regional or local level. Despite this gloomy situation, with significant on museums' budgets, only about half of the respondents looked for alternative founding. This figures museums' budgets, only about half of the respondents looked for alternative founding. This figures suggest a less flexible management, lack of initiative and innovative approaches for pressing managerial and financial aspects. More flexibility was documented in the case of online public activities - for 40% of museum the staff received new responsibilities related to the online activity of the museum. Most of the large museums increased their digital capacity, while less than half of the smaller museums could manage this.

NEMO also offered examples of digital initiatives undertaken in March 2020. Some museums proved very innovative and agile moving online in a significant way. Even during the pandemic, and with no onsite visits, museums continued to focus on their missions in various ways. They offered online educational programs and increased Web-based access to collections. Some museums have started documenting the corona pandemic for future research and understanding (Fondazione Sandretto Re Rebaudengo in Italy, Museum Europäischer Kulturen in Germany, Minnen in Sweden, as well as other museums in the US, Denmark, Slovenia, etc.—see also Abend 2020; Cascone 2020; Passy 2020), or donating materials to hospitals (Rijksmuseum in the Netherlands, MOMA in the US, the National Museum of Finland, and several museums in the UK). Another example of solidarity and support is the campaign *#ARTWORKS 2.0 per Brescia* developed by the Brescia Musei Foundation. This is an innovative digital project which aims to generate funding for the city by presenting 3D animation of Brescian heritage (https://www.bresciamusei.com/). Another significant involvement of museums is the case of the Smithsonian in the US. It offers updates and provides access to relevant and correct information on the pandemic and the virus (Sholts 2020).

A new museum was launched on Instagram in March 2020 by three advertisers from Barcelona, Spain: The Covid Art Museum. It presents various artistic visions of how life and even values have changed during the pandemic (Latorre 2020).

Since museums have been gravely affected by the pandemic, besides trying to continue serving online their audience, they are increasingly concerned about their survival after the pandemic. The Metropolitan Museum of Art in New York, for instance, has launched a lobbying campaign entitled #CongressSaveCulture. The Romanian Network of Museums is also holding active discussions with the Parliament, the Government, and other public bodies regarding the initiation of measures for supporting, in a significant way, cultural organizations, which are among the most affected parties. Various professional associations have been offering support and resources to cope with the pandemic disruption. Added to this, certain initiatives are debating the future of museums and looking for practical approaches to be more effective and inclusive, even in difficult times, such as pandemic crises. An example in this regard is an initiative by Musei Civici Fiorentini which has been launched on Instagram and other social media platforms every week there are live debates on redefining museums (Firenze Today 2020).

Since museums face many difficulties, in some countries public and/or private schemes of support have been proposed. For instance, in the UK and the US, special lines of funding have been instated. In Italy, the Ministry of Culture and Tourism initiated a weekly campaign to raise the visibility of interest in, but also compassion towards museums (or tourism objectives and places), to show that culture is still relevant. The campaign, called #ArtYouReady?, consisted of a digital flashmob initiated in March (Agostino et al. 2020). Other forms of public support emerged as crowdfunding campaigns, auctions for the benefit of museums, or artists-lead initiatives (see for instance Cook 2020; Kendall Adams 2020; Smee 2020; Valentine 2020). The great and unexpected financial toll of the pandemic determined some museums to sell part of their collections, as in the case of the Brooklyn Museum (Pogrebin 2020).

Nevertheless, some museums tried to contribute directly to ease the impact of COVID on society, on the medical system or on the artists—who have been also hardly hit by the pandemic. For instance, some museums donated to hospitals medical supplies (Harper 2020). Other museums bought arts or initiated projects to support the local artists. Such is the case, for instance, of the National Museum of Contemporary Art in Romania—see RFI 2020), or the Baltimore Museum of Art (see Artforum 2020).

Some museums aimed, besides developing their audiences and remaining relevant in times of COVID-19, to support the entire sector in coping during the pandemic. For instance, the Metropolitan Museum of Arts proposed #MuseumCrushModay blog entries every Monday to show support for museums around the globe (Christiansen 2020). Some other museums followed suit, informing on other museums' initiatives or challenges, but not in a systematic way. For instance, in Romania, this was the case for the Romanian National Museum of Art and the National Museums of Maps and Old Books.

Some museums tried to be empathic with their visitors, while also attempting to present their collections and offer support through culture. An example in this regard is the short-films cycle entitled "Antico Presente" posted on Facebook by the Museo Archeologico Nazionale di Napoli, which explores in the context of the famous Neapolitan museum quite common emotions the public face today: desire, fear, inspiration, etc. (Culturefor 2020).

5 The Public's Perception of Museums as Active Members of Society in Italy and Romania

5.1 Methodology

The main aim of the present study is to map how the general public perceives the way museums are active in addressing the main concerns of contemporary society and in contributing to the public discussion on topics such as minorities, migration, global warming, social inclusion, and education. We focused on those issues based on the fact that they represent concerns held by contemporary European society, and are also hot topics in media across Europe.

Global warming has now been a public concern for several years in Europe and conversations on social media are continually increasing (Ellis 2019), which gives rise to the idea that museums have a reasonable amount of time to approach—if they choose to do so—this topic in an adequate manner, as demonstrated by the remarkable initiative of the Natural History Museum (Stephens 2020). Migration and its consequences are felt daily by large portions of Italian, Romanian, and Moldavian societies. However, this occurs differently in each country, with Italy being a country of destinations for economic migrants coming from Eastern European countries as well as migrants from African countries or war zones in the Middle East (Varella 2020). Both Romania (OECD 2019) and the Republic of Moldova are facing many social and economic consequences related to an exodus of the workforce, comprising both high skilled and low skilled labour. We see social inclusion and minorities as being intimately interconnected with the issue of migration, as well as being high on the European agenda (Atkinson et al. 2017). Education is another aspect that is critical for the development of society. Educated citizens contribute to the sustainable development of their communities. At the same time, an increased number of voices are discussing the need to change the education paradigm. Therefore, we chose to include education as an independent sub-dimension in our study.

Special attention is given to the way museums are perceived as reacting in times of the COVID-19 pandemic, a period of great uncertainty and concern for many European countries, especially Italy. Romania, although less hit by the virus at the time of the current research, reacted strongly from the first signs of the pandemic.

In the second half of March 2020, we conducted an online survey using a convenience sample method. We targeted respondents from three countries: Italy, Romania, and the Republic of Moldova. While aware that neither Western Europe nor Eastern Europe are compact blocks and, therefore, differences within their respective nations and societies are considerable, our intention was, nevertheless to come with a comparative perspective on people's perceptions of museums and their degree of connection to topics of interest across society. Italy might be considered a representative of Western Europe, while Romania is an example of Eastern Europe, and the Republic of Moldova, a partner country, is representative of the non-EU European countries.

In terms of cultural profile, considering the six dimensions of national culture, there are strong similarities between Romania and the Republic of Moldova (which is a Romanian-speaking country, with a long shared history). Between Romania and Italy are large differences, especially in terms of power distance, individualism, and masculinity (Hofstede Insights). In terms of museums and cultural infrastructure, there exist essential differences between the three countries, with Italy presenting the best situation. Concerning cultural consumption, including interest for museums, Romania shows some of the lowest figures in the EU, while Italy is close to the EU average (EC 2017; Eurostat 2016). In terms of attitude towards cultural heritage and its importance, Italy and Romania present similar scores, with figures slightly higher in the case of Romanians (EC 2017, p. 32).

We also included a comparison between the perception of museums abroad and that of museums at home. This allows for a glimpse into the preferences of visitors and how satisfied museum-goers are with their experience since the museum image is connected to the level of satisfaction and gives a competitive advantage. It also suggests a wider view of the satisfaction experienced in museums at home compared to museums abroad, since it is influenced by the overall image (Moreno Gil and Ritchie 2009).

We were aware that our prospective respondents must have been experiencing a considerable degree of screen-fatigue induced by many factors, including an increased number of solicitations to participate in all kinds of pandemic-related online surveys. Therefore, we proposed an 11-items, easy to fill-in questionnaire, with simple questions that, in our opinion, did not require a lot of effort to process. At the beginning of the form, we included a paragraph describing the kinds of activities a museum organises when it chooses to become involved in the current

Table 1 Structure of the initial sample

Country	Respondents working within a museum	Lay respondents
Italy	5	207
Romania	47	220
The Republic of Moldova	10	59

concerns of a community or the society at large. To further ensure a broad panel of respondents, we translated the questionnaire into Romanian and Italian respectively.

The survey was promoted on Facebook (mainly through sharing in various Facebook groups but also by being promoted via Facebook ads with texts addressing a wide audience, and not only those interested in culture/museums). It is worth noting that the number of responses obtained as a result of the Facebook ads was significantly higher in Italy than in Romania. Besides, the Facebook ads proved to be almost ineffective in the case of Romania and the Republic of Moldova. They were, therefore, abandoned as a method for increasing the rate of response. The lowest rate of responses was registered for questionnaires coming from the Republic of Moldova ($N = 69$). In Romania, the total number of valid questionnaires was 267, while in Italy it was 212 (Table 1).

Our initial intention was to investigate the concepts from a broader perspective, by also looking into how museum professionals are considering the notion of participation—when applied to their field of activity. As a comparatively higher number of respondents were affiliated with museums in Romania than in Italy (47 and 5), we decided to eliminate their responses. Finally, for the current analysis, we took into consideration only answers coming from people that were not affiliated to a museum in a professional capacity.

5.2 *Respondents Profile: A Comprehensive View*

The disparity between the number of female and male respondents was higher for the Romanian respondents ($M = 33.8\%$, $F = 66.2\%$) and Moldovian ($M = 35.6$ and $F = 64.4$) respondents when compared to the Italians ($M = 50.2\%$, $F = 49.8\%$).

Regarding the distribution of respondents by age group, the best-represented category for the Italian respondents was in people aged 35–50 years (37.7%), while for Romania the highest number of questionnaires was equally filled in by people under 25 and by people aged 35–50 (32.%). Respondents in the Republic of Moldova aged 35–50 were also the best-represented category, with a higher percentage than in Romania and Italy (50.8). The lowest number of respondents by age group for Romanian respondents was registered within the category of senior citizens (over 61 years old, 2.3%), and the situation was the same in the Republic of Moldova (6.8%). However, the Italian senior citizens, although less numerous than other age groups, accounted for 14% of valid responses, which is approximately 4 times higher than the number of questionnaires filled in by older people in

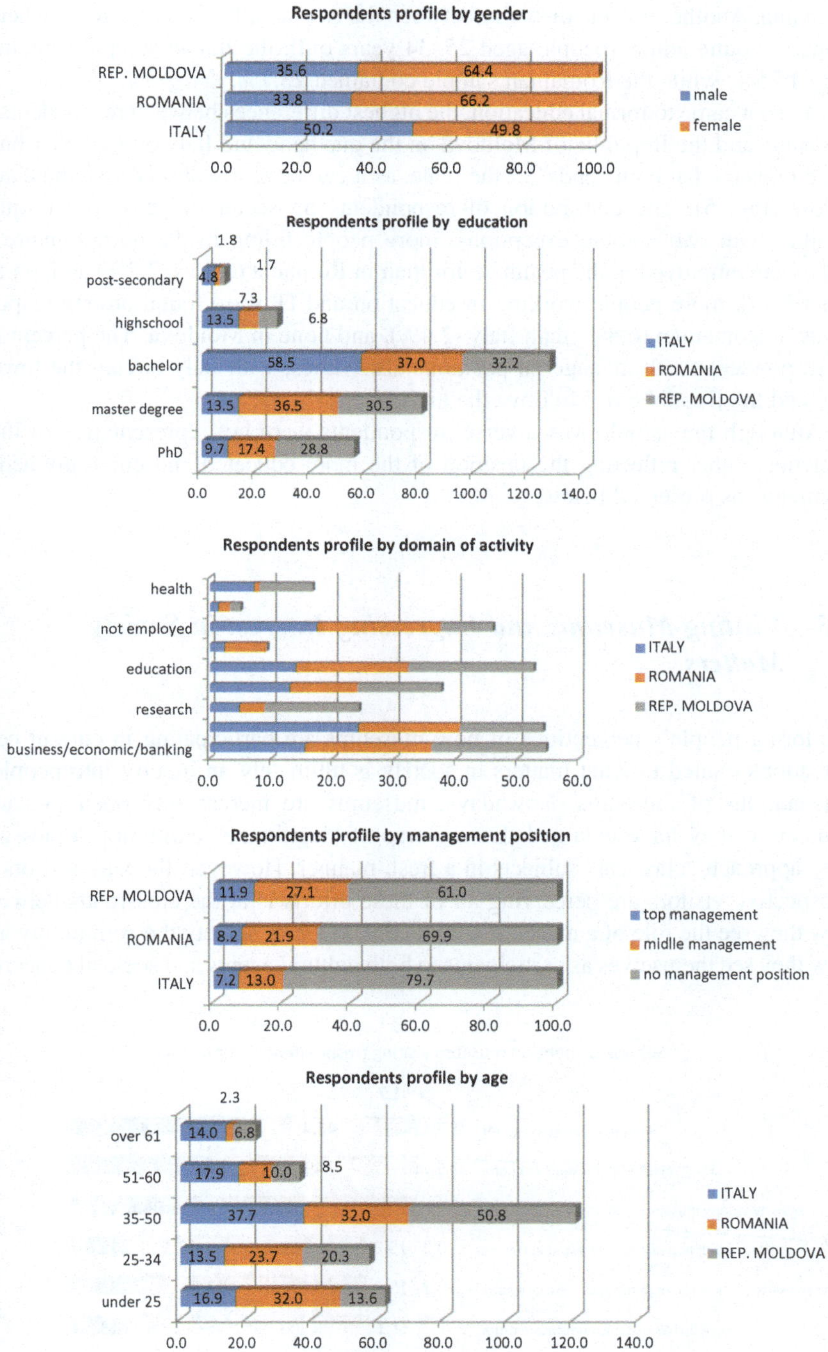

Fig. 5 Respondents' profile

Romania. Another notable disparity between the two samples was registered when it came to young adults (people aged 25–34 years old): the Italian sample contained only 13.5%, while the Romanian sample contained 23.7%.

When it came to formal education, the highest differences between respondents in Romania and the Republic of Moldova on the one hand and Italy on the other hand were registered at both "ends" of the scale, as it can be easily observed in the charts below (Fig. 5). The distribution of respondents by sector of activity was quite similar, with two notable exceptions: more people filling in the questionnaire in Italy were employed in the health sector than in Romania (0.9 vs. 7.2%) and, on the other hand, more people working in education and IT were found among respondents in Romanian (6.8%) than Italy (2.4%), and none in Moldova. The percentage of respondents with managerial positions also varies, with Italy having the lowest rate and the Republic of Moldova the highest.

Although the sample was diverse, respondents were not representative of their societies, rather reflecting the position of the more educated and culturally-active segments, as presented below.

5.3 *Visiting Museums and Expressing Interest in Society Matters*

Exploring people's perceptions of how museums are participating in current conversations related to what matters in society is ultimately an inquiry into people's expectations of museums. Nowadays, museums are increasingly open to wider audiences: they have enlarged the universe of "legitimate" exhibition topics and they approach "classical" subjects in a fresh manner. However, the way visitors or prospective visitors are perceiving all of these efforts is at the crossroads between how they see the role of a museum within contemporary society/the community and how they see themselves as participants to both cultural events and societal concerns.

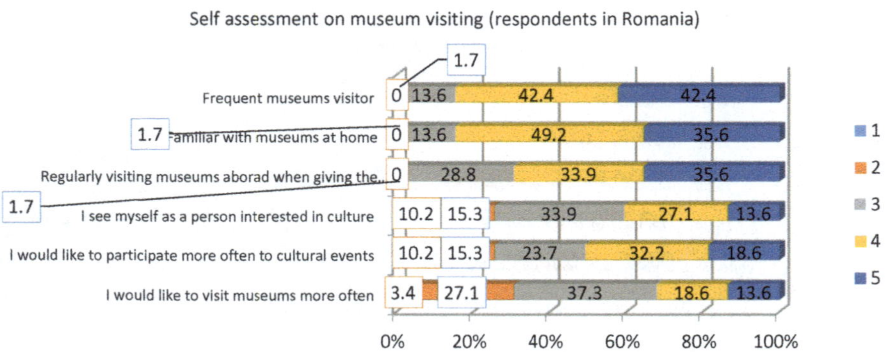

Fig. 6 Self-assessment of museum visiting—Romanian respondents

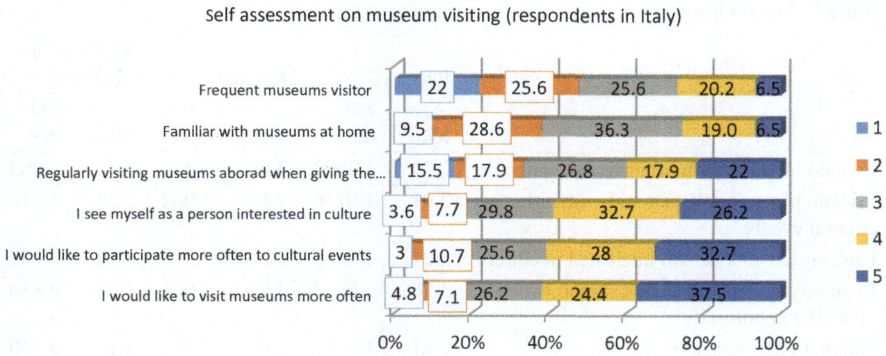

Fig. 7 Self-assessment of museum visiting—Italian respondents

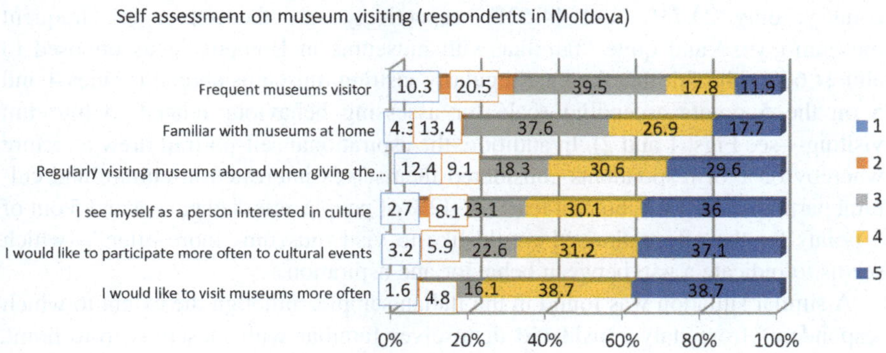

Fig. 8 Self-assessment of museum visiting—respondents from the Republic of Moldova

Therefore, to address our main research question, we devised two sets of questions that we will discuss below.

The first battery of questions aimed to draw a self-portrait of our respondents, by exploring how they see themselves along the general lines of "museum-goers" as well as regarding their interest in participation in cultural activities. This profile helped us to understand better their attitude towards museums and their level of expectation.

The respondents were asked to assess with grades from 1 to 5 (*To what extent do you consider that ... 1 = no; 5 = to a very high degree*) a series of 5 affirmations that would characterise their perceived behaviour and their attitude towards cultural events (including visiting museums at home and abroad). We intended to explore the perceived exposure of our respondents to current exhibitions and recent cultural events, as well as the general overview of their role as participants in cultural events (Fig. 6 for Romanian respondents, Fig. 7 for Italian respondents, and Fig. 8 for the respondents from the Republic of Moldova).

When it came to their perceived behaviour, the Romanian respondents felt that they are more inclined to visit museums while being abroad than when in their own

Table 2 Main statistics

	Italy		Romania		Republic of Moldova	
	Mean	Std. dev.	Mean	Std. dev.	Mean	Std. dev.
I would like to visit museums more often	2.75	1.1567	3.078	1.240	3.12	1.068
I would like to participate more often in cultural events	2.91	1.0716	3.457	1.062	3.34	1.240
I see myself as a person interested in culture	3.21	1.2936	3.635	1.349	3.19	1.167
Regularly visiting museums abroad, when given the opportunity	3.76	1.0818	3.936	1.055	4.03	0.850
Familiar with museums in Italy	3.81	1.0791	3.973	1.111	4.19	0.730
Frequent museums visitor	3.87	0.9418	4.110	1.118	4.25	0.756

country: only 29.73% and 44.62% respectively saw themselves as "frequent museum-goers" and quite "familiar with museums in Romania"—as opposed to almost 60% who felt that they are regularly visiting museums abroad (grades 4 and 5 on the 5 points ascending scale for assessing behaviour related to museum visiting—see Figs. 1 and 2). In addition, the aspirational self-portrait drew a picture whereby 65% of respondents considered themselves interested in culture and cultural participation. Last, but not least, 77.42% of people assessed gave 4 and 5 out of 5 points for the affirmation: "I would like to visit museums more often"—which seems to indicate a gap between behavior and aspiration.

A similar situation was found in the Italian sample, although the extent to which respondents from Italy considered themselves familiar with museums from home was less than for their Romanian counterparts (26.8% frequent visitors and significantly smaller at 25.6% for being "familiar with museums in Italy"). When it came to aspirations related to cultural consumption, the Italians were slightly more moderate: only 60.7% (Italy) and 61.9% (Romania) wished they had more time for cultural participation. (see Fig. 7).

For the respondents in the Republic of Moldova (see Fig. 8), a larger portion considered themselves frequent museum visitors, and more culturally involved and concerned. Nevertheless, 85% stated that they would like to visit museums more often, to a wider degree than in the other two countries. The only item where they scored lower than the Italian or Romanian respondents was related to visiting museums abroad.

To obtain an overall idea of how respondents from Italy, Romania, and the Republic of Moldova see themselves as active museum-goers, we also calculated the mean (see Table 2).

The other two affirmations aimed to examine the (perceived) level of one's civic participation. We have explored perceptions related to self-representation as an *active citizen*, to further consider a possible correlation between paying attention to societal concerns and interest in exhibitions on current "hot" topics on the global agenda. Therefore, respondents were asked, like in the previous example, to assess on a Likert scale (from 1 to 5) to what extent they considered that the affirmations: "I

Table 3 Correlations between self-assessment of respondents as active citizens and museum participation in conversations relevant for society. Respondents from the Republic of Moldova

		Assessment of museum participation at home	Assessment of museum participation abroad	Self-assessment regarding interest in matters of society	Self-assessment regarding being an active citizen
Assessment of museum participation at home	Pearson correlation	**1**	0.220	0.027	0.066
	Sig. (2-tailed)		0.121	0.840	0.614
	N		51	60	60
Assessment of museum participation abroad	Pearson correlation		**1**	0.203	−0.060
	Sig. (2-tailed)			0.154	0.676
	N			51	51
Self-assessment regarding interest in matters of society	Pearson correlation			**1**	**0.613****
	Sig. (2-tailed)				0.000
	N				60
Self-assessment regarding being an active citizen	Pearson correlation				**1**
	Sig. (2-tailed)				
	N				

**Correlation is significant at the 0.01 level (2-tailed)

am paying attention to what is happening in society" and, respectively, "I am an active citizen" suits them. As shown in the tables below (Tables 3, 4, and 5), there are several significant correlations between respondents' assessment of their interest in matters of society and the way they perceived museums as being involved in conversations that are relevant for society. For respondents from the Republic of Moldova, there is no significant correlation between their perception of museums being participative institutions and the perception of their civic engagement (Table 3). However, for respondents in Italy and Romania, there are several significant correlations between how people are seeing museums as being participative and how they see themselves as active citizens.

Table 4 shows that the more respondents in Italy consider museums at home to be participative, the more they consider museums abroad to also be participative (Pearson correlation: 0.482 at 0.01 level). In addition, a high perceived level of participation when assessing museums abroad implies a higher self-appreciation when it comes to self-assessment regarding features such as "active citizen" and "paying attention to what is happening in society" (Table 4 shows a low significant correlation at 0.318 and 0.275 respectively). Table 5 presents the same types of

Table 4 Correlations between self-assessment of respondents as active citizens and museum participation in conversations relevant for society. Respondents from Italy

		Assessment of museum participation at home	Assessment of museum participation abroad	Self-assessment about interest in matters of society	Self-assessment about being an active citizen
Assessment of museum participation at home	Pearson correlation	1	**0.482****	0.136	0.096
	Sig. (2-tailed)		0.000	0.054	0.176
	N		184	202	202
Assessment of museum participation abroad	Pearson correlation		1	**0.318****	**0.275****
	Sig. (2-tailed)			0.000	0.000
	N			184	184
Self-assessment regarding interest in matters of society	Pearson correlation			1	**0.630****
	Sig. (2-tailed)				0.000
	N				207
Self-assessment regarding being an active citizen	Pearson correlation				1
	Sig. (2-tailed)				
	N				

**Correlation is significant at the 0.01 level (2-tailed)

correlations for respondents in Romania, but at other values. To sum up, there is a significant (although low) correlation between how respondents see museums abroad and their involvement in conversations relevant to society, and how they see themselves as citizens interested in matters of society. The more they think about themselves as "active citizens", the more they tend to consider museums abroad as equally involved with topics which are important for society. When it comes to museums at home, there is no such correlation, in none of the categories of respondents (Moldova, Italy, Romania). This finding is consistent with the rest of the study, as museums at home are constantly receiving "lower grades" compared to their counterparts from abroad when they are evaluated in terms of participation in conversations relevant to society.

The second set of questions focused on uncovering whether respondents associate with museum participation with specific themes and revealing their perception of the performances of museums at home vs. museums abroad (Table 6). The themes explored were assessed by respondents against an evaluation scale with 5 degrees, where 1 was the lowest grade and 5 the highest. We looked into topics on the European agenda (e.g. migration, education) on specific topics such as minorities but

Table 5 Correlations between self-assessment of respondents as active citizens and museum participation in conversations relevant for society. Respondents from Romania

		Assessment of museum participation at home	Assessment of museum participation abroad	Self-assessment regarding interest in matters of society	Self-assessment regarding being an active citizen
Assessment of museum participation at home	Pearson correlation	1	**0.457****	−0.025	−0.069
	Sig. (2-tailed)		0.000	0.719	0.318
	N		197	211	211
Assessment of museum participation abroad	Pearson correlation		1	**0.281****	**0.214****
	Sig. (2-tailed)			0.000	0.002
	N			199	199
Self-assessment regarding interest in matters of society	Pearson correlation			1	**0.627****
	Sig. (2-tailed)				0.000
	N				219
Self-assessment regarding being an active citizen	Pearson correlation				1
	Sig. (2-tailed)				
	N				

**Correlation is significant at the 0.01 level (2-tailed)

also into topics on the world agenda, such as global warming and the current pandemic, as well as how museums are adapting to it.

Generally speaking, the respondents in Romania had the feeling that museums at home are addressing global issues to a lesser degree than museums abroad. The highest difference in perception was found in addressing global warming (1.27% points of difference), while the lowest perceived difference was found in education (0.59% points). There was a significant difference in general perception, referring to the overall rating of museum participation in conversations on relevant topics (2.72 museums at home vs. 3.90 museums abroad). The participatory treat of a museum refers to determine the public to engage in dialogue with the museum, to participate in the development and delivery of the museum's offer. It generally refers to a museum that is part of discussions of interest to the society in which it operates.

Respondents in Italy tended to consider that museums at home are closer to what "museums abroad" are currently doing in terms of programs and exhibitions: they also gave lower points for Italian museums, but the differences were small, at less than 0.50%, and sometimes barely noticeable. It is worth mentioning that when assessing museums' perceived behaviour related to the current pandemic, Italian

Table 6 Comparison between the means calculated for the respondents from the three countries and their assessments of museums participating in current conversation within society

	Participation		Minorities		Migration		Global warming		Education		Social inclusion		Others		Pandemic	
	In	Out	In	Out	In	Out	In	Out	In	Out	In	Out	In	Out	In	Out
RO	2.72	3.90	2.39	3.20	2.13	3.10	1.90	3.17	3.29	4.00	2.38	3.33	2.83	3.57	2.82	3.66
IT	3.05	3.45	2.36	2.76	2.30	2.72	2.25	2.67	3.10	3.13	2.59	2.86	2.51	2.62	3.20	3.07
RM	2.77	3.88	2.55	2.01	1.84	3.14	2.37	2.67	3.13	2.81	3.15	3.75	3.22	3.48	2.06	2.92

Table 7 Correlations between the perception of the respondents from Romania over one's cultural consumption and the assessed level of pandemic. A view on museums from home-country and from abroad

	Frequent museums visitor	Familiar with museums at home	Regularly visiting museums abroad	Interested in culture	I would like to participate more often in cultural events	I would like to visit museums more often	Interested in society matters	Active citizen	Museums at home involved in pandemic	Museums abroad involved in pandemic
Frequent museums visitor	1	0.700**	0.589**	0.664**	0.564**	0.502**	0.375**	0.481**	0.128	0.306**
Familiar with museums at home		1	0.564**	0.611**	0.475**	0.509**	0.483**	0.460**	0.072	0.209**
Regularly visiting museums abroad			1	0.609**	0.505**	0.553**	0.395**	0.383**	0.015	0.320**
Interested in culture				1	0.725**	0.718**	0.533**	0.549**	0.085	0.374**
I would like to participate more often to cultural events					1	0.793**	0.512**	0.485**	0.153*	0.379**
I would like to visit museums more often						1	0.518**	0.492**	0.167*	0.406**
Interested in society matters							1	0.627**	0.130	0.314**
Active citizen								1	0.083	0.294**
Museums at home involved in pandemic									1	0.620**
Museums abroad involved in pandemic										1

**Correlation is significant at the 0.01 level (2-tailed)

Table 8 Correlations between perception of the respondents from Italy over one's cultural consumption and the assessed level of museums' involvement in current effort related to the pandemic. Perception about museums from home-country and from abroad

	Frequent museums visitor	Familiar with museums in Italy	Regularly visiting museums abroad	Interested in culture	I would like to participate more often in cultural events	I would like to visit museums more often	Interested in society matters	Active citizen	Museums at home involved in pandemic	Museums abroad involved in pandemic
Frequent museums visitor	1	0.739**	0.676**	0.611**	0.507**	0.445**	0.415**	0.335**	0.191**	0.189**
Familiar with museums in Italy		1	0.613**	0.661**	0.520**	0.477**	0.472**	0.420**	0.189**	0.217**
Regularly visiting museums abroad			1	0.602**	0.462**	0.466**	0.464**	0.410**	0.074	0.139
Interested in culture				1	0.752**	0.669**	0.689**	0.558**	0.201**	0.253**
I would like to participate more often to cultural events					1	0.839**	0.618**	0.598**	0.178*	0.285**
I would like to visit museums more often						1	0.631**	0.502**	0.238**	0.312**
Interested in society matters							1	0.630**	0.138	0.224**
Active citizen								1	0.049	0.155*
Museums at home involved in pandemic									1	0.733**
Museums abroad involved in pandemic										1

**Correlation is significant at the 0.01 level (2-tailed)

Table 9 Correlations between the perception of the respondents from Moldova on one's cultural consumption and the assessed level of museums' involvement in current effort related to the pandemic. Perception about museums from home-country and abroad

	Frequent museums visitor	Familiar with museums in Italy	Regularly visiting museums abroad	Interested in culture	I would like to participate more often in cultural events	I would like to visit museums more often	Interested in society matters	Active citizen	Museums at home involved in pandemic	Museums abroad involved in pandemic
Frequent museums visitor	1	0.456**	0.437**	0.455**	0.461**	0.277*	0.110	0.256*	0.310*	0.032
Familiar with museums in Italy		1	0.354**	0.452**	0.427**	0.358**	0.254	0.388**	0.078	0.106
Regularly visiting museums abroad			1	0.338**	0.211	0.343**	0.255*	0.144	0.049	0.066
Interested in culture				1	0.772**	0.716**	0.541**	0.488**	0.269	0.149
I would like to participate more often to cultural events					1	0.707**	0.430**	0.400**	0.275*	0.229
I would like to visit museums more often						1	0.605**	0.446**	0.063	0.154
Interested in society matters							1	0.613**	0.022	0.125
Active citizen								1	0.095	0.133
Museums at home involved in pandemic									1	0.623**
Museums abroad involved in pandemic										1

**Correlation is significant at the 0.01 level (2-tailed)

Table 10 Perception of the respondents from Italy on the museums at home and from abroad. Correlations between the assessed level of museums' involvement in current effort related to the pandemic and the overall perceived level of museum participation in conversations about social concerns

Correlations—respondents from Italy		Museums at home involved in efforts related to the current pandemic	Museums abroad involved in efforts related to the current pandemic	Museums at home generally involved in conversations on society concerns	Museums abroad generally involved in conversations on society concerns
Museums at home involved in efforts related to the current pandemic	Pearson correlation	1	0.733**	0.310**	0.104
	Sig. (2-tailed)		0.000	0.000	0.167
	N	197	168	194	178
Museums abroad involved in efforts related to the current pandemic	Pearson correlation	0.733**	1	0.249**	0.257**
	Sig. (2-tailed)	0.000		0.001	0.001
	N	168	168	168	160
Museums at home generally involved in conversations on society concerns	Pearson correlation	0.310**	0.249**	1	0.482**
	Sig. (2-tailed)	0.000	0.001		0.000
	N	194	168	202	184
Museums abroad generally involved in conversations on society concerns	Pearson correlation	0.104	0.257**	0.482**	1
	Sig. (2-tailed)	0.167	0.001	0.000	
	N	178	160	184	184

**Correlation is significant at the 0.01 level (2-tailed)

respondents appreciated the efforts made by museums at home and felt they were more consistent than those made by museums abroad (see Table 6).

The respondents from the Republic of Moldova (see Table 6) presented a different attitude towards museums when evaluating how present they are in connection with society's concerns. They felt that minorities and education are of greater concern for the museums at home. Migration was perceived as being almost of no

Table 11 Perception of the respondents from Romania on museums at home and from abroad. Correlations between the assessed level of museums' involvement in current effort related to the pandemic and the overall perceived level of museum participation in conversations about social concerns

Correlations—respondents from Romania

		Museums at home involved in efforts related to the current pandemic	Museums abroad involved in efforts related to the current pandemic	Museums at home generally involved in conversations on society concerns	Museums abroad generally involved in conversations on society concerns
Museums at home involved in efforts related to the current pandemic	Pearson correlation	1	0.457**	0.441**	0.125
	Sig. (2-tailed)		0.000	0.000	0.087
	N	211	197	200	189
Museums abroad involved in efforts related to the current pandemic	Pearson correlation	0.457**	1	0.303**	0.375**
	Sig. (2-tailed)	0.000		0.000	0.000
	N	197	199	189	184
Museums at home generally involved in conversations on society concerns	Pearson correlation	0.441**	0.303**	1	0.620**
	Sig. (2-tailed)	0.000	0.000		0.000
	N	200	189	206	191
Museums abroad generally involved in conversations on society concerns	Pearson correlation	0.125	0.375**	0.620**	1
	Sig. (2-tailed)	0.087	0.000	0.000	
	N	1	0.457**	0.441**	0.125

**Correlation is significant at the 0.01 level (2-tailed)

concern for museums in the Republic of Moldova. They also believed that museums are generally poorly engaged in pandemic-related issues, especially those at home.

As seen in Table 7, there was a moderate positive correlation between aspirational cultural consumption behaviour when it came to museum visiting and general cultural consumption (0.718). There were also several other moderate correlations between being a frequent museum visitor and being familiar with museums. Data for Romania suggested that, to some degree, those who are interested in culture and active in cultural activities, also tend to be concerned citizens. Additionally, those

Table 12 Correlations between the assessed level of museums' involvement in current effort related to the pandemic and the overall perceived level of museum participation in conversations about social concerns

Correlations—respondents from Moldova					
		Museums at home involved in efforts related to the current pandemic	Museums abroad involved in efforts related to the current pandemic	Museums at home generally involved in conversations on society concerns	Museums abroad generally involved in conversations on society concerns
Museums at home involved in efforts related to the current pandemic	Pearson correlation	1	0.220	0.227	−0.032
	Sig. (2-tailed)		0.121	0.103	0.820
	N	60	51	53	53
Museums abroad involved in efforts related to the current pandemic	Pearson correlation	0.220	1	0.276	0.231
	Sig. (2-tailed)	0.121		0.063	0.114
	N	51	51	46	48
Museums at home generally involved in conversations on society concerns	Pearson correlation	0.227	0.276	1	0.623**
	Sig. (2-tailed)	0.103	0.063		0.000
	N	53	46	53	48
Museums abroad generally involved in conversations on society concerns	Pearson correlation	−0.032	0.231	0.623**	1
	Sig. (2-tailed)	0.820	0.114	0.000	
	N	1	0.220	0.227	−0.032

**Correlation is significant at the 0.01 level (2-tailed)

more interested in culture and active in this field tend to perceive museums abroad as being more involved in the pandemic challenges. The same expectation is not projected onto the Romanian museums.

A similar situation was found for respondents in Italy, as shown by the table below (Table 8), illustrating a very low correlation between the level of perceived museum participation in the current pandemic and people's assumed behaviour in terms of cultural participation. In the case of the Republic of Moldova (Table 9), these aspects were not documented, except in the case of frequent visitors, who gave some credit to local museums as being concerned with society's challenges in times of the pandemic.

We also tested for correlations between the overall perception of museums as being involved in conversations of general concern and the perception of their involvement in the current pandemic situation. There were low correlations between those aspects in Italy and Romania (Tables 10 and 11).

In the case of Italy, the data suggested that the general perception of museums influences both the way museums at home and those abroad are perceived, especially in the case of the involvement in times of the pandemic. A somewhat similar situation was seen with the data for Romania (Table 8) while in the case of the Republic of Moldova was a relationship documented only in the case of civic involvement (Table 12).

6 Conclusions and Discussions

The entire museum community agrees that museums are dynamic organisations, changing from keepers of heritage and old-fashioned educators into active members of their communities. This shift has transformed museums into organisations that actively address critical concerns of their audience and are supporters of society. The aforementioned means that museums offer more than interactive exhibitions; they are relevant for their public in a more significant way. They create content while not only bearing in mind their communities but also involving them directly. This evolution is tightly connected with the changing role(s) of museums. From researchers and guardians of heritage, they have become both educators and members of their communities. They not only create experiences but also meaning for their public. As well as being presenters of history and cultural heritage, they are also active citizens, and part of the debates of interest for society, helping to shape the public space.

Key to museums' success, i.e. being able to attract more visitors and have a significant social and educational impact on their audience, is how they are perceived by their public. This sets the levels of interest and expectations, as well as the openness to museums' discourse and also levels of interaction on the social framework. Nevertheless, the subject is more complex. For instance, the Eurobarometer on culture (EC 2017) indicates that, at least for some countries, there is a gap between the attitude towards culture and the actual cultural consumption. Still, developing a better understanding of visitors, and the way they perceive museums helps the museum management and professionals to better design their strategies and offers, and to communicate more effectively.

The wider public in Italy, Romania, and the Republic of Moldova feel that museums are participatory to some degree, but not substantially. The Romanians prove to be more critical of their museums and to appreciate, to the highest degree, the gap between museums at home and those abroad. Moreover, museums are perceived as not being so involved in presenting and being part of discussions on the subjects of concern in society. The public feels that museums pay the most attention to education and the least to global warming and migration. Considering that these two are the hottest topics of discussion, the study reveals that the public believes that museums are not so connected to the concerns of contemporary society. The research also illustrates that audiences interested in museums feel that museums abroad are more participative and connected with the actual concerns of society than museums at home. Bearing in mind that an affective image influences the overall image more than the cognitive image (Moreno Gil and Ritchie 2009), museums should invest more in developing this affective image among their communities. Museums should also act to close the (perception) gap between their image and those of museums abroad, to increase their appeal, as well as the level of satisfaction of their visitors. This would also create conditions for a more significant impact on visitors and communities. This aspect would probably become extremely relevant in the following 2–3 years when, due to the COVID-19 pandemic, museums will cater more to local and national audiences, which will have more difficulties accessing museums abroad.

From a theoretical perspective, more attention could be given to understanding the extent to which "activism" is a drive for museum visits. Previous research concentrated more on educational and recreational aspects—"light consumption" (Brida et al. 2016a). Since more frequent museum visits are connected more to "heavier" motivations than the topics associated with museum activism are, a developed image in this vein might lead to increased visitation.

The reaction of museums to the COVID-19 pandemic is similar in Europe and the US. In March 2020, after closing their doors for on-site visitation, many museums stressed on online visits—they enhanced collection presentations and proposed different types of online guided tours or virtual museums. The online educational materials increased in terms of numbers and typology. Many museums adapted and even proposed new educational programs for children. Educators, families and children seemed the most common online visitors of museums in times of pandemic, but some museums also developed special online programs for adults. Certain museums have more significant involvement, such as donating medical materials and equipment, raising awareness of COVID-19 related aspects, or documenting the times for future presentation and understanding, as well as by supporting independent artists or cultural organizations. Especially museums considered more dynamic and promoters of good practices, that tend to be more connected to the societal framework seem to have been more solidary, both with other museums and with society affected in so many ways by the COVID-19 pandemic.

Among the respondents of the present research, Italians are those who most appreciate museums' reactions during the pandemic, while in the Republic of Moldova people observe the lowest involvement both at home and abroad. The study shows that museums are not perceived as being too involved in the pandemic, and therefore they do not seem to be a voice during these times. Generally, the public, at least those more culturally-involved and concerned, do not perceive museums at home as being very involved in debates relating to contemporary society, whereas museums abroad seem to be more concerned and connected to their communities.

The overall findings reveal that even if in the field of museum studies, as well as in the museum practice, social support, inclusion, and even activism are the current assumed perspectives, the wider public perceives them in relative terms. To some extent, museums are considered participatory, but not so active in terms of being active members of society. The data suggests that even those more interested and active when it comes to museums and culture have this perception. Therefore, museums should communicate more regarding their contribution to the debates concerning contemporary society.

References

Abend, L. (2020, March 31). Museums scramble to document the pandemic, even as it unfolds. *The New York Times*. Retrieved from https://www.nytimes.com/2020/03/31/arts/design/museums-coronavirus-pandemic-artifacts.html

Addis, M. (2005). New technologies and cultural consumption–edutainment is born! *European Journal of Marketing, 39*(7–8), 729–736. Retrieved from https://cursa.ihmc.us/rid=1234517332312_632271524_3467/%EE%80%80edutainment%EE%80%81.pdf.

Agostino, D., Arnaboldi, M., & Lema, M. D. (2020). New development: COVID-19 as an accelerator of digital transformation in public service delivery. *Public Money and Management*, 1–4. https://doi.org/10.1080/09540962.2020.1764206.

Antón, C., Camarero, C., & Garrido, M. J. (2018). Exploring the experience value of museum visitors as a co-creation process. *Current Issues in Tourism, 21*(12), 1406–1425.

Artforum. (2020, May 28). *Baltimore museum of art to provide financial relief to local artists and businesses*. Retrieved from https://www.artforum.com/news/baltimore-museum-of-artto-provide-financial-relief-to-local-artists-and-businesses-83129.

Atkinson, A. B., Guio, A. C., & Marlier, E. (Eds.). (2017). *Monitoring social inclusion in Europe*. Luxembourg: Publications Office of the European Union.

Bagdadli, S. (1997). *Il museo come azienda. Management e organizzazione al servizio della cultura*. Milano: Etas.

Bandelli, A., & Konijn, E. A. (2015). Public participation and scientific citizenship in the science museum in London: Visitors' perceptions of the museum as a broker. *Visitor Studies, 18*(2), 131–149.

Black, G. (2005). *The engaging museum: Developing museums for visitor involvement*. London: Routledge.

Bonacini, E. (2012). Il museo partecipativo sul web: forme di partecipazione dell'utente alla produzione culturale e alla creazione di valore culturale. *Il capitale culturale. Studies on the Value of Cultural Heritage, 5*, 93–125.

Brida, J. G., Dalle Nogare, C., & Scuderi, R. (2016a). Frequency of museum attendance: Motivation matters. *Journal of Cultural Economics, 40*(3), 261–283.

Brida, J. C., Meleddu, M., & Pulina, M. (2016b). Understanding museum visitor's experience: A comparative study. *Journal of Cultural Heritage Management and Sustainable Development, 6* (1), 47–71.

Brown, K., & Mairesse, F. (2018). The definition of the museum through its social role. *Curator: The Museum Journal, 61*(4), 525–539.

Burton, C., & Scott, C. (2007). Museums. Challenges for the 21st century. In R. Sandell & R. R. Janes (Eds.), *Museum management and marketing* (pp. 56–68). London: Routledge.

Cascone, S. (2020, April 8). Masks and other ephemera from the coronavirus pandemic to document history as it unfolds. *Artnet.com*. Retrieved from https://news.artnet.com/art-world/museums-starting-coronavirus-collections-1827606

Casey, D. (2007). Museums as agents for social and political change. In S. Watson (Ed.), *Museums and their communities* (pp. 292–299). London: Routledge.
Christiansen, K. (2020). The Met and the COVID crisis. *Museum Management and Curatorship, 35* (3), 221–224.
Cook, C. (2020, July 7). France: Rodin Museum sells works amid COVID-19 losses. *AA*. Retrieved from https://www.aa.com.tr/en/europe/france-rodin-museum-sells-works-amid-covid-19-losses/1903874
Culturefor. (2020). #CultureForDigital—MANN Napoli. *Culturefor.com*. Retrieved from https://culturefor.com/2020/04/18/culturefordigital-mann-napoli/?fbclid=IwAR2lhCdilBd_N9kb-NVh67CTiEoJHrpGXRFLYGhNE9VONr6f3D3GhOgxCKE
EC. (2017). *Special Eurobarometer 466*. Cultural Heritage Report. Retrieved from file:///D:/articole,%20lucrari%20etc/2020/Springer%20-%20Culture/ebs_466_en_final.pdf
Ellis, K. (2019). How social media is driving the climate change conversation. *NewsWhip*. Retrieved from https://www.newswhip.com/2019/09/social-media-is-driving-the-climate-change-conversation/
Emery, A. R. (2001). The integrated museum: A meaningful role in society*? Curator: The Museum Journal, 44*(1), 69–82.
Eurostat. (2016). *Culture statistics*. Luxembourg: Publications Office of the European Union. Retrieved from https://ec.europa.eu/eurostat/documents/3217494/7551543/KS-04-15-737-EN-N.pdf/648072f3-63c4-47d8-905a-6fdc742b8605.
Falk, J. (2016). Museum audiences: A visitor-centered perspective. *Society and Leisure, 39*(3), 357–370.
Firenze Today. (2020, April 6). Coronavirus, su Instagram per ripensare i musei del future. *Firenze Today*. Retrieved from https://www.firenzetoday.it/eventi/coronavirus-su-instagram-per-ripensare-i-musei-del-futuro.html
Harper, D. (2020, March 26). Museums and art schools donated supplies to hospitals overwhelmed with COVID-19 cases. *Artsy*. Retrieved from https://www.artsy.net/news/artsy-editorial-museums-art-schools-donated-supplies-hospitals-overwhelmed-covid-19-cases
Hede, A. M., Garma, R., Josiassen, A., & Thyne, M. (2014). Perceived authenticity of the visitor experience in museums. *European Journal of Marketing, 48*(7–8), 1395–1412.
Hofstede Insights. *Country comparison*. Retrieved from https://www.hofstede-insights.com/country-comparison/italy,moldova,romania/
Hooper-Greenhill, E. (1995). Museums and communication: An introductory essay. In E. Hooper-Greenhill (Ed.), *Museum, media, message* (pp. 1–12). London: Routledge.
Janes, R. R., & Sandell, R. (2019). Posterity has arrived. The necessary emergence of museum activism. In R. R. Janes & R. Sandell (Eds.), *Museum activism* (pp. 1–21). London: Routledge.
Karp, I. (1992). Introduction: Museums and communities: The politics of public culture. In I. Karp & C. Mullen Kreame (Eds.), *Museums and communities* (pp. 1–17). Washington, DC: Smithsonian Institution Press.
Kelly, L. (2010). How web 2.0 is changing the nature of museum work. *Curator, 53*(4), 405–410.
Kendall Adams, G. (2020, May 1). *Grassroots fundraising efforts take off amid Covid-19 crisis*. The Museums Association. Retrieved from https://www.museumsassociation.org/museums-journal/news/2020/05/01052020-grassroots-fundraising-efforts-take-off/#
Kim Lian Chan, J. (2009). The consumption of museum service experiences: Benefits and value of museum experiences. *Journal of Hospitality Marketing and Management, 18*(2–3), 173–196.
Knell, S. (2019). *The contemporary museum*. London: Routledge.
Komarac, T., Ozretic-Dosen, D., & Skare, V. (2020). Managing edutainment and perceived authenticity of museum visitor experience: Insights from qualitative study. *Museum Management and Curatorship, 35*(2), 160–181.
Kopke, J. (2011). The Denver community museum. *Curator: The Museum Journal, 54*(4), 399–402.
Kotler, N., & Kotler, P. (2007). Can museums be all things to all people? Mission, goals, and marketing's role. *Museums Management and Curatorship, 18*(3), 271–287.

Latorre, A. (2020, April 3). Arte, visibilidad y apoyo moral: nace el Covid Art Museum, el primer museo inspirado en la crisis del coronavirus. *20minutos.es*. Retrieved from https://www.20minutos.es/noticia/4214976/0/nace-el-covid-art-museum-el-primer-museo-virtual-inspirado-en-el-coronavirus/

Legget, J. (2018). Shared heritage, shared authority, shared accountability? Co-generating museum performance criteria as a means of embedding 'shared authority'. *Internationl Journal of Heritage Studies, 24*(7), 732–742.

Lord, B. (2006). Foucault's museum: Difference, representation, and genealogy. *Museum and Society, 4*(1), 1–14.

Lynch, B. T. (2011). Custom-made reflective practice: Can museums realise their capabilities in helping others realise theirs? *Museum Management and Curatorship, 26*(5), 441–458.

Manna, R., & Palumbo, R. (2018). What makes a museum attractive to young people? Evidence from Italy. *International Journal of Tourism Research, 20*(4), 508–517.

Mears, H., & Modest, W. (2013). Museums, African collections and social justice. In R. Sandell & E. Nightingale (Eds.), *Museums, equality and social justice* (pp. 294–309). London: Routledge.

Moreno Gil, S., & Ritchie, J. B. (2009). Understanding the museum image formation process: A comparison of residents and tourists. *Journal of Travel Research, 47*(4), 480–493.

NEMO. (2021, January 2021). *Follow-up-survey on the impact of the COVID-19 pandemic on museums in Europe Final Report*. Retrieved from https://www.nemo.org/fileadmin/Dateien/public/NEMO_documents/NEMO_COVID19-FollowUpReport_11.1.2021.pdf

Nowacki, M., & Kruczek, Z. (2020). Experience marketing at polish museums and visitor attractions: The co-creation of visitor experiences, emotions and satisfaction. *Museum Management and Curatorship*, 1–20. https://doi.org/10.1080/09647775.2020.1730228.

OECD. (2019). *Talent Abroad, a review of Romanian emigrants*. Retrieved from https://www.oecd.org/migration/talent-abroad-a-review-of-romanian-emigrants-bac53150-en.htm

Packer, J. (2008). Beyond learning: Exploring visitors' perceptions of the value and benefits of museum experiences. *Curator: The Museum Journal, 51*(1), 33–54.

Packer, J., & Ballantyne, R. (2002). Motivational factors and the visitor experience: A comparison of three sites. *Curator: The Museum Journal, 45*(3), 183–198.

Passy, Ch. (2020, April 6). History in the moment: Museums begin chronicling coronavirus pandemic. *Wall Street Journal*. Retrieved from https://www.wsj.com/articles/history-in-the-moment-museums-begin-chronicling-coronavirus-pandemic-11586210478

Pogrebin, R. (2020, September 16). Brooklyn museum to sell 12 works as pandemic changes the rules. *New York Times*. Retrieved from https://www.nytimes.com/2020/09/16/arts/design/brooklyn-museum-sale-christies-coronavirus.html

Potts, T. (2020). The J. Paul Getty museum during the coronavirus crisis. *Museum Management and Curatorship, 35*(3), 217–220.

Rentschler, R. (2007). Museum marketing. Understanding different types of audiences. In R. Sandelli & R. R. Janes (Eds.), *Museum management and marketing* (pp. 345–365). London: Routledge.

RFI. (2020, August 11). MNAC: Prima procedură de achiziții publice de artă contemporană din ultimii 12 ani. *RFI*. Retrieved from https://www.rfi.ro/cultura-123895-mnac-prima-procedura-achizitii-publice-arta-contemporana

Sandell, R. (1998). Museums as agents of social inclusion. *Museum Management and Curatorship, 17*(4), 401–418.

Sandell, R. (2002). Museums and the combating of social inequality: Roles, responsibilities, resistance. In R. Sandell (Ed.), *Museums, society inequality* (pp. 3–23). London: Routledge.

Sandell, R. (2003). Social inclusion, the museum and the dynamics of sectoral change. *Museum and Society, 1*(1), 45–62.

Sandell, R. (2007). Museums and the combating of social inequality: Roles, responsibilities, resistance. In S. Watson (Ed.), *Museums and their communities* (pp. 95–113). London: Routledge.

Scott, C. (2010). Museums, the public, and public value. *Journal of Museum Education, 35*(1), 33–42.
Sholts, S. (2020, February 26). How museums can help the public make sense of pandemics. *Smithsonian Magazine.* Retrieved from https://www.smithsonianmag.com/smithsonian-institution/how-museums-can-help-public-make-sense-pandemics-180974281/
Simon, N. (2010). The participatory museum. *Museum 2.0.* Retrieved from http://www.participatorymuseum.org
Smee, S. (2020, April 20). Museum association relaxes rules on selling art. *Washington Post.* Retrieved from https://www.washingtonpost.com/entertainment/museums/this-is-how-bad-things-are-for-museums-they-now-have-a-green-light-to-sell-off-their-art/2020/04/29/b5492a5e-899e-11ea-8ac1-bfb250876b7a_story.html
Stephens, S. (2020). *Natural History Museum declares climate emergency. Museum launches strategy for tackling the planetary crisis.* Retrieved from https://www.museumsassociation.org/museums-journal/news/2020/01/23012020-nhm-declares-climate-emergency/#
Stylianou-Lambert, T. (2009). Perceiving the art museum. *Museum Management and Curatorship, 24*(2), 139–158.
Taylor, C. (2017). From systemic exclusion to systemic inclusion: A critical look at museums. *Journal of Museum Education, 42*(2), 155–162.
Tully, G. (2020). Are we living the future? Museums in the time of Covid-19. In F. Burini (Ed), *Tourism facing a pandemic: From crisis to recovery* (pp. 229–242). Università degli Studi di Bergamo.
Valentine, V. (2020, April 19). In wake of COVID-19, artists are uniting to support the Museum of the African Diaspora San Francisco with an online benefit auction. *Culture Type.* Retrieved from https://www.culturetype.com/2020/04/17/in-wake-of-covid-19-artists-are-uniting-to-support-museum-of-the-african-diaspora-san-francisco-with-online-benefit-auction/
Varella, S. (2020). Migrants arrivals in Italy, 2014–2020. *Statista.* Retrieved from https://www.statista.com/statistics/623514/migrant-arrivals-to-italy/
Verdon, M. T. (2020). Art in time of Corona virus. *Museum Management and Curatorship, 35*(3), 215–216.
Watson, S. (2007). Museums and their communities. In S. Watson (Ed.), *Museums and their communities* (pp. 1–23). London: Routledge.
Zardini Lacedelli, S. (2020, February 27). La risposta Dei musei italiani al coronavirus: Diffondiamo cultura e positività attraverso il web [The reaction of Italian museums to covid-19: Spreading beauty and culture on the web]. *Muzeo Dolom.* Retrieved from http://www.museodolom.it/home/it/2020/02/27/la-risposta-dei-musei-italiani-al-coronavirus-diffondiamo-cultura-e-positivita-attraverso-il-web/
Zbuchea, A. (2015). *Marketing muzeal pentru nonmarketeri* [Museum marketing for nonmarketers]. Bucharest, RO: Tritonic.

Alexandra Zbuchea is Associate Professor and Vice-Dean of the Faculty of Management at the National University of Political Studies and Public Administration, Bucharest, Romania. She is the Executive Manager of the Center for the Study of Responsible Organizations from the same faculty, a member of several organizing and scientific boards of conferences and academic events, as well as a board member for several academic journals. She is part of professional associations such as the Academy of Marketing Science or the International Association of Knowledge Management. Also, she is active in an NGO promoting education for children through access to heritage and culture—Da'DeCe. Until 2018 she was a member of the Board of the National Network of Museums in Romania. Since 2006, she is a consultant for museums in management and marketing, as well as trainer in cultural marketing and visitor studies. She published several books and many articles in marketing for museums, promotion, knowledge management, cultural tourism, etc. She was twice Fulbright Scholar, at Columbia University and New York University, in the field of museum studies.

Mauro Romanelli is Assistant Professor of organisation studies and human resource management at Parthenope University of Naples, Department of Business and Economics, Napoli, Italy, where he teaches courses related to Organization of international businesses, Human resource management, Ethics and public organizations. He is member of international academic and professional associations in management and organization. Main research areas in organisation studies concern sustainability, change, management and ethics of public organisations, information technology and public administration, museums, technology and cultural heritage, smart and sustainable cities, communities and ecosystems.

Monica Bîră is senior lecturer at the Faculty of Communication and Public Relations (within the National University of Political Studies and Public Administration, Bucharest, Romania) where she teaches courses related to PR and Communication (e.g. PR measurement, Communication in Creative Industries). Her research interests generally revolve around the construction of social representation. How museums are contributing to shaping communities and to forge identities has been a constant topic of her publishing activity, starting with her PhD thesis on the national antiquities and their impact in the national construction in Romania during the nineteenth century. As a member of the Research lab on Image and Identity within the Faculty of Communication, she also contributed to several books and published articles on museums and their.

The Leadership Dance in a Performing Arts Organization

Anna Chiara Scapolan and Martina Gianecchini

Abstract The leadership structure of cultural organizations has been a central issue in cultural management studies for many years. However, extant research tends to characterize leadership structure either as unitary or dual, employing a static approach that fails to explain the process underlying its evolution when the cultural organization changes and develops. Drawing on these limitations, we aim to answer the following research question: how do leadership forms change over time in cultural organizations?

We use a qualitative methodology based on an exploratory case study carried out between 2011 and 2018 on the Italian performing arts organization "Fondazione Nazionale della Danza Aterballetto."

Our findings show that different plural leadership forms may emerge during an organizational change. In addition, we show how various factors and dynamics at multiple levels (individual, organizational, and environmental) come into play. Our study suggests that cultural organizations aimed at establishing a stable leadership structure should consider not only their internal strategic demands and organizational conditions but also external factors. In addition, our study demonstrates that both informal and formal rules influence leader's expected behaviors.

Keywords Cultural organizations · Performing arts · Plural leadership · Dual structure · Leadership change

A. C. Scapolan (✉)
Department of Communication and Economics, University of Modena and Reggio Emilia, Reggio Emilia, Italy
e-mail: annachiara.scapolan@unimore.it

M. Gianecchini
Department of Economics and Management "M. Fanno", University of Padova, Padova, Italy
e-mail: martina.gianecchini@unipd.it

P. Demartini et al. (eds.), *Cultural Initiatives for Sustainable Development*, Contributions to Management Science,
https://doi.org/10.1007/978-3-030-65687-4_5

1 Introduction

Management can be viewed as an art rather than as a science. Dancing, playing music, and painting can be viewed as sciences rather than as arts. Nonetheless, very often management and arts are depicted as two separate and contrasting domains, each one characterized by opposite goals and logics (e.g., Caves 2002; Delmestri et al. 2005; Lampel et al. 2000). By echoing an ancient philosophical tradition that defines a trade-off between economic rationality, i.e., reason, and artistic expression, i.e., passion (see, for instance, DeFillippi et al. 2007), this duality is reflected in the leadership forms of cultural organizations. In fact, to manage the economic and artistic domains and their often conflicting core objectives, many cultural organizations adopt, instead of a solo leader, a dual leadership structure (Denis et al. 2012) in which two leaders—an artistic director and an executive or managing director—take the responsibility for the two complementary but separate functions of the company that are, respectively, economic administration and artistic work (Cray et al. 2007).

Even though many studies have underlined the benefits of the dual structure versus a solo leader who struggles in possessing all the skills and competences needed by the management and artistic domains (e.g., Bhansing et al. 2012), lately scholars are starting to question the polarization of this dualism and call for research that explores the intersection of these two leadership spheres. For instance, those studies that focused on the dual leadership structure highlighted issues related to the relationship between the two leaders, such as conflicts (Reid and Karambayya 2009), problems of trust (Reid and Karambayya 2016), and differences in the leaders' perceptions and orientations (Bhansing et al. 2012, 2016). In addition, some authors have recently suggested the existence of leadership forms where the dual leaders, also in cultural settings such as film-making, are supported by a third role as mediator (Ebbers and Wijnberg 2017). As a result, trios or even larger executive constellations may informally emerge or be adopted formally to share the leadership at the top (Alvarez and Svejenova 2005). These studies identify diverse strategies and mechanisms of role crafting, through which the leaders themselves may contribute to define the leadership form, and suggest contextual factors that may impact on the design of leadership structure. However, they fail in capturing the emergence of different forms of leadership in cultural organizations (e.g., unitary, dual, and more extensive plural) and in identifying those factors and dynamics that support the transition from a leadership structure to another over time.

Hence, drawing on the literature of leadership forms of cultural organizations, we aim to answer the following question: how do leadership forms change over time in cultural organizations? Based on a case study of Fondazione Nazionale della Danza *Aterballetto* that is the only Italian fully public dance production center, this chapter contributes to the extant literature in three main ways. First, we contribute to overcome the strict functional duality concerning artistic and management purposes, offering a more nuanced perspective about the leadership structure traditionally attributed to cultural organization (i.e., unitary or dual) showing that different plural leadership forms may emerge. Second, we show how various factors and dynamics at multiple levels (individual, organizational, and environmental) come into play in connection with the role (re)definition of leaders during a change in the leadership structure. In doing so, we add to the current literature which mainly focuses on the individual characteristics of leaders (e.g., Bendixen 2000; Mumford et al. 2002) and the relationships between leaders (e.g., Bhansing et al. 2012; Ebbers and Wijnberg 2017; Reid and Karambayya 2009), overlooking other factors that may affect the adoption of a specific leadership structure and the underlying process which may lead from a collective leadership form to another one (Denis et al. 2012). Finally, as studies on plural leadership have predominantly focused on theoretical development and practical application (e.g., Alvarez and Svejenova 2005), by exploring how leadership structures are introduced and may change over time, we add to the few empirical studies on this topic (Denis et al. 2001) specifically in the cultural setting. The chapter is organized as follows: the next section outlines our theoretical framework, basically reviewing extant literature on leadership structures in cultural organizations; the following section describes the setting and our research method. Then we illustrate the main findings of our empirical study that are discussed in the Sect. 5. Finally, the section "Conclusions" remarks the theoretical contribution, the managerial implications, and the main limitations of our study.

2 Theoretical Framework

Cultural organizations are a very specific setting for studying leadership, as they are examples of pluralistic organizations characterized by multiple logics, values, and core objectives, where power relationships are diffuse (Denis et al. 2012). More specifically, extant research has highlighted that cultural organizations face the challenge of organizing creativity (DeFillippi et al. 2007; Lampel et al. 2000), struggling in managing together art and business, exploration and exploitation, and identities and values of artists and non-artistic personnel (e.g., Caves 2002; Glynn 2000; Knight and Harvey 2015). Such multiple domains are often divergent, sometimes even mutually exclusive (Delmestri et al. 2005). For instance, being too openly concerned with commercial success may be detrimental to an organization operating in the cultural field. Similarly, the exploitation for efficiency and profitability may undermine the efforts of exploration for artistic excellence and innovation whose returns are uncertain (DeFillippi et al. 2007), thus suggesting a possible trade-off also between short-term operational decisions and long-term, strategic decisions.

Moreover, the coexistence within cultural organizations of artistic, technical, and managerial-administrative professional areas and workers (Castañer and Campos 2002; Glynn 2000) may create problems in terms of coordination and integration (Knight and Harvey 2015).

As a result, leading a cultural organization requires the capacity to balance contradictory forces, which often translate into reconciling artistic goals with the economic constraints (Caves 2002). Moreover, leadership in cultural organizations is strategic since leaders, besides being responsible for the internal management, represent the image of the organization, which, in turn, affects resource attraction and consensus from external audiences (Montanari 2017).

Since a single individual (i.e., a solo leader) rarely possesses the education, the professional background, and the skills necessary to manage the multiplicity of domains and objectives of a cultural organization, cultural organizations often adopt forms of plural leadership (Alvarez and Svejenova 2005). These forms are characterized by a collective leadership where multiple leadership roles are attributed to two, three, or more people and linked to the main organizational domains and objectives (Denis et al. 2012). Among these forms, the simplest one, which is very common in museums, theatres, and the film-making industry (Reid and Karambayya 2009), is the dual leadership structure (Cray et al. 2007) or diarchy (Montanari 2017). In a dual leadership structure, the two leaders, sitting at the top of the hierarchy, have a clear division of responsibilities and goals (Ebbers and Wijnberg 2017). The two leaders are hierarchical equivalent executives—usually an artistic director and an executive/managing director—who lead the organization together (Bhansing et al. 2012; Reid and Karambayya 2009).

However, since the activities performed by the two leaders may present strong interdependencies, risks of conflicts (Reid and Karambayya 2009) and issues related to trust (Reid and Karambayya 2016) could arise, especially when their responsibilities and goals are in opposition. Such problems may spread into the whole organization causing, for instance, deepened conflicts or tensions among supporting roles and units (Ebbers and Wijnberg 2017). Hence, questioning the traditional sharp division between artistic and managerial-administrative leadership, some scholars suggest the inclusion of a third member in the leadership structure (Alvarez and Svejenova 2005). The third member can either legitimate one of the other two leaders or plays as integrator, mediator, and/or arbiter between them.

Such a third role may be formalized in a trio or triad, i.e., a structure where the leadership is formally shared among three top executives, but it can also emerge by informal processes where, for instance, either higher-level (e.g., board members) or lower-level (e.g., subordinates) actors facilitate the functioning of the dual structure, eventually determining de facto a trio or even a larger executives constellation (Denis et al. 2012). For instance, in a study on the film industry, Ebbers and Wijnberg (2017) showed how the collaboration between the director and the producer is mediated by a third actor (i.e., the assistant director) and how the three actors craft their leadership roles. Moreover, in accordance with Alvarez and Svejenova (2005) who identified the different abovementioned strategies by the third leader, the authors eventually assert that “when dual leaders have been brought together by others, the role defining dynamics could play out differently” (Ebbers and Wijnberg 2017, p. 1361).

Extant literature on leadership in cultural management has mainly focused on the skills, attitudes, and orientations of cultural leaders regardless of whether they are solo or dual leader (Bendixen 2000; Mumford et al. 2002; Scapolan et al. 2017). Moreover, extant studies investigated the relationship between dual leaders (Reid and Karambayya 2009; Bhansing et al. 2012), and their role definition (Denis et al. 2012; Ebbers and Wijnberg 2017). Only some scholars have examined the boundary conditions of leadership forms and the factors influencing the choice between different leadership structures (mainly between the solo leader and the duo). These studies highlighted the role played by the stage of development of the cultural organization, since "managing a phase of growth involves very different problems than a phase of rationalization or consolidation" (Montanari 2017, p. 280). Similarly, Alvarez and Svejenova (2005) identified the strategic uncertainty as one of the external, environmental factors that are relevant to the design of the leadership structure. They also mentioned normative pressures, the internal organizational context (i.e., the need of multiple perspective in decision-making; the need to manage internal political relationships), and the needs and concerns of leaders themselves. Thus, for instance, strategic uncertainty in the external environment makes plural leadership (meaning not only duos but also trios or more) more relevant, but normative pressures for clear accountability may push toward a solo leader. The need of leaders for psychological security and support may favor plural leadership, while leaders' concerns for career development are likely to promote unitary leadership. However, how these contingent factors may lead to choose a specific form of plural leadership (a duo or a trio) instead of another (e.g., a constellation) has been overlooked. Moreover, even though such studies provide also useful insights on which factors and dynamics may potentially lead to the evolution of the leadership structure over time, rare is empirical research on how plural leadership form and reform as leaders challenge to reconcile internal interpersonal tensions and organizational context with the external environment (Denis et al. 2012). To the best of our knowledge, the few extant research was conducted by Denis et al. (2001) who made the effort of integrating the different (external, internal, interpersonal) factors that may contribute to a change in the leadership form also with change tactics performed by leaders during the change. However, even though they focused on pluralistic organizations, unfortunately they did not take into consideration cultural organizations and their specificities. As a result, there is a gap in the empirical research concerning how to combine contingent factors which may affect leadership change with the abovementioned formal and informal dynamics and strategies of role defining to describe and interpret holistically the change process of leadership in cultural organizations.

To fill this literature gap, we propose a qualitative study aimed at exploring the process of change in the leadership structure of a cultural organization and specifically the factors and dynamics which affected the transition between different forms of plural leadership structure in a performing arts organization.

3 Methods

3.1 Research Setting

We conducted a case study on Fondazione Nazionale della Danza (National Foundation of Dance) *Aterballetto* (hereafter FND *Aterballetto)*, an Italian cultural institution located in Reggio Emilia, a medium-sized city in the North of Italy.

FND *Aterballetto* is the only Italian fully public dance production center, whose core activities are promoting and distributing dance performances and providing educational projects on dance language.

We adopted a case-based methodology since case studies allow to investigate new research areas and are particularly well suited to answer "how? and why? questions" (Eisenhardt and Graebner 2007; Yin 2013).

FND *Aterballetto* was founded in 1979 when the dance company *Aterballetto* was established as the first Italian permanent touring company aimed at performing artworks and producing original shows. In 1991 *Aterballetto* became the dance company of the Centro Regionale della Danza (Regional Dance Centre) which 12 years after was transformed in FND *Aterballetto* whose founding members are the Emilia-Romagna Region and the Municipality of Reggio Emilia. Under this governance structure, FND *Aterballetto* continues to carry out its main, core activities—i.e., producing and distributing dance performances with the brand *Aterballetto*—and it has been invested in consolidating other activities related to its mission (e.g., master training of dancers and teachers). Besides its founding members, *Aterballetto* is financed also by the MIBAC (Italian Ministry of Cultural Heritage and Activities), and its activities are supported by the Italian Institutes of Culture, the Italian embassies, and the Ministry of Foreign Affairs.

FND *Aterballetto* has 33 employees, and it is organized as follows: the Board of Directors has four members who are appointed for a renewable term of 3 years by the two founding members; these latter entrust the role of President to one of the board's member. The Board is in charge of monitoring the General Director's (GD) work, who is recruited by means of a public announcement of selection and then appointed for a renewable term of 4 years by the same Board. The GD supervises the management and administration area, being in charge of the following organizational units: production and tour management, sales and distribution, communication and public relations, special projects and education, and accounting. The GD supervises also the technical and artistic area, which consists of the dance company *Aterballetto*. More specifically, this area includes the dancers (16 people), one *maître de ballet*, one artistic coach, the lighting director, four stage operators, and a costumer. Until 2018, the artistic area was assigned to an Artistic Director (AD) appointed for a renewable term of 3 years by the Board of Directors. Between the end of 2016 and the beginning of 2017, when the Board of Directors was at the end of its term, the GD and the AD were also at the end of their (already renewed) mandate, a recruitment and selection process was carried out, and a new GD and AD were selected by the expiring Board. They were formally appointed on July 2017, as soon as the new Board of Directors was established. However, at the beginning of 2018, only after few months, the new AD left FND *Aterballetto*. At that point, one of the ballet masters of that time was internally selected to perform the role of Artistic Coordinator, a role which has never existed in the organization before. She reports directly to the GD, and together with him and his Executive Assistant, she is a member of the Direction team. As a consequence of the creation of the artistic coordinator role, the GD changed his job title in "Managing and Programming

Director." We collected data and performed data analysis in order to understand the mechanisms involved in the leadership structure change described above.

3.2 Data Collection

The present study builds on earlier research projects for which we collected the data between November 2009 and July 2012 and, again, between April and September 2014, using multiple sources: direct observation, interviews, and documents. These projects resulted in two articles (Montanari et al. 2016; Scapolan et al. 2017). More specifically, for the first project—which was focused on the relationship between the Resident Choreographer of the dance company and FND *Aterballetto*—we conducted direct observation, visiting FND *Aterballetto* twice a month, observing rehearsals and shows, and attending staff and management meetings. In addition, we conducted 16 in-depth semi-structured interviews with the GD (twice), AD (twice), ballet masters (two people), general secretary, tour manager, and dancers (five people). We also engaged in informal conversations with the President and four industry experts. Finally, we gathered documental information from newspaper and magazine articles, FND *Aterballetto*'s website, official documents, and press archives. For the second research project—which was focused on the soft skills of managerial roles in arts organizations and organizational performance—we performed three structured interviews with the same President, GD, and AD we had already interviewed.

The qualitative approach we used in the abovementioned research projects allowed us to collect rich empirical evidence about the leadership structure of FND *Aterballetto* in the period between 2009 and 2014. On this topic, we collected additional evidence between December 2016 and September 2018, when the new GD and AD were selected and the new President was nominated. We used both documental sources and semi-structured interviews. More specifically, we interviewed the former President (two times) and the new President; the new GD; the Artistic Coordinator; and the Executive Assistant. The interviews lasted on average about 1 h and were audio recorded and transcribed. Interviews addressed the following main topics: (1) personal and professional background of respondents; (2) recruitment and selection process of the new GD and AD; (3) exit of the new AD and its impact on the organization; (4) interviewees' roles (tasks, duties, and responsibilities) in FND *Aterballetto*; and (5) interviewees' relationships in FND *Aterballetto*. In addition, we search LexisNexis® Academic for newspaper and magazine articles that contained both (at least one of) the names of the people involved in the leadership of FND *Aterballetto* and the word "Aterballetto" and that were published between January 2009 and September 2018. We found 27 relevant articles that provided useful data to reconstruct the changes in the leadership roles, the selection of the new GD and his activities in FND *Aterballetto*, the exit of the new AD, and the selection of the Artistic Coordinator. Finally, we also gathered data from *Aterballetto*'s website, official documents (e.g., public selection announcements, statutes), and press archives, focusing on articles published in the press before or after leadership structure changes.

Although this additional evidence forms the core data of this article, our findings have been supplemented by evidence from our earlier projects, especially those concerning tasks, duties, and responsibilities of the President, the GD, and the AD and the relationship among them.

3.3 Data Analysis

The data analysis follows the sequence proposed by Langley (1999). The initial stage of the analysis involved the compilation of an "event history" to document "who did what, and when" (Garud and Rappa 1994) and a narrative account of how the leadership structure of FND *Aterballetto* have changed over time. Next, we conducted an inductive analysis of the qualitative data we collected, triangulating the data we obtained by multiple sources. More specifically, first we independently read the data, and then we shared and discussed our impressions and interpretations until we reached a common understanding and agreement. Finally, we coded our data following a grounded theory method with an interpretive approach (Gioia et al. 2013). As we were interested in depicting the change in the leadership form over time, from the interviews and other documents, we identified how the structure of the leadership roles changed depending on two dimensions: (1) the domain of the leadership role (artistic vs. business) and (2) the focus of the leadership role ("strategic/long-term" vs. "operational/short-term"). Moreover, we coded emergent themes concerning both the elements that affected the design of the leadership form (we label these elements as "factors") and the processes (labeled as "dynamics") by which these factors contributed to modify the structure of the leadership roles. We found out that both factors and dynamics operate at different levels: individual, organizational, and environmental.

Table 1 provides a summary of the main findings, showing in the first column the three levels at which the factors and dynamics operate and in the second and third columns, respectively, the factors and the dynamics through empirical evidence (e.g., illustrative quotes). In the next section, we will elaborate our findings in detail.

Table 1 Summary of findings and illustrative quotes

Level	Factors	Dynamics
Individual	The former President, who is a management professor, has brought a strong attention to the economic-financial aspects. I can bring—and I think this is one of the reasons for my presidency—institutional relations. (*Interview with the new President, June 6, 2018*) The GD is one of the most renowned professionals in the field of management and organization of dance. A "big name" in festivals' organization and artistic planning. Thus, entrusting to him both the managerial and the artistic activities was not surprising at all. (*Interview with the Executive Assistant, September 25, 2018*) Currently, I am also the Director of API Reggio Emilia. As Director I am involved in all the managerial activities. (*Interview with the new President, June 6, 2018*)	I strongly believe that through the delegation would improve the abilities of my subordinates to take their own responsibilities and also their professional development. (*Interview with the new GD, July 2, 2018*) Even if it is clear that the final decision is mine, she [the Artistic Coordinator] fully participates to the whole decision-making process; she fully realizes the added value of me keeping both the responsibilities. She is happy with her role and she really likes it. So, it works really well and we're both very happy. (*Interview with the new GD, July 2, 2018*) I do not go in the hall with dancers and choreographers, that is a responsibility of the Artistic Coordinator (*Interview with the new GD, July 2, 2018*) I talk a lot with the Artistic Coordinator and we are slowly developing a high level of understanding. (*Interview with the new GD, July 2, 2018*)
Organizational	We decided to assign the artistic responsibilities not to a "traditional" artistic director, but to an artistic coordinator that acts as an intermediary between the company and the choreographer. (*Interview with the new President, June 6, 2018*) I coordinate the choice of choreographers and shows to host with the management and planning. (*Interview with the new GD, July 2, 2018*) The President has a guaranteed role, while the GD (who is not a legal representative) is entrusted with all the executive aspects: he basically manages the organization. (*Interview with the new President, June 6, 2018*) My role experienced a radical transformation, not only formally in the job title, but also substantially: in the previous governance system I was in staff to the General and Artistic Direction, now I am an Executive Assistant with higher responsibility and more tasks. (*Interview with the Executive*	The new GD is revisiting the internal division of labor, paying attention to the sharing of information among not only the managers of offices but among all employees, by means of weekly and other regular meetings. (*Interview with the Executive Assistant, September 25, 2018*) Since the GD has an influence on the artistic decisions the integration between the artistic and the business activities is higher, and the exchange of information is easier. There is a risk of conflict, but the same—and even higher!—was present when the artistic and business directions were appointed to two different people. (*Interview with the new President, June 6, 2018*) The company used to have a GD and an AD, then, as a result of an unfortunate experience [the AD leaving], we decided to change. (*Interview with the new President, June 6, 2018*) When the new AD left, we were in the

(continued)

Table 1 (continued)

Level	Factors	Dynamics
	Assistant, September 25, 2018) The new GD is involved in the strategic decision even if he is not seated on the Board. The GD should be an executive role, but he participates to the decision-making process. (*Interview with the new President, June 6, 2018*) As far as the new formal division of labor between the President and GD is concerned, basically the new Statute formalized what already happened informally. (*Interview with the Executive Assistant, September 25, 2018*)	middle of the artistic season [. . .] When something like this happens a company tries to handle the situation until the new season. In addition, we wanted to preserve our public image, so we choose a temporary solution: the artistic coordination of the company was entrusted to one of the maître. [. . .] However, this solution is still going on because it works very well! (*Interview with the Executive Assistant, September 25, 2018*) While in the past, the business relationships were created by the GD, and then partially managed by the AD, now I involved three other people—besides me—in creating and managing relationships with new partners. (*Interview with the new GD, July 2, 2018*) The former GD at the end of his mandate actively contributed to the selection of the new GD. Also the new GD, after his official appointment, will participate as an external member in the process of selection of the new artistic director. (*Interview with the new President, June 6, 2018*)
Environmental	All these institutions were changing their directors and there was the willingness to choose people who would willing to create inter-organizational collaboration, supporting the development of a local cultural system. (*Interview with the Executive Assistant, September 25, 2018*) The mandate for the new GD is to further strengthen the position of FND Aterballetto within the local, national and international cultural networks. The aim is to make the Foundation "the" Dance Center and not "a" Dance Center: Aterballetto should become the most important company in Italy. (*Interview with the new President, June 6, 2018*) Nowadays many cultural organizations do not have separated roles, especially in big arts organizations, such as the Opera of Rome, where one would	This was my priority: position FND Aterballetto within a national and international network. (Interview with the new GD, July 2, 2018) Here in FND Aterballetto I can dedicate myself to the activity I prefer, the lobbying activity. (*Interview with the new President, June 6, 2018*)

(continued)

Table 1 (continued)

Level	Factors	Dynamics
	expect a greater multiplication of roles. This is done in order to more effective, to control the costs and to optimize the planning. (*Interview with the new GD, July 2, 2018*)	

4 Findings

4.1 The Leadership Structure in FND Aterballetto *Between 2011 and 2017*

From 2011, when the President of the Board was appointed for his first term, the GD and the AD had been just renewed by the Board, so they could sit on the top management team of FND *Aterballetto* for the next 6 years together.

The AD and the GD have separate tasks: the AD was in charge of the artistic development of ballet company, and the GD was in charge of the business activities. According to the organizational structure of the time, the GD was in a hierarchically higher position compared to the AD. For instance, it was GD who established a 3-year collaboration, starting from May 2012 (and then renewed in 2014), between FND *Aterballetto* and Piccolo Teatro of Milan (the first permanent theatre in Italy and internationally renowned): that agreement granted to *Aterballetto* the opportunity to spend each year 3 weeks in Milan, presenting its performances to a large national and international audience. In the same period, the GD was also in charge of defining a contractual agreement with the labor unions with the aim of "reducing the fixed costs" (*Il Resto del Carlino*, September 27, 2011). Regardless of their distinct tasks and hierarchical roles, the GD and the AD collaborated on many important decisions related to the development of the company. For example, when the Resident Choreographer, who collaborated with FND *Aterballetto* until 2012, was going to leave the company, also the GD was involved in "exploring the best of contemporary dance, to identify a new resident choreographer" (*Il Resto del Carlino*, March 12, 2014).

The duo worked very closely with the President "Between the President and the GD, beyond the formal meetings of the Board, there was a very frequent exchange of opinions and a continuous discussion on different issues. [...] very often the President was also physically present here in FND *Aterballetto*" (interview with the Executive Assistant, September 25, 2018). The President, according to the Statute, was entrusted with the supervision of the administrative and financial activity of FND *Aterballetto* and the management of human resources (Article 11). The Statute did not include the roles and the responsibilities of the AD and of the GD. Therefore, even if the President was formally appointed to take all the relevant decisions in the organization, "in reality, I [the President] decided together with the Board the strategies and then the GD was entrusted to translate them in operational activities. Obviously, the GD and the AD discussed with me the artistic decisions, but they were the ultimate decision-makers on those aspects" (interview with the former President, May 30, 2018).

4.2 The Change in the Leadership Form

At the end of the second term, the President, together with the outgoing GD and AD, drafted two public announcements in October 2016, for a new GD selection, and in December 2016, for a new AD. In doing so, they took into consideration the desire of the two shareholders (represented by the Board of Directors) to promote a renovation in continuity with the past. More specifically, "they wanted to strengthen the excellent work made so far on the development of the dance company, which is certainly our core activity, but also to promote a change in the management of other important activities, such as the education projects and the activities related to the production center [i.e., planning and hosting performances of other dance companies]" (interview with the Executive Assistant, September 25, 2018).

Consistently, the responsibilities and the tasks described in the public announcement for the selection of a new GD reflected the leadership structure that had characterized the organization in the previous 6 years:

> The Director [GD] works according to the guidelines and within the terms established by the Board of Directors. In this context, the Director [GD] is responsible for planning and managing all the activities of the Foundation. To this end s/he proposes to the Board of Directors a three-year artistic plan (in agreement with the Artistic Director), s/he is responsible of the financial statements, s/he directs the staff of the Foundation, s/he executes the Board of Directors' directions. [*Excerpt from the job description included in the advertisement of public selection for the position of Managing Director, October 30th, 2016*]

The desire for a continuity in the artistic identity of the dance company was also clearly expressed by the outgoing AD who, since the exit of the Resident Choreographer, worked for the development of a new repertoire of contemporary dance pieces created by both internationally renowned choreographers and young promising Italian choreographers: "The new GD will not change the artistic direction of the company [. . .]. We designed it in the last 3 years, producing over 20 ballets: do you think that a single person could sponge off such endowment?" (interview with the former AD, press agency *Ansa*, June 6, 2017). To favor such continuity, the new GD and AD were expected to work together with the outgoing directors for a brief period, to manage the transition "without any shock" (*Ansa*, June 6, 2017). However, as a consequence of the indications included in the new Statute of FND *Aterballetto* approved in July 2017, some changes in the duties and the responsibilities of the President and the GD were introduced. In particular, the new current Statute explicitly defines the role and the responsibilities of the GD and, at the same time, limits the power of the President:

> The tasks of the GD, who is responsible for managing the Foundation, are: a) collaborating in the definition of the general guidelines and the operational strategies of the Foundation;

> [. . .] c) directing and coordinating, [. . .] the technical-administrative, business-financial, artistic-cultural activities of the Foundation; [. . .] e) preparing the activity programs to be submitted to the Foundation, considering the AD's opinion; [. . .] i) managing the Foundation staff. [*Excerpt from the Article 14 of FND Aterballetto Statute*]

The selection process of the GD was articulated in two steps: an initial screening was based on the candidates' curriculum vitae; in the second step, the selected people were required to present a 3-year plan including an artistic and business development proposal of FND *Aterballetto*. Out of 12 candidates, 4 were invited to send their projects. The four plans were analyzed by a commission composed by the members of the Board of Directors, the outgoing GD, and some representatives of the founding members. The project presented by the winner was appreciated by the commissioners because "it contained some ideas about how to improve networking and increase the internationalization of *Aterballetto*" (interview with the former President, May 30, 2018). A change in the strategy of FND *Aterballetto*, toward an increased collaboration with other cultural organizations at local, national, and international level, was indeed encouraged in that period by the new management of the most relevant cultural institutions operating in the city of Reggio Emilia and in the Emilia Romagna region (e.g., Fondazione Palazzo Magnani). According to the Executive Assistant "FND *Ateballetto* was part of a broader renewal project which involved all the main cultural organizations of the city [of Reggio Emilia]" (interview, September 25, 2018).

After having appointed the new GD, FND *Aterballetto* proceeded with the selection of the AD. The company received 14 curriculum vitae. After an initial screening based on prior artistic and managerial experience, a few candidates were invited for an interview with the members of the Board. The interview concerned professional experience, motivation, knowledge of the national and international dance industry, and artistic projects. In order to take the final decision, the selection committee "involved also the new GD" (interview with the former President, May 30, 2018).

4.3 The New Leadership Structure

For the previous 15 years, the new GD worked at the Teatro Stabile of Turin (Italy) where he directed "Torinodanza," a relevant international festival of dance. Holding a graduation in contemporary history and further specialization in culture management, he has directed permanent festivals in Italy and collaborated with several Italian theatres, during his entire professional career. The new GD was appointed with the specific aim of positioning FND *Aterballetto* as a point of reference for national and international institutions and organizations. This goal clearly appeared in his first interviews "In the future, *Aterballetto* will no longer simply be a ballet company [. . .]. We will open our artistic development to a plurality of arts ranging from theater to music" (*Corriere della Sera*, April 8, 2018). In addition, some commentators noted that he was well-known in the ballet industry as a skilled "weaver" of institutional relations; this orientation was also evident in his words "we will try to create partnerships with important European organizations. Our goal is to fill an unprecedented role in our country, accelerating the processes of internationalization of the entire Italian dance sector. This means that we will immediately seek a conversation with all the institutions and in particular with the Ministry and the Region" (the new GD interviewed by Silvia Poletti, *Delteatro.it*, March 31, 2017).

The new GD had clear ideas about both the development of the dance company and about his role: he started to involve his direct subordinates in his activities through delegation, stimulating participation and sharing his decision-making power. As a result, although the Statute assigned him a strong formal power over all the decisional domains, he promoted the empowerment of his Executive Assistant (a long-term employee of FND *Aterballetto* as she worked there from 2005), who was included in the Direction team:

> I have a clear idea about leadership. People in *Aterballetto* were used to a fragmented decision decision-making process: now I introduced a shared process, where decisions are quick, and everybody is aware of their own responsibilities. Without clear responsibilities, as it was in the past, every decision can be easily questioned. (Interview with the new GD, July 2, 2018)

The new AD had never covered a similar role before, but she had a long international experience as dancer and teacher. She worked with Mats Ek of Cullberg Ballet of Stockholm for 24 years, having the possibility to work with very important choreographers (e.g., Jiri Kylian and Carolyn Carlson). Therefore, her long, international artistic experience represented an element which favored her appointment. In addition, the GD and the new AD knew professionally each other as, after leaving Stockholm, she established in Turin, where the GD was very active in the dance scenario. Nonetheless, 5 months later, she left FND *Aterballetto*. "The GD and I did not know each other personally [. . .]. When we started working together we discovered that our personalities were not compatible and, as we all know if the GD and the AD are not in complete harmony nothing can happen" (the new AD interviewed by *Sipario*, April 2018). This exit was unexpected because the relationship was apparently running smoothly: as a matter of fact, when a newspaper printed the news (Zapparata, *La Repubblica*, November 15, 2017), other journalists immediately reacted saying it was a fake news (Ratti, *Dance&Culture*, November 16, 2017). However, some people in the organization—and particularly in the dance company—were not totally surprised since they perceived that she was not comfortable in her role. "[The new AD] was a brilliant professional but she had never managed a ballet company before: she had substantial experiences in teaching choreographies, but nothing more. This, in my opinion, was a problem" (interview with the Artistic Coordinator, July 12, 2018). Also, the new GD maintained "The relationship with the AD failed because it was based on a premise that couldn't work well: she expected to behave as a traditional AD, but I had something else in mind" (interview, with the new GD, July 2, 2018).

To face this unexpected situation, the GD together with the Board had to decide whether to search for another AD, bearing the cost and the time of a new selection process, or to cover this vacancy in a different way:

> If I had selected a new AD, I would have said to her 'you will offer advice, provide suggestions, manage the relationships in the studio, but the final decisions are mine'. Moreover, I discovered that nowadays young people move directly from dancing to creating. They start to create when they are in their 40s. However, 'real' ADs are usually 70 years old and they are impossible to find. (Interview with the new GD, July 2, 2018)

As suggested by the new President, even if "initially we intended to search for another AD, then, looking at the new job description and at the professional background of the GD, we opted for a different solution. The new GD would be required to cover also some artistic responsibilities formerly assigned to the AD" (interview with the new President, June 6, 2018). Knowing that "I could have covered both the roles, since I am basically an artistic person but I had developed a lot of experience on management issue" (interview with the new GD, July 2, 2018), the GD decided to look among the people already collaborating with FND *Aterballetto*. He identified a *maître de ballet* as a possible Artistic Coordinator, since "I had not any experience on managing relationships with dancers and choreographers" (interview with the new GD, July 2, 2018). She collaborated with *Aterballetto* during two periods: between 1984 and 1993 as a dancer and from 2004 as a *maître de ballet* and *maître repetiteur*. She had the competencies for both managing the most operative aspects of the dance company (e.g., teach daily ballet class, rehearse the *corps de ballet* and character parts, advise on proper techniques, and assist with auditions) and for contributing to the definition of the artistic strategy of the company together with the GD (e.g., choice of choreographers, occasional *guest teachers*, and new dancers). As suggested by the Artistic Coordinator herself "I know *Aterballetto* since I was very young: I started to dance here when I was 19 years old and then I have been in touch with this company for the last 30 years. [. . .] *Aterballetto* is a peculiar context as for outsiders it is not easy to be accepted by the dancers and the staff. In my case, I have covered different artistic roles and thus it was easier for me to manage the complexity of such a company" (interview with the Artistic Coordinator, July 12, 2018).

Eventually, the GD proposed to the Board to promote the *maître de ballet* as Artistic Coordinator, assuming himself the responsibility for the planning of the artistic development of FND *Aterballetto*. The Artistic Coordinator role is not defined in the Statute, and it encompasses operative artistic activities (as described above) and the definition of the artistic strategy of the company in collaboration with the GD. In other words, this role is less autonomous than the one of a traditional AD. As explained by the Artistic Coordinator herself "on a practical level I do the same things of an Artistic Director: coordination of the dancers, preparation of the group for the performances, supervision of the rehearsals, maintenance of the relationships with the choreographers. But I have a little less responsibility than an AD: when the GD and I discuss and decide about artistic issues he has the 'last word' on every decision" (interview with the Artistic Coordinator, July 12, 2018). The

GD's proposal was accepted by the Board, and since January 2018, the GD is supported by the Artistic Coordinator.

In the meanwhile, the renewal of the leadership roles within FND *Aterballetto* was completed with the appointment of a new President. He is a management consultant for public administrations and public businesses, who used to work as General Manager for multiutility companies. In the last years, he covered different roles in the local Association of Small and Medium Enterprises (API Reggio Emilia), eventually becoming Director of the association. He does not have an artistic background "I am not a dance expert: however, I appreciate dance and I often go to see dance performances with a lot of pleasure" (interview with the new President, June 6, 2018). While the new President was Director of API Reggio Emilia, he knew the former President of FND *Aterballetto*, and the new President's association sponsored many public events where *Aterballetto* dancers were performing. According to him, this previous knowledge was a base for his appointment as President of FND *Aterballetto* "I think the Major called me also on the advice of the outgoing President" (interview with the new President, June 6, 2018).

In line with his background and with the definition of the job as prescribed by the new Statute, the current President has mainly a representative role: he coordinates the activities of the Board, he legally represents FND *Aterballetto*, and he manages the relationship with the two public founders and main shareholders of organization. One of his major tasks concerns fundraising from (public and private) funders. Because in the last years public funding to arts and culture have been reduced in Italy, many arts organizations are searching for new sources of funds "to reduce the financial dependence on public funders, we are working on increasing private sponsorships" (interview with the new President, June 6, 2018). Regarding the redistribution of tasks, duties, and responsibilities between the new President and the new GD, the former asserts "I am the legal representative of FND *Aterballetto*. I am not responsible for the day-to-day management of the organization: I do not handle the artistic part, neither the business nor human resources. There is a GD for this!" (interview with the new President, June 6, 2018).

5 Discussion

Our findings show how leadership forms change over time in cultural organizations. We can represent the roles constituting a leadership structure in a bidimensional space (see Fig. 1). The vertical axis relates to the leadership focus and captures the dialectic between long-term (strategic) decisions and short-term (operational) decisions. Since leadership roles are important for the definition and execution of the organizational strategy, identifying both who the leaders are and what their responsibilities and activities consist of are two key elements for supporting the long-term development of a cultural organization. But the leadership form may be also affected by short-term organizational needs, which relates to the day-by-day management of the business and artistic activities. The horizontal axis relates the leadership domain

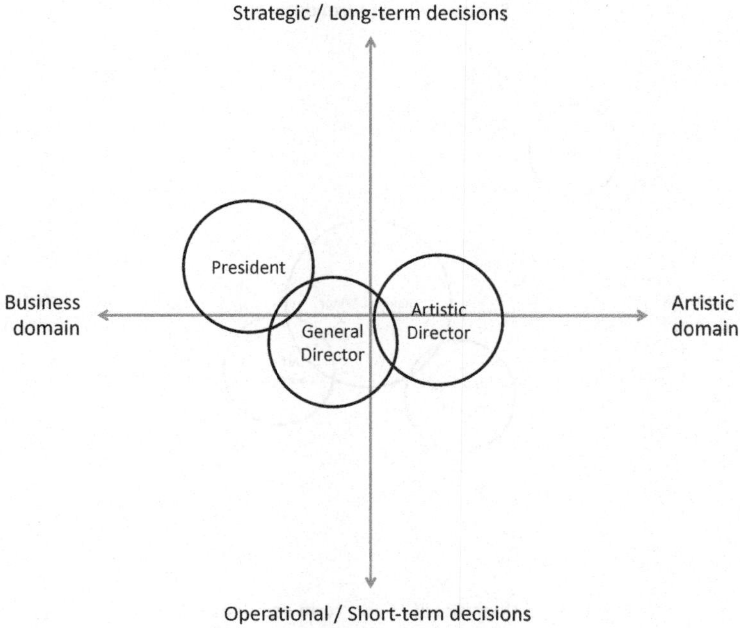

Fig. 1 Leadership structure of FND Aterballetto between 2011 and 2017

and represents the dialectic between the two complementary but separate functions of any cultural organization that are, respectively, business and artistic work.

In particular, as FND *Aterballetto* is concerned, between 2011 and 2017, the formal leadership structure mainly resembles the characteristics of a dual leadership, since the GD managed all the managerial and administrative issues, while the AD was in charge of the artistic development of the dance company. An additional role in the leadership structure is represented by the President, who was primarily involved in the business-related decisions and secondarily in the relationship between the GD and the AD. Hence, we may assert that the de facto leadership structure was indeed a trio (Fig. 1).

After 2017, the urgent organizational need of covering an unexpected job vacancy, when the new AD suddenly left the company, contributed to a redesign of the extant leadership structure. When FND *Aterballetto* renewed its leadership roles, different driving forces such as reduction of the public support to the cultural sector, limited pool of candidates in the leaders' selection process, time constraints in the transition between the old and the new leaders, and internal labor market rules that affect the turnover of some key figures created the conditions for the emergence of a different leadership structure (Fig. 2).

In this new structure, the GD consolidated his role (Alvarez and Svejenova 2005) acquiring both the responsibility for the business activities and the supervision of the

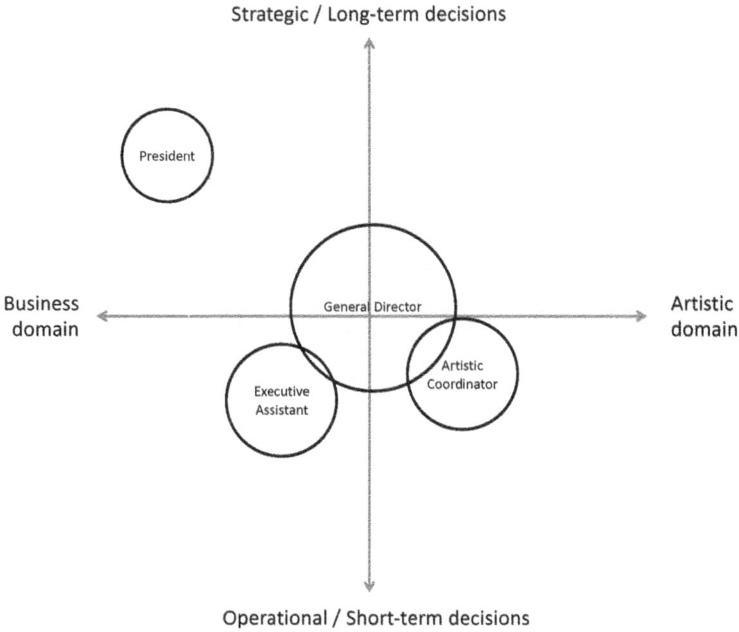

Fig. 2 Leadership structure of FND *Aterballetto* after 2017

artistic domains, whereas the President limited his previous role reducing his participation to the management of the ordinary business activities. At the same time, the new plural leadership structure is characterized by role creation (Alvarez and Svejenova 2005). In particular, the GD created the role of the Artistic Coordinator, to receive support for the artistic decisions and for the management of some operative activities related with the *corps de ballet*, and he coopted in the leadership team the Executive Assistant with the task to support him in managing ordinary business-related activities. In doing so, the GD addressed the business and artistic contingencies emerging during his mandate and actually contributed to modify the leadership structure as formally defined by the Statute.

Our analysis reveals that the emergence of this new plural leadership structure was affected by *factors* and *dynamics* operating at three different levels: individual, organizational, and environmental. By *factors* we mean characteristics of the individual, the organization, and the environment that shape the tasks and responsibilities of the leadership roles within an established leadership structure. By *dynamics* we mean behaviors and actions of the individual, the organization, and the environment that drive the emergence of a new leadership structure.

5.1 Individual Level

Individual factors concern individual resources, such as knowledge, (lack of) experiences, personal and professional relationships, and personality, of the individuals who are in charge of different leadership positions. These resources influence how they define their roles either reducing or expanding their activities in relationship with the other roles composing the leadership structure of the organization. For instance, the experiences and skills of the leaders can affect the design of the leadership structure and the relationships between the leaders. In our case study, the Presidents contributed to their roles in two different ways: the former President introduced a strong focus on managerial aspects, and the new President brought his prior experience in managing relationships with institutional stakeholders. Moreover, their expertise supported the creation of different relationships between them: whereas the former President was involved, at least with a consultative role, in all the relevant artistic and business-related decisions of the company, the new President willingly adopts a representative role delegating all the decisions to the GD. Similarly, the Artistic Coordinator's experience as *maître de ballet* facilitated her collaboration with the GD. In addition, the leaders brought, as a resource, their personal and professional relationships: for instance, the former President knew the new President, the new President has a strong reputation in the network of institutional relationships of the local context where the FND *Aterballetto* is located; the GD knew the AD. All these relationships represent relevant resources for accessing to the leadership roles and defining its structure.

The change of the leadership structure is influenced at individual level by dynamics that concerns the behaviors of individuals, while they act their leadership roles. Leaders adopt these behaviors with the aim to align their role with their own preferences, motives, and passions. In the case of plural leadership, the expected tasks, duties, and responsibilities are less defined than in traditional unitary leadership, and therefore there is a larger room for the individual interpretation of one own's role. In our case, different leaders acted their roles according to their preferences, for example, the new President, who is passionate about the lobbying activity, and the new GD, who is a believer of a participative management style.

5.2 Organizational Level

Organizational factors concern rules and formalized processes that contribute to defining the expected leadership behaviors, such as job descriptions, management practices (e.g., recruitment and selection, on-boarding), organizational charts, the Statute, and employment agreements. For instance, in the case of FND *Aterballetto,* our findings show that the new leadership structure was in line with the formal separation of responsibilities between the President and the GD, as reported in the new Statute.

Organizational dynamics concern processes modifying the formal leadership structure as the leaders execute their expected tasks and responsibilities. Indeed, the boundaries separating individual responsibilities are blurred when it comes to their execution, because of the involvement of other actors into the activities formally assigned to one specific actor. The distribution of roles within the leadership structure may be dynamically shaped by the interaction between formal and informal drivers. For instance, even if the sharing of tasks and duties is favored where the responsibilities are more broadly defined—as in the former leadership structure of FND *Aterballetto* when all the trio actors were involved in many artistic and business-related activities—also in the case of a leadership structure where the separated responsibilities are well-defined, a leader can decide to involve other people in the execution of his/her leadership. More specifically, our findings show that notwithstanding the separation of roles and the different hierarchical level, the GD and the Artistic Coordinator share all the relevant artistic decisions, so their common artistic knowledge combined with a strong communication reduces the likelihood of conflicts and improve their collaboration. Similarly, in the management of international collaborations, the GD involved other people of FND *Aterballetto* staff (i.e., the Artistic Coordinator, the Production Manager, the Marketing Manager), and, in some business-related decisions, he included other subordinates, such as his Executive Assistant, in the decision-making process. The GD has a strong formal power, and he is informally involved in all the other relevant decisions (e.g., the selection of the new AD). However, he decided to involve different actors in his decision-making processes and activities, and therefore a new plural leadership structure has emerged in the organization. Consequently, new actors were empowered, as they are delegated to execute some tasks and/or are involved in duties which were originally assigned exclusively to the GD.

5.3 Environmental Level

The environmental factors represent institutional elements, in the form of rules, norms, and cognitive schemas that, operating at a local and national level, exert isomorphic pressures on the organization, therefore influencing the leadership roles. For instance, in the case of FND *Aterballetto*, local communities' demands played a role as the members of local political institutions expressed the desire that the new leadership of the organization would create a local cultural network by means of strong inter-organizational collaborations. In addition, many Italian cultural organizations were changing their former dual leadership structure toward unitary forms of leadership.

As a consequence of these pressures, environmental dynamics drive the change in the leadership structure. This change is decided by the organizational actors in order to respond to the external pressures, and it is aimed not only to improve the organizational performance but also to attain external legitimacy in the industry and thus survival. For instance, in order to satisfy the local pressures for the creation of a network with other local cultural organizations, FND *Aterballetto* selected a GD with a relevant national and international experience and a President who has already a leadership role in a local trade union. In addition, in order to align its leadership structure to the trend of the Italian cultural industry, FND *Aterballetto* formally centralized the power in the hands of the GD, who then informally decided to involve other members in the Direction team creating de facto a plural leadership.

6 Conclusions

Our study illustrates the characteristics of the process through which a leadership structure changes over time in cultural organizations. We look beyond the traditional dichotomy between unitary and dual leadership and recognize the existence of different structures of plural leadership, which are shaped by individual, organizational, and environmental factors and dynamics. By mapping the leadership structure drawing on two dimensions (i.e., the domain and the focus of the leadership), we show how change in plural leadership forms may involve the formal and/or informal (re)definition of leadership roles (e.g., role creation or consolidation, Alvarez and Svejenova 2005) either along one of these dimensions or both. As for the factors and the dynamics driving change, whereas some of them were already discussed in the literature, we provide some original insights, showing, for instance, (1) how at the individual level not only skills, attitudes, needs, or concerns of the leader(s) but also their prior experiences and other personal resources (e.g., relational resources) may affect the leader's role (re)definition and (2) how at the organizational level the interplay between formal and informal processes may shape the leadership structure and its change over time.

In terms of managerial implications, our study suggests that leadership structure in cultural organization are prone to many institutional pressures; therefore companies aimed at establishing a stable leadership structure should consider not only their internal strategic demands and organizational conditions but also should be aware that external factors may impact on their decisions. In addition, our study demonstrates that the creation of formal rules for the execution of the leadership, such as job descriptions and contractual arrangements, provides only limited indications about the "actual" leaders' actions. Therefore, cultural organization should consider both informal and formal drivers to influence expected behaviors.

The limitations of this study may open avenues for further research. First, even if our study considers a long period of time, it illustrates only a specific change in the leadership structure. Future analysis of the evolution of FND *Aterballetto* may expand the current findings showing how future environmental, organizational, and individual factors will support the emergence of a different leadership arrangement. For instance, it could be interesting focusing on the relational dimension of leadership, analyzing the evolution of the personal and professional social network of the leaders and how this evolution may affect the leadership structure. Second, some characteristics of our setting (e.g., the national culture, the public ownership of the company, the industry) might limit the generalizability of the findings. Third, it may be interesting to explore the change in the leadership structure adopting a different point of view, for instance, the artists working in the organization (e.g., dancers and choreographers) or external partners. Finally, so far we have not collected enough empirical evidence on the relationship between the leadership structure change and the organizational performance, which represents, at the same time, a limitation of our study but also a promising future line of research with relevant managerial implications.

References

Alvarez, J. L., & Svejenova, S. (2005). *Sharing executive power: Roles and relationships at the top*. Cambridge: Cambridge University Press.
Bendixen, P. (2000). Skills and roles: Concepts of modern arts management. *International Journal of Arts Management, 2*(3), 59–73.
Bhansing, P. V., Leenders, M. A., & Wijnberg, N. M. (2012). Performance effects of cognitive heterogeneity in dual leadership structures in the arts: The role of selection system orientations. *European Management Journal, 30*(6), 523–534.
Bhansing, P. V., Leenders, M. A., & Wijnberg, N. M. (2016). Selection system orientations as an explanation for the differences between dual leaders of the same organization in their perception of organizational performance. *Journal of Management and Governance, 20*(4), 907–933.
Castañer, X., & Campos, L. (2002). The determinants of artistic innovation: Bringing in the role of organizations. *Journal of Cultural Economics, 26*(1), 29–52.
Caves, R. E. (2002). *Creative industries: Contracts between art and commerce*. Cambridge, MA: Harvard University Press.
Cray, D., Inglis, L., & Freeman, S. (2007). Managing the arts: Leadership and decision making under dual rationalities. *The Journal of Arts Management, Law, and Society, 36*(4), 295–313.
DeFillippi, R., Grabher, G., & Jones, C. (2007). Introduction to paradoxes of creativity: Managerial and organizational challenges in the cultural economy. *Journal of Organizational Behavior, 28* (5), 511–521.
Delmestri, G., Montanari, F., & Usai, A. (2005). Reputation and strength of ties in predicting commercial success and artistic merit of independents in the Italian feature film industry. *Journal of Management Studies, 42*(5), 975–1002.
Denis, J. L., Lamothe, L., & Langley, A. (2001). The dynamics of collective leadership and strategic change in pluralistic organizations. *Academy of Management Journal, 44*(4), 809–837.
Denis, J. L., Langley, A., & Sergi, V. (2012). Leadership in the plural. *Academy of Management Annals, 6*(1), 211–283.
Ebbers, J. J., & Wijnberg, N. M. (2017). Betwixt and between: Role conflict, role ambiguity and role definition in project-based dual-leadership structures. *Human Relations, 70*(11), 1342–1365.
Eisenhardt, K. M., & Graebner, M. E. (2007). Theory building from cases: Opportunities and challenges. *Academy of Management Journal, 50*(1), 25–32.
Garud, R., & Rappa, M. A. (1994). A socio-cognitive model of technology evolution: The case of cochlear implants. *Organization Science, 5*(3), 344–362.
Gioia, D. A., Corley, K. G., & Hamilton, A. L. (2013). Seeking qualitative rigor in inductive research: Notes on the methodology. *Organizational Research Methods, 16*(1), 15–31.
Glynn, M. A. (2000). When cymbals become symbols: Conflict over organizational identity within a symphony orchestra. *Organization Science, 11*(3), 285–298.

Knight, E., & Harvey, W. (2015). Managing exploration and exploitation paradoxes in creative organisations. *Management Decision, 53*(4), 809–827.

Lampel, J., Lant, T., & Shamsie, J. (2000). Balancing act: Learning from organizing practices in cultural industries. *Organization Science, 11*(3), 263–269.

Langley, A. (1999). Strategies for theorizing from process data. *Academy of Management Review, 24*(4), 691–710.

Montanari, F., Scapolan, A., & Gianecchini, M. (2016). Absolutely free? The role of relational work in sustaining artistic innovation. *Organization Studies, 37*(6), 797–821.

Montanari, F. (2017). Organizational design and people management. In F. Cirrincione, P. Dubini, & F. Montanari (Eds.), *Management of cultural firms* (pp. 257–291). Milan: Bocconi University Press.

Mumford, M. D., Scott, G. M., Gaddis, B., & Strange, J. M. (2002). Leading creative people: Orchestrating expertise and relationships. *The Leadership Quarterly, 13*(6), 705–750.

Reid, W., & Karambayya, R. (2009). Impact of dual executive leadership dynamics in creative organizations. *Human Relations, 62*(7), 1073–1112.

Reid, W., & Karambayya, R. (2016). The shadow of history: Situated dynamics of trust in dual executive leadership. *Leadership, 12*(5), 609–631.

Scapolan, A., Montanari, F., Bonesso, S., Gerli, F., & Mizzau, L. (2017). Behavioural competencies and organizational performance in Italian performing arts. *Academia Revista Latinoamericana de Administración, 30*(2), 192–214.

Yin, R. K. (2013). *Case study research: Design and methods*. Thousand Oaks, CA: Sage.

Anna Chiara Scapolan (PhD in Management, Ca' Foscari University of Venice) is Associate Professor at the University of Modena and Reggio Emilia (Italy). Her main teaching and research activities concern organization design, human resource management, and organizational behaviors. Recently, her studies have focused on organizational solutions and people management for creativity and innovation. She published articles in international journals, such as *European Journal of Innovation Management, Organization Studies, International Journal of Human Resource Management, European Management Journal*, and *International Journal of Arts Management.*

Martina Gianecchini (PhD, Business Management, University of Udine) is an Associate Professor of Human Resource Management at the Department of Economics and Management at the University of Padova, Italy. She is Scientific Director of the Executive Master in Human Resource Management at CUOA Business School. Her main research interests regard career management and labor market dynamics. She is member of the international research group 5C (Cross-Cultural Collaboration on Contemporary Careers) which aims at understanding meanings and determinants of career success around the world.

Interpretive Innovation in the Performing Arts: The Role of Organizations

Paola Trevisan

Abstract This chapter provides an original analysis of innovation in the performing arts, giving centrality to the practices that lead to original interpretations of the works of art that are organized and performed every day by performing arts organizations. In particular, the study examines interpretive innovation as emerging from evaluative practices in which multiple principles and performance criteria concur in the definition of value. Stark's notions of heterarchy, lateral accountability, and creative friction are applied to analyze the development of an original interpretation in an Italian opera house. The analysis demonstrated how innovation emerges from the interplay of multiple performance criteria. These criteria are based not only on the different technical and discipline-specific requirements involved in the production of live performances (industrial principles) but also on a number of individual sensitivities (the inspiration principle). The analysis also demonstrated that a meaningful organization of diversity generates frictions that are productive of novel and unexpected interpretations. The chapter contributes to our understanding of innovation in the performing arts by highlighting the role of performing arts organizations in setting up complex evaluative processes that lead to a re-generation of cultural heritage and sustain its transmission.

Keywords Artistic innovation · Interpretive innovation · Performing arts · Heterarchy · Creative frictions · Sustainable heritage

1 Introduction

> Our world lies at the borders of archeology. Almost the entire standard repertoire is made up of works from the nineteenth century ... they are part of our cultural heritage. Under these

P. Trevisan (✉)
Department of Operations Management, Copenhagen Business School, Frederiksberg, Denmark
e-mail: pt.om@cbs.dk

P. Demartini et al. (eds.), *Cultural Initiatives for Sustainable Development*, Contributions to Management Science,
https://doi.org/10.1007/978-3-030-65687-4_6

> circumstances, we must determine if there is something we can do to make the interpretation innovative, to make the opera suitable for our times. In this sector, innovation cannot be anything other than this .... Maybe you don't make opera innovative as a genre, but this leads to innovation in all the applied arts that it involves. Then it becomes creative. (Stage Director)

The performing arts, such as music, dance, opera, and prose, constitute a form of cultural heritage that requires continuous re-creation in order to be experienced, diffused, and transmitted (Diettrich 2015; UNESCO 2018; the stage director quoted above). In order to make this heritage viable for future generations, and thus to establish sustainable transmission mechanisms, a certain degree of innovation is necessary (Chan 2018). From such a perspective, cultural repertoires such as musical archives and play scripts—the inherited tangible elements of the performing arts—play an important role in that they provide resources that stimulate the creativity needed to re-create the cultural heritage, making it contemporary and living and further enriching the art field (Pereira Roders and Van Oers 2011; UNESCO 2018). Performing arts organizations are the custodians of such cultural heritage because it is through their work that performing arts are renewed and transmitted and new cultural assets are created. However, evidence of growing conformism in repertoires worldwide (Dimaggio and Stenberg 1985; Dowd et al. 2002; Heilbrun 2001; Martorella 1977) has led to a critique of the lack of innovative ability in the performing arts field (Heilbrun 2001) and to a questionable association of performing arts organizations with museums (Heilbrun 1993) as if their role in heritage management was limited to preservation and display of works of art. The critique is based on an assumption that relates performing arts organizations' programming decisions to artistic innovation. This perspective has a number of limitations (Castañer 2014; Castañer and Campos 2002), including its failure to recognize the innovation found in musical, visual, and dramatic interpretations of works of art. As a result, it discredits the artistic value and daily work of orchestra conductors, stage directors, actors, scenographers, and other artistic professionals and questions the innovative abilities of the performing arts organizations that stage artistic works. In addition, the literature on the drivers and mechanisms of artistic innovation has mainly focused on analysis at the art field level and on the radical forms of innovation that are capable of disrupting established artistic conventions (Becker 1974; Delacour and Leca 2017; Patriotta and Hirsch 2016; Wijnberg and Gemser 2000). This literature has rarely explored the characteristics of incremental forms of innovation in the arts or the role of arts organizations in offering forms of novelty within familiar and traditional artistic fields.

This chapter addresses such limitations by providing a conceptualization of innovation in the performing arts as *interpretive innovation* and analyzing the role of performing arts organizations in producing and staging novel interpretations of past works. The main goal is to understand how performing arts organizations support continuous processes of incremental innovation. It does this by addressing the following questions: What are the characteristics and dynamics of innovation processes in the performing arts? What management models can be applied to support their organization and coordination? To answer these questions, a processual view of innovation is adopted (Moeran and Christensen 2013), which sees innovation as emerging from a set of evaluative practices that are characterized by the interplay of different notions and criteria for defining what is "good" (Stark 2009). Stark's notion of *heterarchy*, together with the related concepts of lateral accountability, creative frictions, and organized dissonance, is employed to understand the mode of organization that makes possible the ongoing search for and production of novelty. When applied to a case study of an opera production in an Italian opera house, the concept of heterarchy helped to identify the various competing performance criteria and investigate how they interacted to make interpretive innovation possible.

The study contributes to the literature on artistic innovation in the performing arts by analyzing the role of organizations in combining different values that lead to original interpretations of works of art that belong to the artistic and cultural heritage. In addition, as a practical implication of the study, the recognition of the plurality of values involved in cultural production, which is seen as a resource for innovation rather than a problem for coordination, suggests that the true challenge for arts managers is not how to settle divergences and differences but how to encourage diversity and make it productive. In this sense, the notion of heterarchy provides a useful model for organizing diversity in arts organizations.

2 Innovation in the Performing Arts: Introducing Interpretive Innovation

In an art field, innovation is seen as a radical departure from the existing conventions defining the artistic canons and the cooperative mechanisms among artists and institutions (Becker 1974; Delacour and Leca 2017; Patriotta and Hirsch 2016), as in the case of avant-garde movements that lead to new artistic genres (Delacour and Leca 2017; Wijnberg and Gemser 2000). Within this definition, the field of the performing arts seems rather stable, especially in the case of symphonic music and opera, which are still dominated by the works and canons of the nineteenth century (Heilbrun 2001). Even when new works are produced, which is more common in theaters and the dance sector (Heilbrun 1993), these are not necessarily innovative in terms of challenging existing conventions (Castañer 2014). However, every art field is rich in "small" innovations, as artistic conventions always leave room for new interpretations and negotiations (Becker 1974). These forms of incremental innovation occur within the existing conventions (Jones et al. 2016) through the production of innovative yet recognizable styles (Montanari et al. 2016), which, rather than introducing radical differences, ensure variation within the field. The mechanisms of artistic incremental innovations, however, have rarely captured academic interest, which has been more directed towards researching the drivers of radical innovations. As a consequence, arts organizations have mainly been considered not as sites in which new artistic productions are made but as repositories of established conventions and therefore as constraints on the innovative ability of artists (Becker 1974).

An exception is represented by Castañer (2014) and Castañer and Campos (2002), who recognize the potential of cultural organizations to produce incremental innovations, namely through "form innovations" (innovation in the ways of presenting and performing a work, such as new ways of interacting with the audience) and "content innovations" (created by combining multiple art forms and disciplines in ways that have not been performed before, as in the case of ballet and prose). With specific reference to the performing arts, we add to these concepts the notion of *interpretive innovation*. This refers to innovation that relies on a novel artistic interpretation of an existing work of art, which, while maintaining the authenticity of the original work, explores it through a novel perspective or theme. It can be innovative in terms of both the form and content of the performance, which are meaningful live expressions of an interpretive frame. Interpretive innovation can be applied in the case of opera, where original librettos can be interpreted in novel ways, with each interpretation opening up multiple possibilities of representation in terms of the dramatic, visual, and musical elements that characterize the live performances (Heilbrun 1993). For instance, stage director Damiano Michieletto's interpretation of Mozart's classic *The Magic Flute* at the Venice opera house in 2015 was that of "an allegory of the forces fighting for control over mankind." He decided to stage this allegory inside a school, where the protagonists experience the conflict between religious and secular education.[1] In this sense, the performing arts can be innovative even within the increasing conformism in terms of repertory (indeed, *The Magic Flute* is one of the most performed operas worldwide). Looked at from a cultural heritage perspective, these forms of innovation are those that allow the re-production and re-generation of heritage and its sustainable transmission, as they are familiar and traditional enough to the audience to be understood and accepted, they preserve the authentic cultural value of the work of arts, while at the same time they introduce novel elements and new associations with contemporary meanings that were not conceivable when the work was originally composed.

While the idea of interpretive innovation, and of incremental innovation in general, has remained in the background of artistic innovation studies, it has assumed a more central role in the literature on creative organizations. In the creative industries, innovation is indeed an ambiguous concept for producers of cultural content, who face the opposing imperatives of constantly seeking and proposing new things and keeping them within the realm of familiarity and tradition so that they can be recognized and accepted by the market (DeFillippi et al. 2007; Lampel et al. 2000; Perretti and Negro 2007; Wijngaarden et al. 2019). People in creative organizations, as in the performing arts, engage every day in practices and processes aimed at continuously searching for new interpretations and re-combinations that will populate television channels, bookstores, clothes stores, and so on. Indeed, this

[1]Taken from the press release dated September 2015 of the opera house La Fenice, available at the website www.teatrolafenice.it

aspect is what makes creative organizations permanently innovative (DeFillippi et al. 2007; Jones et al. 2016; Lampel et al. 2000). To understand how creative organizations can ensure continuous innovations, some authors have suggested moving beyond the conceptualization of innovation as a "thing" that can be assessed ex-post (by the market, critics, or other agents), seeing it instead as a process (Moeran and Christensen 2013) whose outcome can only be understood "by telling the story of how it came out that way" (Becker 2013, p. xiv). For instance, in the study by Islam et al. (2016), the authors retraced the development process of a new perfume, showing how the actors used analogies to connect abstract cultural schemas with material forms and link visual and olfactory representations. To analyze artistic innovation as a process, we need to refer to works of art such as live performances as the result of a collective action involving not only artists but also the support of humdrum personnel (Becker 1974; Caves 2003). As more people, disciplines, and interests are involved, the production process requires different people to engage with the product during the whole process in order to meet several—sometimes competing—criteria and navigate the variety of visions of what the end result should be. Therefore, it is important to see the innovation process as a set of evaluative practices that entail discussions, negotiations, and compromises (Moeran and Christensen 2013). Such continuous evaluative practices occur during the production process and can be defined as formative assessments, as they enable learning and are distinct from the more formalized summative evaluations that occur at the end of the process (Sawyer 2013).

3 Innovation Through Evaluation and the Organization of Dissonance

In the study of interpretive innovation as a process characterized by multiple evaluative practices, economic sociology provides useful theoretical perspectives and constructs. In particular, the main theoretical reference is David Stark's work, *The sense of dissonance: Accounts of worth in economic life* (2009). Stark's framework has the peculiarity of linking valuation studies, a discipline developed within the realm of sociology, with organizational innovation, emphasizing how the presence of different valuation principles can generate productive frictions that support search and innovation. As a result, Stark's framework provides an original theoretical lens for the analysis of the artistic process as a set of different, sometimes conflicting, evaluative practices (Crepaz et al. 2016).

Stark, drawing mainly from John Dewey's theory of valuation (1939) and Luc Boltanski and Laurent Thévenot's economies of worth (2006), states that the interplay of diverse valuation principles generates perplexing and troubling situations that prompt search and innovation. Valuation principles, according to Boltanski and Thévenot (2006), are organizing mechanisms that allow individuals to assess and distribute value in real situations around a notion of the common good.

In particular, Boltanski and Thévenot identify six such principles: the market principle, the industrial principle, the civic principle, the domestic principle, the fame principle, and the inspiration principle. However, as different notions of the common good exist, people can resort to different and incompatible principles to organize, interpret, and evaluate the same situation. Stark sees the frictions between values as having a generative power. It is the disagreement about values—about what counts—that produces the perplexing situations that, following Dewey, support the search for something novel and unknown.

This theoretical approach can be applied by organizations. According to Stark, "instead of avoiding perplexing situations, organizations can embrace them. Even more radically, organizations can take the next step: if perplexing situations provoke innovative inquiry, then why not build organizations that generate such situations?" (p. 5). Therefore, Stark defines entrepreneurship as "the ability to keep multiple evaluative principles in play and exploit the resulting friction of their interplay" (p. 15). Moreover, he links innovation with valuation practices that bring together incompatible principles in an unpredictable way, finding new combinations by continuously questioning values.

Stark refers to the organizational form that produces perplexing situations as a *heterarchy*. Instead of enforcing a single principle of evaluation, hierarchically ordering principles, or separating them, heterarchies recognize that it is legitimate to articulate alternative conceptions of what is valuable. For example, in their ethnography of a new-media startup, Stark and Girard (2009) show how the frictions that enabled multidisciplinary teams to come up with new solutions regarding website construction were caused by the interplay of diverse regimes of valuation or, as they call them, logics. These comprised the logic of programming (valuing efficiency and accuracy), the logic of design (valuing creativity, experience, and imagination), the logic of information architecture (valuing clarity and usability), and the logic of merchandising (valuing market desires). Sawyer, reviewing a collection of case studies on different creative productions, identified similar evaluative regimes: the manufacturing regime (valuing industrial and economic knowledge of processes); the aesthetic regime (valuing aesthetic taste, history, and tradition); the craft/professional regime (valuing professional standards); and the brand/genre regime (valuing market and brand image) (Sawyer 2013). However, the studies reviewed by Sawyer usually describe cases of conflicts in which the "creative" (an artist or a designer) proposes something and the pragmatic counterpart (a manager or a technician) refuses because of the lack of resources or because the proposal does not meet other technical requirements (Becker 2013). In contrast, the notion of creative frictions developed by Stark is open to the possibility that the tension between different values can be generative of new ideas and solutions.

To exploit the creative potential of the frictions, companies rely on *lateral accountability*, meaning that all units are involved in the innovation process and are held accountable to each other based on different principles of evaluation or performance criteria. Innovation thus looks like an emergent process distributed across many actors who collaborate to figure out how different values can fit together (Stark and Girard 2009). A heterarchy is therefore a decentralized form of organization characterized by strong interdependences, requiring a lateral (rather than vertical) form of accountability and multiple performance criteria.

Innovating through disagreement is not a harmonious process. However, as Stark advises us, nor is it random or chaotic. Heterarchies are *organized forms of dissonance* in which disagreement is *principled*, meaning that valuations and judgements are always justifiable according to a general principle of worth or an accepted *compromise* between principles (Boltanski and Thévenot 2006). Lateral accountability, coupled with principled judgements, enables actors to debate values and temporarily make settlements to accomplish the project (what Stark calls *discursive pragmatism*). Therefore, it is not enough to place together people with different disciplinary backgrounds or different sensitivities and let them interact. Organizations need to ensure that relevant principles are mobilized and that people are disciplined towards organizational goals. It is necessary to organize diversity in a meaningful way, keeping the demarcation and separation of values (thus avoiding convergence, which would lead to the reproduction of sameness) and, at the same time, allowing interactions through proximity (Stark and Beunza 2009).

4 Artistic Innovation in Performing Arts: The Case of an Opera Production

This section analyzes a case study of an opera production in an Italian opera house (La Fenice in Venice). The case represents a form of interpretive innovation in the sense described above, as it refers to the production of an operatic work that was first performed at the beginning of the nineteenth century and that has since been reproduced by numerous opera houses around the world. The case site too staged the same production 2 years earlier and at the time of this study was preparing the first revival. The opera produced was one of the first operatic farces written by composer Gioachino Rossini and librettist Giuseppe Maria Foppa.

This study is part of a larger research project on arts management (Trevisan 2017). The data were collected during 1 year of fieldwork from April 2014 until March 2015. The fieldwork included a 6-month period of full-time observations in the organization. However, the narrative concerning the artistic production, which is the core of this study, has been elaborated mainly from notes taken during 2 weeks of observations conducted during the rehearsal of Rossini's opera. Having assisted in the entire process—from the first rehearsal to the first performance—it was possible to provide a complete account of the process. The description of the process emphasizes the interactions between individuals with diverse disciplinary backgrounds and diverse visions. In addition, I conducted interviews with six managers (covering all departments), four employees, and, specifically for the Rossini opera, the stage director (who was also the opera house's director of production and planning) and two contracted singers. The interviews focused on a variety of topics that were the object of the research project, in particular the role of management thinking in an arts organization. For the specific purpose of the present study, I relied on the interviews with the personnel involved firsthand in the artistic production (the stage director, singers, a stage technician, and a musician), with whom I discussed the most relevant aspects emerging from the analysis of the observations conducted during the rehearsals, especially the relations between different professional profiles and the tensions between different visions and interests. During the fieldwork, I assisted in the rehearsals of various productions, and I commented on my observations in several informal conversations with employees and managers. Therefore, the analysis of the specific Rossini production was able to benefit from discussions and observations regarding the opera house's production system in general. The data analysis enabled the identification of the different value frames and the disciplined practices of making settlements.

4.1 The Opera House: Programs as Compromises Between Principles

The opera house's mission is to spread the culture of opera through the production of live performances. In pursuing this mission, the opera house states that it operates according to the principles of enterprise and efficiency. Therefore, it responds to cultural and artistic principles as well as industrial and market ones. To keep together such diverse principles, the opera house follows a production strategy according to which, every season, it presents operas that span from baroque to contemporary music, including the more or less well-known classics of the nineteenth century. When selecting the operas, the managers can choose those that are already part of the opera house's repertory or they can opt to set up a new production. For the realization of the operas, the opera house hires specialist artists with different degrees of expertise and reputation. Thus, in the programming phase, the managers play with these variables (artistic genres, names of works/composers, and artists/interpreters) with the aim of simultaneously reaching a balanced budget, cultural variety, and artistic novelty.

While financial, cultural, and originality principles respond to different and often conflicting logics, the managers build their programs by exploring the possible synergies between them to find an acceptable compromise. For example, the revival of a big classic such as *La Traviata*, interpreted by renowned artists, leads to good results in terms of ticket revenues and a positive margin thanks to the economies of scale deriving from the earlier investment in a previous production. Even if it can hardly be judged as giving artistic novelty, it generates resources that can be invested in other, more original productions. These new productions can be investments in the classical repertory and provide long-lasting interpretations, particularly when attached to names of interpreters with a high artistic reputation. Such productions introduce elements of novelty and variety into the program and then become part of the repertory. New productions can also be made in the case of less popular operas or to invest in the artistic potential of less-experienced artists. These productions usually have a minor budget and—even if they become part of the repertory—are less likely to be revived in the near future, as they are not highly popular with the audience. Yet, they add novelty and variety to the programs and, in addition, contribute to the artistic development of young artists and the cultural education of the public.

The following words of the general manager summarize how the compromise between artistic, cultural, and financial principles works at the programming level and provide an idea of the narrative used to sustain it:

> Am I criticized because we have 50 performances of *La Traviata* in one season? Well, there are another 17 works this year. The resources generated from revivals are reinvested in new productions . . . . This year, we have Maestro X [a well-known Verdi interpreter]. We cannot always afford him, but we created the conditions to hire him: *La Traviata* pays for him.

As emerges from the way in which the general manager justifies—more than explains—the programming decisions, the compromise between the financial, cultural, and originality values is not exempt from critique. Both internally and externally, the management of the opera house has been criticized for being more oriented towards market and financial principles than artistic and cultural ones. The critiques have related, in particular, to the increasing number of revivals of *La Traviata* every year and to a general decrease in rehearsal time. This observation is in line with Boltanski and Thévenot's (2006) conceptualization of compromises as inherently fragile.

The Rossini opera production analyzed below is among the first works written by the famous composer and one of his least performed operas in theaters. The main interpreters were young artists, and the scenography, costume, and light design were designed by students of the local fine arts academy. The stage director was also the opera house's director of production and planning, and he did not ask for additional salary. Therefore, this was a "low budget" production compared to those involving more—and more famous—artists. Moreover, it required less rehearsal time than when it was first produced 2 years earlier. It was also a "small" production, which, thanks to its small stage set and limited number of orchestra players, was suitable for rehearsal and performance during the same period in which the opera house staged *La Traviata* and another opera. In addition, the first performance was broadcast on national radio, thus increasing the opera house's contribution to the promotion and diffusion of culture. This, together with the involvement of young artists, helped fulfill the opera house's mission and public-service vocation of supporting arts and culture. In itself, the production was thus already the manifestation of a compromise between financial and productivity principles, the values of promoting culture and culture education, and the values of originality and innovativeness that artists bring to an old work.

4.2 The Microdynamics of Interpretive Innovation in the Production Process

The cast of the Rossini opera was composed of the orchestra conductor, the stage director and his assistant, and four lead singers, all specifically contracted for the production. The orchestra comprised a number of musicians and instruments suitable for playing the musical score; and there was also the technical corps responsible for the stage set, the lighting, the recording system, the costumes, and the stage props. The stage coordinator was also present to ensure everything was ready and in place for each rehearsal.

The performance was produced in collaboration with students from the arts academy, which provided the scenographer, the lighting designer, and the costume designer. The stage director chose the set design among those received by students through an open call. He chose a scenography depicting a boat rather than the traditional mining setting that the opera would suggest. As the stage director commented, "I love working with students of the academy ... I am always impressed by the originality and freshness that they can bring to old works." The boat-inspired set was constructed in the opera house workshop by carpenters, and the costumes, also approved by the stage director, were sewn and tailored in the opera house's costume workshop. From the interviews with the chief of the stage technicians, it emerged that the collaboration between set designers, technicians, and carpenters is important for the development of a project that meets the artistic vision and at the same time is feasible and maneuverable:

> They [scenographers] send us the design and then we meet, and we always have a discussion because scenographers ... they are artists. They want James Bond's car! So I, the most pragmatic man on heart, have to bring things back on track. Then, mediation is important—because I never say no to a project—that would be stupid, but we have to work on it because it [the translation of the scenographer's ideas] is not easy .... Then it goes to the workshop, and there, well, if it is wooden material, no problem, but if we have to work with some particular resins or plastics, we need more contacts with the scenographer, and we learn a lot.

The director's vision, already nurtured and shaped by the design of the scenographer, was developed further during the rehearsals by interactions with the singers, seeing what they did on stage, trying out variations, and listening to their suggestions. Scene by scene, the director watched the singers acting and intervened with suggestions. Often, he climbed up on stage to show the moves himself. The stage director wanted to emphasize the comical aspects of the farce. However, while one singer was evidently at ease with the comical acting and even suggested some moves, the director noticed that another singer had difficulties following his vision: she tended to follow with her body the melody of the romantic aria she was singing, giving a more classical, serious interpretation than a comical one. The director and his assistant politely stopped the rehearsal when this happened and encouraged the singer to move differently. When the singer departed from her "traditional" interpretation with an unexpected clumsy move, the director and his assistant, sitting in the first row, laughed amusingly at this surprising occurrence.

The director is ultimately responsible for a coherent dramatic representation and guides its development, but how the direction develops is the result of collaborative work. To explain what happens in the production, the director used the word *commozione*, whose English translation is "to get emotional." As he explained,

> the etymological meaning is "to move together." This is what happens—there is this moving together that eventually will also involve the audience . . . I always want them [the singers] to put something of their own into the production. Everyone should bring their own brick along the way.

In fact, the director took inspiration from the various people involved, from the scenographer to the singers, and this inspiration ranged from their understanding and vision of the opera to their general notion of what was worthwhile. A singer, during the interview, confirmed the collaborative work of setting up an operatic production and, in particular, the collaborative relationship between the director and the singers:

> Usually, directors want first to see you—see what you do—and then they tell you what to keep or what to change. Some directly tell us what we should do. But even then, there are no two people who will do the same character in the same way. Everyone includes his personal signature. Because of how we are, our appearance, our way of acting, etc.

Another singer stated the following:

> Regarding the character, we develop our idea when we study the opera. But then we meet with the director, who maybe has a different vision, and then we discuss. That is the beauty of the work on the stage. It is through the development of the opera that we really understand, more and more clearly, what we are making. And this is especially so in the encounter with the director, as he can clarify better some aspects. And then we work all together to make it better. It may be that I don't always agree with the director's visions, but that does not matter that much because he signs the direction, not us, the singers.

The collaborative work also concerned the use and selection of the props, which, together with the scenery and costumes, define the visual aspects of the performance. For instance, at a certain point, while rehearsing and talking, the team came up with the idea of adding a scene where an imaginary train was leaving an imaginary station. The director asked the prop technician—who also participated in the rehearsals and was responsible for arranging and handling the stage props—to search for an oil lantern similar to those used in railways in the 1800s. He added, "I don't think we have the budget; see what you can do." On the next day, the prop technician came to the rehearsal with the right object, without having asked for an increased budget. A walk-on was then asked to go on stage holding the lantern and wearing a train driver's hat (also expressly searched for). The prop technician told me that he was a collector of antiquities and that he knew where to search for the right prop, whether in the opera house's warehouse (as in this case), his own collections, or elsewhere; or he could even create props, as happened on a different day following the request by the director for a fake newspaper with the picture of the starring singer on the front page. The stage director, this time linking the incidents to his role as manager of the technical corps, commented as follows:

> Technicians, props people, electricians . . . they are not electricians working for the municipality to change the bulbs on the street. At a certain point, each one of them contributes

> through something, something that only he can give—and only if he decides to. This positive individual dynamic is essential for the energy of the opera house.

The production continued to develop also from the musical point of view. During the musical rehearsals, the singers sat in a semicircle in front of the piano, while the conductor, next to the pianist, guided the development of their individual and collective interpretations. While in the rehearsals with the director, the singers kept their voices low in order to preserve them, but in the musical rehearsals, the singing was the main focus. The singers sang their parts as they had studied and thought about them. They discussed issues with the conductor, telling him how they had interpreted the piece and how they thought of singing it. The conductor corrected their inflection and pronunciation and determined, for example, how much *piano* and how much *forte* they should sing and how long a pause there should be. He made reference to the emotions that the music should transmit (e.g., anger or sweetness) for each part, based on the lyrics and the plot, which he described according to his interpretation of it. Similar to what the stage director did, the conductor used his voice to give a better idea of how to model the melody and the voice. Some singers recorded the rehearsal to perfect their interpretation through additional individual study, and the pianist took notes on the score sheets. Before the singers and the orchestra met, the conductor had separate sessions with the orchestra as well as sub-sessions with the orchestra's instrumental groups. Speaking in general about the role of the orchestra conductor during an interview, a cello player stated the following:

> It is like reading a poem. The words are always the same, but the result can be very different. It is a matter of sensitivity and professionalism. . . . A conductor is a good professional when he can communicate, even without saying a word, just with his hands and eyes. . . . it is a matter of sensitivity—it is difficult to explain if you are not a musician.

So far, the visual (e.g., set, props, and costumes), musical (e.g., playing and singing), and dramatic (acting) aspects of the opera performance had been kept separate. Later in the rehearsal process, the various elements were gradually put together, increasing the number and complexity of the interdependences. In particular, it emerged how acting and singing are not two independent practices. For instance, during a stage rehearsal, the stage director suggested to the lead singer a movement to make on stage to emphasize the meaning of a specific part of the libretto. She replied that with that movement she would be prevented from looking at the orchestra conductor in the orchestra pit. She stressed that, given the complexity of the aria she had to sing at that point in the performance, it was important that she could see the conductor's instructions. During the *assieme* rehearsals, where both stage director and orchestra conductor were present, the team made specific settlements in this and other cases, finding ad hoc solutions. During the interview with the singer, she narrated how it was common for stage directors to ask for acting efforts that contrasted with the singing:

> Once a director wanted me to sing a very complex aria while lying down, with a quite heavy man almost sitting on me. We tried, and I managed to do it almost as he wanted. Sometimes it is good that they push you to try. But at other times it would be better if they didn't.

By contrast, the director emphasized the use of collaborative and interactive ways of finding a suitable solution:

> Having experience in stage direction, I know it is easier to get to the singers if you show them how to do it. Often they believe they cannot make a move, but if I show them, it is easier. Then I can be accused of staging many "mini-mes," but that is how I work.

4.3 *The First Night and Ex-post Evaluations*

When the night of the first performance arrived, everyone backstage was observing the performance and working to ensure the timing and coordination of the lighting, the set moves, and the scene entrances. The very last scene was a kiss between the two main characters. When the two singers kissed outside the lit circle projected on the stage from the spotlight above, the stage director reacted unbelievingly to what he saw: "Come on, the final scene! How is that even possible!" The incident was soon forgotten and, after the final bow, everyone congratulated each other and left to celebrate, while the stage director and the orchestra conductor proudly gave an interview to the national radio station that broadcasted the entire performance. As the singer said during the interview, "when I get the bow from the audience, that is the recognition that I did well—that I did something to the audience—that I reached them somehow."

Despite these celebrations, the stage coordinator quietly said the following to me:

> This [the mistake in the final scene] should have happened during the rehearsals. These things happen when there is less time for rehearsal. Everything is rushed, things get mixed up, and the quality suffers ... but today nobody cares about quality; everything is about numbers.

Indeed, for the staging of Rossini's work, the time was scarce, the rehearsals were short, and perhaps the artists were not sufficiently professionally experienced. The team working on the production, however, did not expect to have more time, a more articulated stage set and more expensive props, or a famous conductor, even if they probably wished for this. As a singer admitted, "I wish we had more time." However, she added, "but I am aware you cannot ask a very efficient and productive house like this one to keep the orchestra busy for 2 entire days only for the staging of one production." When asked about how it felt to rehearse on stage with the assembled set for the first time only the day before the premiere, the singer commented as follows:

> This opera house is very well organized. True, we saw the stage set only the day before the premiere, but we did it two years ago—we knew how it was. This is a situation in which there are three productions at the same time. It is understandable ... that we cannot have the stage set from the first rehearsal day.

The team seemed to understand and accept the reasons why the production was valuable for the opera house. For instance, a singer stated the following:

> These productions, made with students of the arts academy and with us, young singers, work. It is an idea that works. It is low cost. We earn almost nothing, and it gives young artists the possibility to work, which is always good.

Not only time and space matter; the operatic material itself can be valued differently by artists. This holds especially for the musicians, who are permanently employed by the opera house. For them, works and composers can be boring, challenging, emotional, etc. A bass player, for instance, explained that the kind of music written by Rossini, which requires light touches, did not match his kind of specialty and technique, which were suitable for music that is more vigorous. As a result, he continued, it was physically painful for his arms, especially when they had two rehearsals in 1 day (as was the case when this conversation occurred). He added, "It is better to play Wagner for five hours than half an hour of this." Yet this did not prevent him from playing with the rest of the orchestra and developing, together with the conductor, the artistic interpretation. The musicians knew that there would be other opportunities during the season with operas and conductors they liked better.

Thus, the valuation process that leads to the programming, budgets, and plans had already prepared the basic elements for the Rossini production: the operatic material was provided, the cast was hired, the budget was assigned, and the time and rooms were booked for the rehearsals. The people involved in the production of Rossini's opera were disciplined with respect to these constraints, which were neither questioned nor challenged, allowing the disciplined practice that Stark has defined as lateral accountability. In fact, rather than being seen as constraints, the operatic material, the performers' abilities, the budget, and the time available were considered as resources that the individuals collaboratively decided how to combine in novel ways by mobilizing technical and inspirational values.

5 Discussion

5.1 The Principles of Interpretive Innovation in the Performing Arts

From the description of the production process, two main sets of principles—or evaluative registers—that guided the rehearsals can be identified, which resound with Boltanski and Thévenot's (2006) definitions of the industrial and inspiration registers. The industrial register includes all the technical criteria that define what is functional and what performs properly. This applies not only to the technical professions, such as carpenters, stage technicians, and tailors, but also to the artistic professions, as both singers and musicians have knowledge of how to sing or play an instrument correctly. Even if different values are emphasized according to the discipline involved (e.g., the technical concerns of carpentry are not the same as those of playing music), the industrial logic values specific knowledge, professionalism, ability, feasibility, and functionality. The second logic is guided by inspiration. This refers to the individual interpretation of the meanings, values, and emotions of the operatic material that are transmitted in the live performance. Inspiration values sentiments, emotions, passions, abstract visions, and everything that can nurture imagination. It applies to all human beings, regardless of their profession. This second frame of value, according to Boltanski and Thévenot (2006), is less equipped than the first in terms of valuation tools and proofs of value. In fact, while there are ways to assess the technical ability of professionals and the technical feasibility of various options, inspirations and visions are unexpected; they are worthy because of their originality, and they cannot be ranked in a non-ambiguous way.

These registers of evaluation resound with the aesthetic/design logic and the professional logic identified by both Stark and Girard (2009) and Sawyer (2013) in creative productions. Differently from them, however, the findings showed that the market/brand and business/manufacturing principles, also found in the literature, were not part of the interpretive search during the production process. Such criteria were instead mobilized in the programming phase, when it was decided which works to stage and which artists to hire. During the production, the opera itself (and related operatic materials such as scores and librettos) and the artists represented the principal sources for the interpretation, and the opera house managers' decisions regarding whether or not to prefer popular works and renowned artists were not discussed. Only later, in the summative, ex-post evaluations, these principles were brought back in to criticize the reduction of the rehearsal time, which had been decided upon for the sake of efficiency and productivity, at the detriment of production values.

5.2 *The Emergent Interpretation: Lateral Accountability and Creative Frictions*

In the rehearsal rooms, the two sets of principles interweaved during dialogues and practices involving different people and the objects they evaluated and valorized (e.g., music, lyrics, and props), making the innovative interpretation an emergent process (Stark 2009). For instance, the stage director did not think about a marine setting at the beginning, as the libretto suggested a mining setting instead. It was thanks to the inspiration of a young scenographer that the original set was conceived. Similarly, it was thanks to conversations during a rehearsal that the idea of suggesting a departure by train was conceived, and, thanks to the creativity of the props technician and his ability to search for and find the right props, this idea was staged. Therefore, even if technical and inspiration values refer to different valuation principles and performance criteria, they can be combined in multiple and unexpected ways. Such collaboration, which the stage director nicely defined as *commozione*, is a form of what Stark calls lateral accountability. It transcends the classical hierarchical lines and departmental divisions to embrace the individual inspirations and personal interpretations of both artists and technicians, who work together to figure out what to make and how to make it. In this fashion, the inspiration and technical realization—or design and implementation (Stark and Girard 2009)—developed together, often resulting in unexpected arrangements.

The more the production process was at an advanced stage, the more complex were the interdependences, making principles clash with each other and generating tensions. In particular, it emerged how the different disciplines involved in the visual, musical, and dramatic elements of operatic performances embody different and conflicting valuation criteria. This can be seen in the example of acting (drama) that could conflict with the singing (music) should particular movements or positions be required, but it could also be the case that props and costumes (visual) might interfere with the acting or singing. Such tensions were handled through discussions and resolved on a case-by-case basis, often leading to learning (cf. Sawyer 2013) (e.g., the singer realizing something new that she could do). These discussions were an additional source of inspiration, one that allowed the search for and definition of settlements, which proves that inspiration and technical realization develop together and influence each other. Unlike conflicts, which are resolved by privileging one criterion over the others, these tensions take the form of creative frictions (Stark and Girard 2009), which are resolved through collaborative dialogues that lead to solutions that the parts had not considered at the beginning, as they had not learned of or envisioned them before.

5.3 *Organized Diversity and the Fragility of Compromises*

In the rehearsal process described above, the tensions never degenerated into conflicts. We only have one account in which a singer recounted the possibility of a stage director's inspirations that were not shared by the singer. In this case, the conflict might have been resolved by returning to the traditional hierarchical method, with the director's vision counting for more than the singer's. In other words, the problem of different values would have been resolved by privileging the industrial principle, according to which those who occupy the higher positions in the hierarchy should take decisions, over the inspiration principle, according to which any creative outpouring is valuable as long as it is original and unique (Boltanski and Thévenot 2006). But in the data here provided, the rivalry between principles, criteria, and visions was instead a source for the creation and acceptance of novel arrangements, as described above.

How did the organization enable such a productive arrangement of diversity, which Stark calls organized dissonance, without incurring unproductive conflicts? It has already been anticipated that the market and business principles, which are often considered in the literature as contrasting with cultural and artistic principles, are not part of the interpretation process. Or, better, they are incorporated in it without being questioned. In fact, by defining the season's program, which contains the names of the works to be performed and the works' interpreters, the managers set up a frame within which the interpretation is developed (see also Maier 2017), allocating the fundamental resources for developing an artistic interpretation. In a season's program, as we have seen, productions are valued differently (e.g., revivals of big classics are worthwhile for their positive margin, while new productions are worthwhile for cultural variety and artistic novelty), and this seems to be recognized and accepted within each production, making people disciplined with respect to the management's decisions (e.g., the musician who preferred playing Wagner recognized and accepted the need to play Rossini). Therefore, there should be a certain coherence in framing the creative space in order to maintain people's accountability, motivation, and discipline and ensure that frictions are generative rather than destructive. For example, allowing only a little time for rehearsal for a very complex opera or teaming up a very famous director with a student of the art academy as scenographer could risk degenerating into conflicts. This suggests that decisions about how to equip productions have an important effect on the dynamics of innovation, as it is through these decisions that diversity is organized in a meaningful and productive way. This confirms the importance of working out acceptable compromises, which, according to Boltanski and Thévenot (2006), have the power to suspend conflicts between principles for the sake of a common interest (which, in our case, was to combine the available resources to set up and stage an original performance).

It was only at the end of the first performance, when the two starring singers kissed away from the stage lighting, that the stage coordinator criticized the managers for preferring the principle of efficiency to the detriment of artistic quality. This judgement relates to the idea that the amount of resources allocated to a given project (and time, in particular, appears to be the most critical resource) is a determinant of a "good" performance. It is thus a critique that stems from industrial principles, according to which a "good" performance is technically correct, with everything functioning well, while a performance is "bad" if a mistake happens on stage. This industrial notion of goodness, however, contrasts with other notions that, even if not expressly articulated, justified the emotions, applause, and celebrations that followed the performance. Therefore, this final critique seems addressed to the management's decisions that lead to a general decrease in rehearsal time for the sake of efficiency and productivity values; it does not relate to the value of the particular production or its interpreters. This demonstrates, in line with Boltanski and Thévenot (2006), how every compromise is fragile and that critique can always be resorted.

6 Conclusions

The sustainable management of performing arts is a form of safeguarding cultural heritage. It requires arts organizations to continuously provide society with live performances that give cultural variety and artistic novelty in a financially sustainable way. In order to present on stage something that is appreciated and suitable for contemporary generations, each performing arts production should meet criteria of originality and artistic novelty that create an emotional effect in the audience (Becker 1974) without betraying the authenticity of the work of art. We have called this *interpretive innovation*. Performing arts organizations, therefore, should organize their productions in such a way that multiple performance criteria are considered and applied to guarantee the generation of artistic interpretations that are feasible and original. A useful way to look at how organizations can achieve this goal is to look at the organization of production as a form of heterarchy (Stark 2009).

The case study analyzed above demonstrated how the daily innovative work of an opera production is performed through valuation practices in which multiple performance criteria are at play (Moeran and Christensen 2013; Sawyer 2013; Stark 2009). These criteria are based not only on the technical principles of functionality and efficiency, which govern the different professional disciplines involved in the production of the musical, dramatic, and visual elements of live performances, but also on several individual sensitivities (inspiration principle). The innovative interpretation emerges gradually through the pragmatic and collaborative engagement of individuals with the objects they evaluate and valorize during the whole process. In particular, the emergence of novel interpretations relies on the frictions that the interdependence of different performance criteria generates and that opens up multiple possible solutions that need to be actively sought. In this way, diversity can become a resource rather than a constraint. Producing a resourceful dissonance requires a form of governance defined as heterarchy and based on lateral accountability (Stark 2009), in which hierarchical and departmental divisions are set aside and where the common responsibility of "getting the job done" keeps various actors accountable to each other, stimulating a pragmatic cooperation. In addition, the study showed that managers can prevent the emergence of unproductive conflicts by allocating resources that are diverse (in terms of disciplines, professionalism, and sensitivities) yet meaningfully organized. The meaningful organization of diversity might be built on a generally accepted compromise between market, cultural, and artistic values (e.g., made in programming decisions) that transcends the single productions but influences each of them.

This chapter makes three main contributions. First, it contributes to the literature on artistic innovation, which has mainly focused on the analysis of radical innovation, by providing a more nuanced understanding of incremental innovation through the conceptualization of interpretive innovation. Interpretive innovation appears as a recombination of new and old elements, styles, and ideas (Lampel et al. 2000; Wijngaarden et al. 2019) that leads to an original visual, dramatic, and musical interpretation in the live performance of operatic works. This holds for new works, new productions, revivals (as was the case in this analysis), and both less-performed works and the big classics. The notion of interpretive innovation enables scholars to understand what is going on in art fields that are apparently stable, as in the performing arts.

Second, as urged by Castañer (2014), this study analyzes the role of organizations in artistic innovation. Organizations, in fact, are of secondary importance in most of the literature on artistic innovation, which has instead focused on artists' agency as the primary innovation driver (Jones et al. 2016; Montanari et al. 2016), limiting the analysis of organizations to their programming decisions (Dimaggio and Stenberg 1985; Heilbrun 2001). This study shows that while programming decisions have a role in distributing resources for innovation (in terms of linking existing works of art to artistic–technical teams), the practices that lead to the development and staging of novel interpretations happen daily during the production process, which supports the view of arts organizations as "permanently innovative organizations" (DeFillippi et al. 2007, p. 513), even within an artistic field characterized by a tendency for repertory conformism. In particular, by adopting Stark's notion of heterarchy (and the related concepts of creative frictions, lateral accountability, and organized diversity), this study contributes to the understanding of innovative processes in arts organizations as sets of valuation practices involving different principles.

Finally, it contributes to the studies on the sustainable development of intangible cultural heritage (Pereira Roders and Van Oers 2011). Interpretive innovations in the performing arts blend contemporary creativity and inherited artistic activities, and thus contribute to the re-generation of cultural heritage. Therefore, heterarchy, as a form of governance that enables interpretive innovations, is a relevant concept for cultural heritage management, as it is a management mechanism through which intangible heritage can be renewed and transmitted.

This is the first study to apply Stark's framework to the performing arts. Future research should deepen our understanding of heterarchies as a management model in the wider cultural and creative industries. For instance, this chapter showed that compromises between market and artistic values were worked out during the programming and budgeting phase, framing the production process in a time–space in which such conflicts were suspended (and yet resurfaced at the end of the production process). Future research could explore how heterarchies generate not only frictions but also compromises, considering in particular that the management techniques and technologies used for programming and budgeting respond traditionally to an industrial logic and are alien to cultural and artistic notions of worth. Regarding the principle of inspiration, another interesting aspect relates to the presence of evaluation principles that are weakly equipped with measures and that escape rankings and comparisons. In the case analyzed here, there were no tensions regarding aesthetic taste or emotional feelings, but we might expect that these could happen. Future research should investigate how claims that are based on the principle of inspiration are made operational, particularly in the presence of more immediately operative criteria based on functionality and efficiency.

References

Becker, H. S. (1974). Art as collective action. *American Sociological Review, 39*(6), 767–776.

Becker, H. S. (2013). Foreword. In B. Moeran & B. T. Christensen (Eds.), *Exploring Creativity. Evaluative practices in innovation, design and the arts* (pp. xiii–xvi). Cambridge: Cambridge University Press.

Boltanski, L., & Thévenot, L. (2006). *On justification: Economies of worth*. Princeton: Princeton University Press.

Castañer, X. (2014). Cultural innovation by cultural organizations. In *Handbook of the economics of art and culture* (Vol. 2, pp. 263–276). Amsterdam: Elsevier B.V.

Castañer, X., & Campos, L. (2002). The determinants of artistic innovation: Bringing in the role of organizations. *Journal of Cultural Economics, 26*(1), 29–52.

Caves, R. E. (2003). Contracts between art and commerce. *The Journal of Economic Perspectives, 17*(2), 73–84.

Chan, C. S. C. (2018). Sustainability of indigenous folk tales, music and cultural heritage through innovation. *Journal of Cultural Heritage Management and Sustainable Development, 8*(3), 342–361.

Crepaz, L., Huber, C., & Scheytt, T. (2016). Governing arts through valuation: The role of the state as network actor in the European Capital of Culture 2010. *Critical Perspectives on Accounting, 37*, 35–50.

DeFillippi, R., Grabher, G., & Jones, C. (2007). Introduction to paradoxes of creativity: Managerial and organizational challenges in the cultural economy. *Journal of Organizational Behavior, 28* (5), 511–521.

Delacour, H., & Leca, B. (2017). The paradox of controversial innovation: Insights from the rise of impressionism. *Organization Studies, 38*(5), 597–618.

Dewey, J. (1939). Theory of valuation. *International Encyclopedia of Unified Science, 3*(4), vii–67.

Diettrich, B. (2015). Performing arts as cultural heritage in the Federated States of Micronesia. *International Journal of Heritage Studies, 21*(7), 660–673.

Dimaggio, P., & Stenberg, K. (1985). Why do some theatres innovate more than others? An empirical analysis. *Poetics, 14*((1–2), 107–122.

Dowd, T. J., Liddle, K., Lupo, K., & Borden, A. (2002). Organizing the musical canon: The repertoires of major U.S. symphony orchestras, 1842 to 1969. *Poetics, 30*(1–2), 35–61.

Heilbrun, J. (1993). Innovation in art, innovation in technology, and the future of the high arts. *Journal of Cultural Economics, 17*(1), 89–98.

Heilbrun, J. (2001). Empirical evidence of a decline in repertory diversity among American Opera companies 1991/92 to 1997/98. *Journal of Cultural Economics, 25*(1), 63–72.

Islam, G., Endrissat, N., & Noppeney, C. (2016). Beyond "the eye" of the beholder: Scent innovation through analogical reconfiguration. *Organization Studies, 37*(6), 769–795.

Jones, C., Svejenova, S., Strandgaard Pedersen, J., & Townley, B. (2016). Misfits, mavericks and mainstreams: Drivers of innovation in the creative industries silviya svejenova. *Organization Studies, 37*(6), 751–768.

Lampel, J., Lant, T., & Shamsie, J. (2000). Balancing act: Learning from organizing practices in cultural industries. *Organization Science, 11*(3), 263–269.

Maier, E. R. (2017). The budget in the aesthetic: The role of calculative practice in the production of popular culture. *Management Accounting Research, 35*, 83–98.

Martorella, R. (1977). The relationship between box office and repertoire: A case study of opera. *The Sociological Quarterly, 18*(3), 354–366.

Moeran, B., & Christensen, B. (Eds.). (2013). *Exploring creativity: Evaluative practices in innovation, design, and the arts*. Cambridge: Cambridge University Press.

Montanari, F., Scapolan, A., & Gianecchini, M. (2016). "Absolutely free"? The role of relational work in sustaining artistic innovation. *Organization Studies, 37*(6), 797–821.

Patriotta, G., & Hirsch, P. M. (2016). Mainstreaming innovation in art worlds: Cooperative links, conventions and amphibious artists. *Organization Studies, 37*(6), 867–887.

Pereira Roders, A., & Van Oers, R. (2011). Editorial: Bridging cultural heritage and sustainable development. *Journal of Cultural Heritage Management and Sustainable Development, 1*(1), 5–14.

Perretti, F., & Negro, G. (2007). Mixing genres and matching people: A study in innovation and team composition in Hollywood. *Journal of Organizational Behavior, 28*, 563–586.

Sawyer, K. (2013). Evaluative practices in the creative industries. In B. Moeran & B. T. Christensen (Eds.), *Exploring creativity. Evaluative practices in innovation, design and the arts* (pp. 278–304). Cambridge: Cambridge University Press.

Stark, D. (2009). *The sense of dissonance*. Princeton: Princeton University Press.
Stark, D., & Beunza, D. (2009). The cognitive ecology of an arbitrage trading room. In D. Stark (Ed.), *The sense of dissonance: Accounts of worth in economic life* (pp. 118–162). Princeton: Princeton University Press.
Stark, D., & Girard, M. (2009). Creative friction in a new-media start-up. In D. Stark (Ed.), *The sense of dissonance: Accounts of worth in economic life* (pp. 81–117). Princeton: Princeton University Press.
Trevisan, P. (2017). *Reshaping opera. A critical analysis on arts management*. Cambridge: Cambridge Scholars.
UNESCO. (2018). *Convention for the safeguarding of the intangible cultural heritage*. Paris: UNESCO.
Wijnberg, N. M., & Gemser, G. (2000). Adding value to innovation: Impressionism and the transformation of the selection system in visual arts. *Organization Science, 11*(3), 323–329.
Wijngaarden, Y., Hitters, E., & Bhansing, P. V. (2019). 'Innovation is a dirty word': Contesting innovation in the creative industries. *International Journal of Cultural Policy, 25*(3), 392–405.

Paola Trevisan is Assistant Professor in Management Accounting at Copenhagen Business School, Department of Operations Management. In 2018, she was granted a Marie Skłodowska-Curie fellowship for the project MACRAME "Management Accounting and Creativity: Analysis of Meanings". She has carried out research on various sectors of the cultural and creative industries—such as performing arts, fashion, advertising, and traditional craftsmanship—investigating the organizational and sociological aspects of handling creativity with management technologies.

'Start Me Up': The Challenge of Sustainable Cultural Entrepreneurship for Young Cultural Workers

Fabrizio Montanari, Lorenzo Mizzau, and Damiano Razzoli

Abstract Cultural entrepreneurship has been seen as a process through which cultural workers organize activities around culture, motivated by their passion and willingness to create cultural value. This chapter provides an account of the actions that cultural entrepreneurs deploy to satisfy the needs related to developing and sustaining their activities. Based on 120 interviews with cultural workers at an early stage of their entrepreneurial activity, the chapter shows that sustainability is a multifaceted concept consisting of three main dimensions: creative, economic, and social. Individuals consider these different dimensions simultaneously while dealing with three main needs: sense of purpose, networking, and professional development. In so doing, they engage in a varied set of actions aimed at figuring out a personal path to sustainability. The chapter contributes to the understanding of how cultural entrepreneurs manage to establish and sustain an entrepreneurial activity, by experiencing the hurdles related to manage the different and often contradictory aspects of it.

Keywords Sustainability · Cultural work · Cultural entrepreneurship · Creative industries · Qualitative study

---

F. Montanari (✉) · D. Razzoli
Department of Communication and Economics, University of Modena and Reggio Emilia, Modena, Italy
e-mail: fabrizio.montanari@unimore.it; damiano.razzoli@unimore.it

L. Mizzau
Department of Economics and Business, School of Social Sciences, University of Genoa, Genoa, Italy
e-mail: lorenzo.mizzau@economia.unige.it

P. Demartini et al. (eds.), *Cultural Initiatives for Sustainable Development*, Contributions to Management Science,
https://doi.org/10.1007/978-3-030-65687-4_7

1 Introduction

In the last decade, many industrialized countries have experienced a significant growth in the segment of the workforce represented by people "loosely connected to organizations or selling directly to the market" (Petriglieri et al. 2019, p. 125). Statistics show that self-employed people and microbusiness owners represent more than one third of the overall workforce (Bureau of Labor Statistics 2016; Eurofund and ILO 2017). Cultural sectors have been no exception to this trend. On the contrary, the proportion of start-up activity, microbusiness and self-employment represent nowadays the norm in cultural sectors (e.g., Hausmann 2010; Merkel 2019).

The reasons for such an enhanced development of autonomous and self-employed activities in cultural sectors could be traced in several factors. Technological advances in information and communication technologies and the emergence of new business models related to the digital economy have certainly facilitated this transformation (e.g., McRobbie 2016; OECD 2018). However, a crucial factor lies in a change in approaches to culture by governments, policymakers, and managers of cultural organizations. Since the end of the last century, cuts in government expenditures have increasingly put pressure on cultural actors to adopt a market-oriented approach (e.g., Lewandowska 2015; Slavich and Montanari 2009). The increasing attention to efficiency and cost reduction has led cultural organizations to adopt managerial practices in their day-by-day operations, as well as reduce the number of permanent labor contracts (Hausmann 2010; Hesmondhalgh and Baker 2010). Furthermore, cultural workers have been encouraged to start up ventures and self-employed activities by a general political climate that has emphasized the potential of cultural industries for contributing to the development of national, regional, and urban economies (DCMS 2001; Flew 2011; Loacker 2013).

In this context, both policy makers and scholars from different disciplines have advocated for the importance of cultural entrepreneurship, which refers to "a management process through which cultural workers seek to support their creativity and autonomy, advance their capacity for adaptability, and create artistic [or cultural] as well as economic and social value" (Chang and Wyszomirski 2015, p. 11). However, cultural entrepreneurship is no easy feat. Indeed, individuals face several challenges in developing and sustaining their activity in such unpredictable and fast-changing settings as the cultural sectors. Indeed, cultural sectors are characterized by high demand uncertainty, high incidence of fixed sunk costs, a lack of formal labor regulations, and a skewed distribution of revenues accruing to the few products and individuals that are able to reach a critical mass of attention and popularity (Alacowska 2018; Caves 2000; Morgan and Nelligan 2015). Besides the risks inherent in any venture, individuals have also to deal with the difficulties related to the need to balance different logics, prominently the artistic and economic ones (Eikhof and Haunschild 2007; Lampel et al. 2000). It has been shown, in fact, that individuals strongly motivated towards artistic and cultural activities struggle in dealing with managerial issues that can conflict with artistic goals (Hausmann 2010;

Lingo and Tepper 2013). Sometimes, cultural workers even espouse a counter-establishment ethos that could represent a hurdle to learning managerial skills and following formalized practices (Lee et al. 2018, p. 16; Townley et al. 2009).

Previous literature on cultural entrepreneurship has provided a framework for understanding the motives, activities, and goals pursued by individuals engaged in entrepreneurial activities in cultural sectors (e.g., Hausmann 2010; Mietzner and Kamprath 2010; Svejenova et al. 2010). These studies have focused mostly on the analysis of skills, competences, and mindsets that cultural workers should develop to ensure economic sustainability to their ventures over time, while also managing potential conflict with their ethos akin to "art for art's sake." However, the concerns and aspirations of cultural workers go beyond the traditional contraposition between artistic and economic logics, embracing a wide array of needs and motives (see also Lee et al. 2018). Indeed, developing and sustaining a cultural activity has several implications on individuals in terms of, for example, emotional well-being, personal life, and the pursuit of self-actualization. Thus, cultural workers need to find a personal solution to deal with the different challenges that a sustainable cultural activity implies. This could be particularly the case of young cultural workers who are urged to confront the high levels of insecurity and uncertainty characterizing the contemporary socio-economic context (for similar considerations, see Lee et al. 2018; Morgan and Nelligan 2015). This chapter, thus, aims to delve into how cultural workers experience the hurdles related to manage the different aspects of developing a sustainable cultural activity.

In doing so, we highlight how sustainability is a multifaceted concept that cultural workers interpret along three main dimensions: creative, economic, and social. Developing and sustaining a cultural activity implies that individuals deal with these different dimensions simultaneously. This stems from the necessity to address a varied range of needs pertaining to different dimensions, which nonetheless individuals perceive all part of the same challenge—that of developing and sustaining a cultural activity. In this sense, individuals seek to find out a personal path to sustainability by deploying proper actions in different realms of activity. By delving into how young cultural entrepreneurs try to figure out a personal solution to the challenges related to sustainability in cultural sectors, this chapter offers a deeper understanding of the processes through which individuals make sense of their cultural ventures, also dealing with different, apparently contradicting logics. From the empirical viewpoint, we present the results of a qualitative study conducted in a medium-sized city in northern Italy, and based on 120 interviews with cultural workers at an early stage of their entrepreneurial activity (i.e., less than 4 years).

The chapter is structured as follows. The first section presents the theoretical background, the second illustrates the research methods, and the third one shows the main findings. The last section discusses the results, also proposing some policy implications and ideas for future research.

2 Theoretical Background

Cultural entrepreneurship has attracted the attention of an interdisciplinary cohort of scholars with different backgrounds, from entrepreneurship, arts management, and organization studies, to cultural studies, sociology, and economic geography (e.g. Eikhof and Haunschild 2006; Konrad 2013; Lazzeretti and Vecco 2018; Scherdin and Zander 2011). Cultural entrepreneurship has been variously defined as a process, a set of practices, a "state of mind" of individuals, and/or a way to define actions deployed by individuals and organizations that operate in cultural sectors and aim at generating value (see Hausmann and Heinze 2016, for a review on definitions). Similarly, cultural entrepreneurs are defined as those individuals who organize cultural, financial, social and human capital, to generate revenue from a cultural activity (e.g., Hausmann 2010; Konrad 2013; Swedberg 2006). Recent changes in the context of cultural sectors (e.g., development of digital technologies, reduction of public funds, a streamlining of employment-based cultural organizations, and a political climate advocating for a managerialization of culture) have emphasized the necessity for individuals operating in these fields to become more entrepreneurial (Ellmeier 2003; Markusen 2013; Morgan and Nelligan 2015). As a result, we have witnessed an increase in the number of cultural workers engaged in developing entrepreneurial activities (Lingo and Tepper 2013).

However, developing and sustaining activities in cultural sectors is difficult, as it requires strong motivation and vocational attitude, a set of varied competences, and the ability to combine different, sometimes diverging goals and expectations from different audiences (e.g., Lee et al. 2018; Stinchfield et al. 2013; Svejenova et al. 2010). In particular, the literature highlights the presence of three main areas of attention (or domains) relevant for cultural entrepreneurs.

A first domain regards the responses to the challenges that individuals face in order to ensure the economic durability of their venture. For example, studies suggest that developing an entrepreneurial activity in cultural sectors requires managing aspects that are functional to assure economic sustainability. In this sense, it is key to acquire competences necessary to develop a business model, set a business plan, and launch a marketing campaign. Similarly, organizational or "coordinative" abilities are useful to coordinate different professionals involved in a project (Svejenova et al. 2010). However, individuals interested in starting up a cultural activity often show a "lack of management expertise (particularly marketing), lack of skills, insufficient information, [and] inadequate planning and lack of organization" (Hausmann 2010, p. 23), and are "acutely aware of their financial vulnerability" (Lee et al. 2018, p. 18). While an educational background in fine arts and an adherence to a bohemian ethos sometimes leads to an adversarial attitude towards managerial competences (Morgan and Nelligan 2015), many cultural workers show "a desire for more commercial awareness or for knowledge about pricing and marketing techniques" (Lee et al. 2018, p. 18).

A second domain regards the responses to the challenges that individuals face in order to introduce (radical or incremental) innovation in their work. Indeed, innovation has a key role in cultural sectors, as "consumers need familiarity to understand what they are offered, but they need novelty to enjoy it" (Lampel et al. 2000, p. 292). However, innovation is highly risky and uncertain, as cultural sectors are characterized by aesthetic canons and conventions that pose constraints to the autonomy of individuals to deviate from them (Becker 1982; Jones et al. 2016). Therefore, individuals need to balance the need to seek novelty by differentiating their products and services with that of creating something accessible and familiar to customers. In this sense, individuals cannot rely only on their artistic talent, but also on the knowledge of markets, and on the ability to develop relations with relevant actors such as peers, organizations, and critics (Lange 2008; Konrad 2013; Montanari et al. 2016).

A third domain regards the challenges implied by high levels of complexity and uncertainty, which require constantly recombining, mobilizing and stretching available resources (Klamer 2011). For example, literature highlights how individuals who engage in entrepreneurial activity should orient themselves to typically entrepreneurial behaviors such as bricolage or effectuation rather than strictly planned behaviors (e.g., Hooi et al. 2016; Rao et al. 2005). More precisely, bricolage consists of "making do with what is at hand" (Baker and Nelson 2005, p. 329), while effectuation is a process that drives a course of action starting from the means one has at hand and discovering the "ends" (outcomes, products, markets, etc.) along the way, rather than from ends to means as in prescriptive planning models (Sarasvathy 2001). In a similar vein, in their study conducted on entrepreneurs from different realms (amongst which arts), Stinchfield et al. (2013) found out that the durability of ventures is achieved mainly through the persistence and resilience of entrepreneurs in the face of difficulties, thus lending support to the importance of adapting to (particularly tough) conditions instead of reaching goals through carefully planned predictions.

All in all, the literature agrees in highlighting that individuals struggle to develop and sustain an activity in cultural sectors as they have to face several challenges and potentially contradictory requirements. In particular, studies highlight the complexity of balancing multiple logics, primarily the artistic and the economic ones (Eikhof and Haunschild 2007). However, researchers need to go beyond the traditional contraposition between artistic and economic imperatives in order to better understand the nuances of the process by which individuals engage in cultural entrepreneurship (for similar considerations, see also Lee et al. 2018). Accordingly, this chapter aims to give an account of (what are) the needs perceived by cultural workers in developing and sustaining their cultural activities, and of how they strive to find a way to satisfy them. In so doing, we first of all offer a fine-grained view of how individuals articulate these needs and develop responses to cope with the struggles and the ambivalent emotions implied by the pursuit of their aspirations. Secondly, we articulate a multifaceted view of sustainability, which we propose consists of three different dimensions that cultural entrepreneurs try to deal simultaneously with.

3 Methodology

In this chapter, we adopted an interpretivist approach using qualitative, in-depth interviews with a sample of cultural workers living and working in Reggio Emilia, a city of about 170,000 inhabitants located in northern Italy. The city represents an interesting setting to investigate cultural entrepreneurship, as local and regional authorities have recently implemented several policies aimed at supporting individuals, associations, and organizations developing activities in cultural sectors (Municipality of Reggio Emilia 2009, 2016). Reggio Emilia is also home to internationally renowned cultural institutions such as Collezione Maramotti—the contemporary art gallery established by local fashion company Max Mara—and Fondazione Nazionale della Danza Aterballetto, Italy's foremost contemporary ballet company. Overall, the city hosts the fourth largest creative cluster in the administrative region of Emilia-Romagna, with the 4% of the local workforce employed in cultural sectors (ERVET 2018).

The data analyzed for this study were collected as part of a wider research project aimed at investigating the creative dynamics of the city. In order to understand how cultural workers experience the hurdles related to developing and sustaining a cultural activity, we focused on self-employed individuals who were in the early stages of their careers, i.e., founders of cultural ventures established by less than 4 years. Indeed, the quest for sustainability practices has increased in recent years and workers at an early stage of career are faced with more complex challenges compared to the past (see also Morgan and Nelligan 2015).

In order to identify which cultural workers to interview, we used a purposeful, non-probabilistic sampling strategy (Zikic et al. 2010). More specifically, we used a mix of databases provided by the Municipality of Reggio Emilia and the Italian Association of Young Italian Artists (GAI), personal contacts, and snowballing techniques (Biernacki and Waldorf 1981). The interviews were conducted in Italian between October 2016 and December 2018, and lasted between 30 and 50 min each, and were recorded and subsequently transcribed. Altogether, 120 subjects were interviewed (72 males, 48 females).[1] Interviews were based on guiding questions aimed to understand the perceptions of respondents about the needs, challenges and aspirations that they had as cultural entrepreneurs. Consistent with the interpretative focus of this study, we asked participants to describe and reflect upon their experience on the following themes: fields and scope of activity, particularly in terms of relationships within and across cultural sectors; the network of clients, suppliers, and peers; the gaps in terms of what is needed for running activities; the ways and means used to fulfill the needs and overcome the perceived hurdles to develop and sustain a cultural activity.

[1]We are aware that our sample is composed predominantly by men. However, this reflects the official data of the regional creative industries (64% males, 36% females—ERVET 2018), and is consistent with a skewed distribution in terms of gender, which has been found to characterize the cultural industries (for example see Bennett and Hennekam 2018).

Table 1 The areas of affiliation of the subjects surveyed

Cultural sectors	Percentage (%)
Music	29
Visual art	11
Architecture/graphics/design	11
Event management	9
Publishing	9
Photography	7
Cinema/video-making	6
Dance	6
Theatre	6
Other	6

Source: Our elaboration based on the data collected during the interviews

Interviewees' ages range between 18 and 35 years (mean: 28 years); they are mostly born in Reggio Emilia (65%) and are highly educated (35% has at least a bachelor degree).[2]

They operate in different cultural sectors, as illustrated in Table 1.

We started to analyze the data while we were still collecting them. More precisely, we first read and coded the transcripts of the interviews independently in light of the main domains suggested by extant literature of cultural entrepreneurship. Then, we had several collective sessions in which we shared our interpretations of the key themes emerged. More specifically, analysis highlighted several needs that were grouped into three main themes: sense of purpose, networking, and professional development. We refined these themes as long as we proceeded to read and code the remaining interviews. In doing so, we focused on the means and actions that interviewees declared to deploy to satisfy the declared needs.

4 Results

The analysis of interviews revealed that cultural workers are aware of the importance of "spending great effort in making [their] activity sustainable" (#53, illustrator, 30 years). In this sense, they bring forward different declinations of the idea of sustainability. All of them agree that carrying out an enduring activity in cultural sectors requires the ability to "innovate, generate new ideas that can be appreciated by the general public and other professionals" (#40, video maker, 25 years). Further, they acknowledge a lack of managerial skills and competences essential for the survival of their activities in the liberal or fine arts programs that they have attended: "After school, I was more naïve and I believed in the strength of my artistic ideas.

[2]The average of graduated people in Italy is 27.8%, as registered by Eurostat, way under the European average that is 40.3% (Eurostat 2020).

Soon after I realized that I need to pay attention also to commercial issues" (#5, graphic designer, 21 years). Interestingly, some interviewees decline sustainability in terms of their impact on the local community as "being appreciated for my effort toward my neighborhood and city motivates me to go on in my activity" (#22, architect, 28 years).

Regardless the interpretation of sustainability, interviewees agree that developing and sustaining activities in cultural sectors implies several challenges in terms of competences, skills, and mindsets. In this sense, they lament that coping with different hurdles causes considerable stress and a continuous necessity to find out "a personal way to solve the problems and pursue my aspirations" (#114, web designer, 29 years). Our analysis shows that cultural workers show three main needs while developing and sustaining a cultural activity: sense of purpose, networking, and professional development. They perceive that satisfying these needs is key to support the sustainability of their activities through actions such as creative exploration, networking, and acquisition of competences, as explained below.

4.1 Sense of Purpose

The first need springs from the "call" to find out a "purposeful career" (#6, architect, 32 years) in a given cultural sector. Indeed interviewees perceive their job as a calling, showing a strong motivation to carry out an activity that contributes to actualizing their selves: "I always strived to follow my path as a dancer, and to establish my dance studio where I could share my love for choreographies with my pupils: this dream has always guided me even when I had to work as waiter to pay the dance school" (#30, dancer, 26 years). In line with such a call, interviewees show the desire to derive a sense of purpose from their activity. Indeed, they perceive their activity as a means to construct their identity as cultural workers. To illustrate, a visual artist noted: "Everybody was looking at me as someone who makes visual installations for hobby: starting this business helped me promote myself as a professional visual artist" (#36, 35 years).

Interviewees decline such a need along two dimensions. The first dimension relates to the desire to perform an activity that sustains self-actualization: "When we started this company, we were looking for something that made sense for us and our audience, something that could help us to do what we like the most and to say ourselves: 'Look at what we have done'" (#57, founder of a theatre company, 33 years). Interviewees want to develop professional outcomes (artworks, products, services, etc.) that represent a true expression of their self. To illustrate, a producer of cultural events claimed: "Setting up performing arts events in public really defines who I am: I love to engage people and other professionals, to search for their curiosity and appreciation" (#35, 26 years).

The second dimension refers to the interest in developing a sustainable career in their sector: "I've always wanted to become a visual artist and to make a living with my art" (#1, visual artist, 27 years). Since high uncertainty makes it difficult to obtain stable sources of income ("I have a project-based activity: I know that I have to go through hard work and precariousness before I can make it"; #67, deejay, 23 years), about two thirds of the interviewees hold multiple jobs as a means to differentiate revenues and face potential periods of low activity (see also Hausmann 2010). Thus, they are strongly interested in making their professional activity a stable source of income, which can allow them to fully dedicate to their vocation: "I want to increase my revenues so that I can just do this" (#116, fashion designer, 34 years). Interviewees do not only want to increase incomes from cultural activities, they also claim that they decided to start an autonomous activity in pursuit of better overall working conditions. Actually, the willingness to continually put oneself to the test typical of creative work goes hand in hand to the desire to make culture the sole working activity: "I can deal much more easily with my business if I find a balance between life quality and the need for an atmosphere that fosters my creativity" (#103, video-maker, 32 years).

Interviewees declare that they work very hard in order to satisfy the need for a sense of purpose. In particular, they claim that dedicating passion, effort, and time to their cultural activity implies going beyond the traditional "9 to 6" work day. To illustrate, one video-maker told us:

> I don't have a regular work time. If I don't need to meet deadlines or I don't have anything urgent to do, I can call a Monday as a day off; but sometimes I work till 2am and, if I have to meet a deadline, I can work round the clock for two or three days even if they are Saturday and Sunday (#34, 33 years).

Interviewees continuously scrutinize the surrounding environment to find out opportunities and resources "that can sustain [their] motivation to commit to [their] cultural ventures" (#51, graphic designer, 27 years). More specifically, they mention the importance of having continuous creative stimuli from the environment: "I need to be inspired by the place where I live and where I work; it should help me to challenge my creativity and provide different features which I can feed my mind with" (#105, photographer, 27 years).

Interviewees also note the relevance of working in a context characterized by a diffused entrepreneurial mindset, which is key "for the development of a venture in a cultural field, since you can feel that your business is legitimated even though it belongs to particular sectors such as the cultural ones" (#118, founder of a creative hub, 34 years). In their attempt to satisfy the need for a sense of purpose, cultural workers declare to be interested in extending their activities to social projects: "Acting in theatre started to become not enough for me: I felt that to give a purpose to my artistic work I needed to engage myself in a real social stage like a neighborhood" (#58, actress, 32 years). Interviewees engage in a wide range of socially engaged activities, which include urban gardening, organizing activities for children of their neighborhoods, and intercultural workshops for migrants, among the others.

4.2 Networking

The second need reported by interviewees refers to their desire of being surrounded by a dense network of relations. Indeed, they claim that having a large set of connections at hand is key to sustain their activities: "No matter whether you're at the beginning of your career and striving for getting paid, or you've got an established activity and need to guarantee a salary to your employees. Networking is always helpful!" (#118, founder of a creative hub, 34 years).

Interviewees decline such a need along two dimensions. The first dimension relates to the desire of having the "right" relationships that could mitigate the entrepreneurial risk related to their activities, as well as being conducive to opportunities for developing new projects. The high level of informality characterizing cultural sectors makes an individual's relations with other actors (customers, other professionals, employees of public administration, etc.) crucial in sustaining entrepreneurial activity. To illustrate, one video-maker noted: "Reggio Emilia is a small city, so having a friendship network is important. It can enable word of mouth and help you find projects to work on" (#113, 34 years). Interviewees also highlight the supportive nature of strong and trustworthy relations, which give them the opportunity to reach out for help and support:

> It is helpful having two of three "colleagues" with whom I can confront my viewpoint about the next trends of the field, as well as getting some sort of support to face the extreme uncertainty of my activity. Sometimes, I am almost scared of the personal and financial commitment that my activity requires (#61, illustrator, 28 years)

However, interviewees think that it is also important to have a diverse set of relations. In this sense, occasional, weak ties can "help you get out of the usual loop" (#39, musician, 23 years), for example, entering in contact with professionals operating in other geographical contexts.

The second dimension refers to the importance of being at the center of a hub of relations in which ideas and information are shared and discussed at a fast pace. Interviewees agree that creativity is key to make an entrepreneurial venture in cultural sectors sustainable. Accordingly, they need to always generate new ideas and projects, which in turn require "being connected with several people who are keen on sharing ideas, perspectives, and information with each other" (#87, visual artist, 25 years). To this regard, interviewees mention how stable and long-lasting ties are beneficial to receive useful feedback for developing ideas and projects:

> If you have few good friends with whom you feel free to share ideas, you can even discuss ideas and projects when they are at early stages of development. Indeed, you do not fear a risk either of opportunism or rejection; you never receive negative or unsupportive feedback (#109, video-maker, 26 years).

To satisfy their need for networking, interviewees dedicate time and effort in cultivating and nurturing a lot of different ties, as "you never know what you can get from a person who is introduced to you by a friend of a friend or whom you know from many years" (#50, graphic designer, 28 years). Thus, they engage in a wide set of networking actions, which are aimed at both creating new ties with previously unknown actors and at managing existing, often long-lasting ones. In particular, interviewees do not "draw lines between professional and personal spheres: you can meet interesting people when you are working for a client or when you are drinking an espresso in a café in the city center" (#108, photographer, 35 years). Thus, they are committed to leveraging all potential occasions for networking. For example, they declare to attend events such as festivals, workshops or exhibition openings because "they are occasions for spontaneous encounters with other creatives" (#9, architect, 28 years). Similarly, they underline the importance of living in a city that hosts prominent cultural institutions, a thick fabric of companies, and higher education institutions. Indeed, this offers opportunities for developing relations with individuals coming from different fields: "During the shows of Aterballetto you always meet other professionals" (#90, dancer, 34 years).

4.3 Professional Development

The third need regards the interviewees' desire of continuously improving their competences and be updated "to the latest trends in the field" (#2, graphic designer, 30 years). It is important to highlight that interviewees admit that they often embark in self-employed initiatives without an adequate training: "I attended a music school where I never had a lecture on business, marketing or other managerial subjects" (#94, musician, 24 years). Thus, they are concerned about compensating a gap in their educational background that can pose severe constrains to the sustainability of their activities.

The need for professional development is declined along two dimensions. The first one relates to the necessity of acquiring the required technical competences to perform their activity. To this regard, interviewees mention those competences that pertain to both creative and managerial issues related to their activity. To the former, interviewees mention those competences required to produce cultural products and services with mastery, and that they need to refresh in order to keep up with the evolutions in customer tastes and competition: "In my field, you need to be updated with new technologies to better position your business in the market and propose the best solution to customers who are more and more competent and demanding" (#110, graphic designer, 28 years). As far as managerial competences are concerned, interviewees remark the importance of being able to deal with different aspects of their activities such as marketing (e.g., capacity to manage social media), business (e.g., capacity to write a business plan), and organization (e.g., capacity to manage teams). To illustrate, one interviewee highlighted: "It is very important to become more skillful in managing the relations with my customers in order to be more responsive to their needs and expectations" (#3, visual artist and event manager, 30 years).

The second dimension regards the emotional competencies, which refer to the ability to recognize, understand, and use emotional information about oneself and others (see also Boyatzis and Goleman 2007). More specifically, interviewees highlight two key competences to support sustainability of their activities: achievement orientation and self-control. The former regards the capacity to achieve a high standard of excellence and to set properly the goals that drive their action: "I like the autonomy that my activity implies; however, it is fundamental that you are organized and goal-oriented, otherwise you can get lost" (#60, video-maker, 27 years). Self-control refers to the capacity to keep a positive approach in difficult moments and maintain a control over one's own emotions in stressful situations, which interviewees declare to experience almost every day.

To satisfy the need for professional development, interviewees are interested in both attending educational programs and getting involved in relevant projects. To the former, interviewees declare that they invest as much resources as they can in attending workshops and courses, rarely being able to attend specialized masters in arts and creativity management. To illustrate, one dancer noted: "I really would like to attend this master's program, but at the moment I cannot afford it. It is not only a matter of money, but also of time, as I've got to work to earn a living" (#85, dancer, 29 years). Such an interest in education is paralleled by a strong emphasis on the expertise that they can acquire from working in challenging projects or with well-reputed professionals (or organizations): "Through practice I can learn the tricks that more experienced individuals have developed over their career" (#92, musician, 25 years).

5 Conclusions

This chapter proposes an account of how cultural workers experience the hurdles and challenges related to developing and sustaining a cultural activity. Besides confirming the well-known difficulties experienced by cultural entrepreneurs (e.g., Lee et al. 2018; Stinchfield et al. 2013), our results shed light on three key needs that cultural workers report in trying to overcome the high levels of uncertainty characterizing the first years of activity, and on the ways by which they try to satisfy them.

In so doing, we show that the sustainability of cultural activities is a multifaceted concept that could be articulated along three main dimensions. First, cultural workers interpret sustainability in creative terms, as they perceive that the essence of their ventures consists of being capable of developing continuously new creative ideas. Since cultural workers conceive their products and services as a true expression of their creative selves, they are eager to show to customers, peers, and the general public that they are able to develop new and breakthrough ideas (for similar considerations, see also Eikhof and Haunschild 2006; Svejenova et al. 2010). Second, cultural workers interpret sustainability in economic terms, as they are aware of the importance of developing stable streams of revenues from their activities. Whereas cultural entrepreneurs sometimes struggle in combining cultural and economic goals, they seem to pragmatically trade their dreams of 'arts for art's sake' with a sober reckoning of how to sustain economically their activities (see also

Morgan and Nelligan 2015). Finally, cultural workers interpret sustainability in social terms as they evaluate the social impact of their work as an important means to derive a sense of purpose and have an impact on their local community (see also Markusen 2013). In this sense, cultural entrepreneurs are keen on giving back to their local community, in many instances engaging in social activities (e.g., art workshops for children, informal exhibitions for their neighborhood) that extend their creative activities.

Such a proposed conceptualization of sustainability in cultural sectors is consistent with previous studies claiming that concerns and aspirations of cultural workers go beyond the traditional contraposition between artistic and economic logics, embracing a wide array of needs that are only apparently contradictory (e.g., Lee et al. 2018; Markusen 2013). In particular, our results lend further evidence to the argument that cultural workers are interested in being "socially engaged in their communities" rather than driven just by the "desire to optimize individual opportunities for individual creativity" (Lingo and Tepper 2013, p. 343).

Cultural workers deal with the different dimensions of sustainability simultaneously, trying to sort out a personal way to develop a sustainable cultural activity. Such a way stems from the attempt to satisfy a varied range of needs that they deem important for them, defining their aspirations as cultural workers. More specifically, cultural workers feel that they have to meet three main needs: sense of purpose, networking, and professional development. The first one regards the cultural workers' desire to find their way in complex and volatile environments such as the cultural sectors. More specifically, cultural workers are interested in developing a venture that provides the opportunity to make a living as well as to express their creative selves. The second kind of need regards the desire of developing a large network of relations with other actors (individuals, organizations, etc.) operating in cultural sectors. Cultural workers perceive such relations as key collaborative mechanisms that help them to mitigate the entrepreneurial risk related to their activities. Indeed, these relations represent key informal channels for job opportunities as well as social conduits through which to exchange and discuss creative ideas (see also Konrad 2013; Montanari et al. 2016). The third need regards the cultural workers' desire for continuously improving their competences. Cultural workers are committed to invest time and resources in enhancing their artistic and managerial competences in order to keep up with constant changes in technology, competition, and customer tastes. Similarly, they are keen on developing emotional competences that can help them navigate complex and multifarious relations, as well as turbulent and fast-changing environments. This lends support to previous studies highlighting that cultural workers are keen on acquiring technical competences but also on receiving emotional support and mentoring (e.g., Hausmann 2010; Lee et al. 2018).

Results highlight that cultural workers continuously scrutinize the surrounding environment to find out any potential opportunity that can satisfy their needs. In particular, they look at the cultural, economic, social, and symbolic resources provided by the city context as means to sustain their activities (see also Montanari et al. 2020). For example, cultural workers acknowledge the importance of operating in a context characterized by the presence of renown educational institutions and organizations operating in cultural (and related) sectors, which increase the availability of job and learning opportunities that stem from enhanced occasions for meeting other professionals and exchanging ideas with them. Finally, cultural workers refer to the importance of perceiving that their entrepreneurial activities are in line with values and mindsets promoting entrepreneurial activities. This result highlights the importance of the alignment between their aspirations and the local culture, as a supportive culture and histories of entrepreneurship "often encoded in histories of past successes . . ., supporting and 'normalizing' entrepreneurial activities, risk taking, and innovation create a breeding ground for other actors to engage in such activities" (Bolzani and Mizzau 2019, pp. 81–82). Since cultural entrepreneurship presents several challenges, individuals are more confident to overcome them when they perceive that their entrepreneurial activities are accepted and sustained in the local context.

Our study confirms the complex challenges associated with developing and sustaining an activity in cultural sectors as young cultural workers need to face several hurdles and a lack of managerial skills and competences (e.g., Hausmann 2010; Lee et al. 2018; Morgan and Nelligan 2015). To cope with such difficulties, individuals are required to become agents of their own professional trajectory. In so doing, they should adopt an approach that is open to adaptation and to emerging possibilities. Indeed, cultural workers aiming at developing and sustaining their activity need to continuously seek for learning opportunities, enhancement of their creative ideas, and new potential business opportunities, wherever and whenever they arise. This lends support to previous studies that depicted the figure of the cultural entrepreneur as an adaptive and resilient individual, mainly driven the passion for her work, rather than a rational and calculative one (e.g., Hooi et al. 2016; Stinchfield et al. 2013).

However, we add to previous studies an account of how cultural workers figure out their personal ways in developing and sustaining a cultural activity. In particular, we show how they strive to find out a way to satisfy different, apparently contradictory challenges by paying attention not only to the artistic/creative and the economic dimensions, but also to a third one, which relates to a social declination of sustainability. While they are aware of the complexity of keeping all these three dimensions together, the efforts to do so generate considerable stress on the psychological side, and that one fundamental way to cope with is to rely on their social context. In this sense, cultural entrepreneurs make sense of their actions together with the actors of—and in relation with the values and mindsets diffused in—the local context, co-constructing the value of their activities and thereby contributing also to the artistic and economic sustainability of their activities. In fact, while previous studies in sociology and economic geography highlighted the relevance of the cultural and symbolic resources provided by the city context (e.g., Drake 2003; Scott 2000), we elucidate how creative workers can draw on these resources (aesthetic stimuli, cultural traditions, signs, etc.) not just as sources of inspiration for their products and services, but also as points of reference against which to evaluate their conduct as cultural entrepreneurs (Montanari et al. 2020; Romanelli and Khessina 2005).

In terms of implications, our study suggests that policy makers and local administrators could deploy relevant actions to sustain cultural entrepreneurship. For example, they can dedicate resources to finance educational programs aimed at developing managerial competences, or collaborate with local art schools so as to offer programs that add managerial competences to artistic or creative ones. Furthermore, results suggest that policy makers can invest in order for their local contexts to become really enabling for cultural entrepreneurs. For example, policy makers can adopt an open and participative approach to sustain collaboration among different sectors, particularly financing initiatives such as cultural events and festivals, and the creation of creative hubs. In this vein, they can support the diffusion of 'third places' (Oldenburg 1989) such as coworking, incubators and maker spaces, which offer occasions to meet other professionals, develop a sense of community and create entrepreneurial opportunities (Merkel 2019).

Our study has limitations that offer opportunities for further research. For example, whereas our qualitative data allowed us to unravel cultural entrepreneurs' needs and articulate their ways to respond to them, it would be interesting to adopt a longitudinal approach to map the relations between exhibiting certain needs, different ways to respond to them, and conceptions of sustainability. Similarly, it would be intriguing to analyse how needs unfold over one's career and how different ways to cope with them can sustain both objective and subjective career achievements.

References

Alacovska, A. (2018). Informal creative labour practices: A relational work perspective. *Human Relations, 71*(12), 1563a1589.

Baker, T., & Nelson, R. E. (2005). Creating something from nothing: Resource construction through entrepreneurial bricolage. *Administrative Science Quarterly, 50*(3), 329–366.

Becker, H. S. (1982). *Art worlds*. Berkeley: University of California Press.

Bennett, D., & Hennekam, S. (2018). Self-authorship and creative industries workers' career decision-making. *Human Relations, 71*(11), 1454–1477.

Biernacki, P., & Waldorf, D. (1981). Snowball sampling: Problems and techniques of chain referral sampling. *Sociological Methods and Research, 10*(2), 141–163.

Bolzani, D., & Mizzau, L. (2019). Supporting migrant entrepreneurs in entrepreneurial ecosystems: Insights from Milan. *Piccola Impresa/Small Business, 3*, 78–105.

Boyatzis, R. E., & Goleman, D. (2007). *Emotional competency inventory*. Boston, MA: Hay Group.

Bureau of Labor Statistics. (2016, June 24). American time use survey—2015 results. *News Release*. Retrieved July 10, 2020, from http://www.bls.gov/news.release/pdf/atus.pdf

Caves, R. E. (2000). *Creative industries. Contracts between arts and commerce*. Cambridge: Harvard University Press.

Chang, W. J., & Wyszomirski, M. (2015). What is arts entrepreneurship? Tracking the development of its definition in scholarly journals. *Artivate: A Journal of Entrepreneurship in the Arts, 4*(2), 11–31.

DCMS—Department of Culture, Media and Sport. (2001). *Creative industries mapping document. Demonstrating the success of our creative industries*. London: HMSO.

Drake, G. (2003). This place gives me space: Place and creativity in the creative industries. *Geoforum, 34*(4), 511–524.

Eikhof, D. R., & Haunschild, A. (2006). Lifestyle meets market: Bohemian entrepreneurs in creative industries. *Creativity and Innovation Management, 15*(3), 234–241.

Eikhof, D. R., & Haunschild, A. (2007). For art's sake! Artistic and economic logics in creative production. *Journal of Organizational Behavior, 28*(5), 523–538.

Ellmeier, A. (2003). Cultural entrepreneurialism: On the changing relationship between the arts, culture, and employment. *The International Journal of Cultural Policy, 9*(1), 3–16.

ERVET, Emilia-Romagna Valorizzazione Economica Territorio. (2018). *Economia Arancione in Emilia-Romagna—Cultura, Creatività, Industria*. Retrieved July 10, 2020, from http://www.ervet.it/wp-content/uploads/downloads/2018/06/1_Economia_Arancione-in-Emilia-Romagna_Parte-I-Mappe-Vol-I.pdf

Eurofound and the International Labour Office. (2017). *Working anytime, anywhere: The effects on the world of work*. Luxembourg: Publications Office of the European Union, The International Labour Office. Retrieved July 10, 2020, from http://eurofound.link/ef1658

Eurostat. (2020). *Eurostat regional yearbook: 2020 Edition*. Luxembourg: Publications Office of the European Union.

Flew, T. (2011). *The creative industries: Culture and policy*. London: Sage.

Hausmann, A. (2010). German artists between bohemian idealism and entrepreneurial dynamics: Reflections on cultural entrepreneurship and the need for start-up management. *International Journal of Arts Management, 12*(2), 17–29.

Hausmann, A., & Heinze, A. (2016). Entrepreneurship in the cultural and creative industries: Insights from an emergent field. *Artivate: A Journal of Entrepreneurship in the Arts, 5*(2), 7–22.

Hesmondhalgh, D., & Baker, S. (2010). 'A very complicated version of freedom': Conditions and experiences of creative labour in three cultural industries. *Poetics, 38*(1), 4–20.

Hooi, H. C., Ahmad, N. H., Amran, A., & Rahman, S. A. (2016). The functional role of entrepreneurial orientation and entrepreneurial bricolage in ensuring sustainable entrepreneurship. *Management Research Review, 39*, 1616–1638.

Jones, C., Svejenova, S., Pedersen, J. S., & Townley, B. (2016). Misfits, mavericks and mainstreams: Drivers of innovation in the creative industries. *Organization Studies, 37*(6), 751–768.

Klamer, A. (2011). Cultural entrepreneurship. *The Review of Austrian Economics, 24*(2), 141–156.

Konrad, E. K. (2013). Cultural entrepreneurship: The impact of social networking on success. *Cultural Entrepreneurship, 22*(3), 307–319.

Lampel, J., Lant, T., & Shamsie, J. (2000). Learning from organizing practices in cultural industries. *Organization Science, 11*(3), 263–269.

Lange, B. (2008). Accessing markets in creative industries: Professionalization and social-spatial strategies of culturepreneurs in Berlin. *Creative Industries Journal, 1*(2), 115–135.

Lazzeretti, L., & Vecco, M. (Eds.). (2018). *Creative industries and entrepreneurship: Paradigms in transition from a global perspective*. Cheltenham: Edward Elgar.

Lee, B., Fraser, I., & Fillis, I. (2018). Creative futures for new contemporary artists: Opportunities and barriers. *International Journal of Arts Management, 20*(2), 9–19.

Lewandowska, K. (2015). From sponsorship to partnership in arts and business relations. *The Journal of Arts Management, Law, and Society, 45*(1), 33–50.

Lingo, E. L., & Tepper, S. J. (2013). Looking back, looking forward: Arts-based careers and creative work. *Work and Occupations, 40*(4), 337–363.

Loacker, B. (2013). Becoming 'culturpreneur': How the 'neoliberal regime of truth' affects and redefines artistic subject positions. *Culture and Organization, 19*(2), 124–145.

Markusen, A. (2013). Artists work everywhere. *Work and Occupations, 40*(4), 481–495.

McRobbie, A. (2016). *Be creative: Making a living in the new culture industries*. Cambridge: Polity Press.

Merkel, J. (2019). 'Freelance isn't free.' Co-working as a critical urban practice to cope with informality in creative labour markets. *Urban Studies, 56*(3), 526–547.

Mietzner, D., & Kamprath, M. (2010). A competence portfolio for professionals in the creative industries. *Creativity and Innovation Management, 22*(13), 280–294.

Montanari, F., Scapolan, A., & Gianecchini, M. (2016). 'Absolutely free'? The role of relational work in sustaining artistic innovation. *Organization Studies, 37*(6), 797–821.
Montanari, F., Mizzau, L., Razzoli, D., & Rodighiero, S. (2020). City context and subjective career success: How does creative workers' need for recognition filter city identity? *Human Relations*. https://doi.org/10.1177/0018726720956700.
Morgan, G., & Nelligan, P. (2015). Labile labour—Gender, flexibility and creative work. *The Sociological Review, 63*, 66–83.
Municipality of Reggio Emilia. (2009). *Reggio Emilia Città Creativa*. Unpublished research report.
Municipality of Reggio Emilia. (2016). *Policies for youth creativity 2017–19*. Unpublished policy document.
OECD. (2018). *The emergence of new forms of work and their implications for labour relations*. 1st meeting of the G20 employment working group, Buenos Aires.
Petriglieri, G., Ashford, S. J., & Wrzesniewski, A. (2019). Agony and ecstasy in the gig economy: Cultivating holding environments for precarious and personalized work identities. *Administrative Science Quarterly, 64*(1), 124–170.
Rao, H., Monin, P., & Durand, R. (2005). Border crossing: Bricolage and the erosion of categorical boundaries in French gastronomy. *American Sociological Review, 70*, 968–991.
Romanelli, E., & Khessina, O. M. (2005). Regional industrial identity: Cluster configurations and economic development. *Organization Science, 16*(4), 344–358.
Sarasvathy, S. D. (2001). Causation and effectuation: Toward a theoretical shift from economic inevitability to entrepreneurial contingency. *Academy of Management Review, 26*(2), 243–263.
Scherdin, M., & Zander, I. (Eds.). (2011). *Art entrepreneurship*. Cheltenham: Edward Elgar.
Scott, A. J. (2000). *The cultural economy of cities*. London: SAGE.
Slavich, B., & Montanari, F. (2009). New trends of managerial roles in performing arts: Empirical evidence from the Italian context. *Cultural Trends, 18*(3), 227–237.
Stinchfield, B. T., Nelson, R. E., & Wood, M. S. (2013). Learning from Levi-Strauss' legacy: Art, craft, engineering, bricolage, and brokerage in entrepreneurship. *Entrepreneurship Theory and Practice, 37*(4), 889–921.
Svejenova, S., Planellas, M., & Vives, L. (2010). An individual business model in the making: A chef's quest for creative freedom. *Long Range Planning, 43*(2–3), 408–430.
Swedberg, R. (2006). The cultural entrepreneur and the creative industries. *Journal of Cultural Economics, 30*(4), 243–261.
Townley, B., Beech, N., & McKinlay, A. (2009). Managing in the creative industries: Managing the motley crew. *Human Relations, 62*(7), 939–962.
Zikic, J., Bonache, J., & Cerdin, J. L. (2010). Crossing national boundaries: A typology of qualified immigrants' career orientations. *Journal of Organizational Behavior, 31*(5), 667–686.

Fabrizio Montanari is Associate Professor at the University of Modena and Reggio Emilia, where he is Scientific Coordinator of OPERA, a Research Unit that specializes in the study of creativity and innovation. His research interests regard creative industries, creativity, social networks, and new collaborative workspaces. He is also Research Affiliate at ASK Research Center, Bocconi University. His work has appeared in international scholarly journals such as *Journal of Economic Geography, Journal of Management Studies, Organization Studies* and *Urban Studies*.

Lorenzo Mizzau is Senior Assistant Professor at the Department of Economics and Business, School of Social Sciences, University of Genoa. He is interested in the management of cultural industries, the role of context in organizational practices (particularly applying spatial perspectives), and the dynamics of creative work. His work has appeared on international edited books and scholarly journals such as *Human Relations*, the *International Journal of Arts Management, Journal of Economic Geography*, and *Urban Studies*.

Damiano Razzoli is Research Fellow and Adjunct Professor at the Department of Communication and Economics of the University of Modena and Reggio Emilia, where he is a member of OPERA, a Research Unit that specializes in the study of creativity and innovation. His research interests include the link between creativity and communication, the role of space in creative industries, and city identity. His work has appeared in international journals such as *Human Relations, Studi Organizzativi* and *Sociologia del Lavoro*.

Organizing Academic Entrepreneurship Drawing on Cultural Knowledge: The puntOorg Experience

Davide Bizjak, Luca Pareschi, and Silvio Ripetta

Abstract This chapter presents the case of puntOorg International Research Network to reflect on the issues of cultural entrepreneurship and academic entrepreneurship, especially in the light of University reforms that reframed the role of professors and researchers working in Academia, as well as the rules of the game of academic publishing. After having described the relationship between academic entrepreneurship and cultural production, and after having introduced the Italian academic context, we present the case of puntOorg, which is described as an illustrative example of a not-ordinary way of being academic entrepreneurs. We describe the *praxis* of puntOorg, as it is crystallized in its intra- and interdisciplinary research practices, and we also describe its intellectual outcome, as reified in different publications.

Keywords Academic entrepreneurship · Cultural entrepreneurship · Publishing · puntOorg

1 Introduction

The relationship between business and cultural activities has drawn growing attention over the years, with a major emphasis on the culturalization of business (Buswick et al. 2004; Boltanski and Chiapello 2005) in terms of creative inspiration and innovativeness. Seemingly, cultural management's scholars are expected to

D. Bizjak (✉)
University of Naples Federico II, Naples, Italy
e-mail: davbiz@unina.it

L. Pareschi
University of Rome Tor Vergata, Rome, Italy
e-mail: luca.pareschi@uniroma2.it

S. Ripetta
University of Naples Federico II, Naples, Italy

P. Demartini et al. (eds.), *Cultural Initiatives for Sustainable Development*, Contributions to Management Science,
https://doi.org/10.1007/978-3-030-65687-4_8

make original contributions to their field of study, doing research on the functioning of cultural organizations. The liaison is not always convenient, and scholars are demanded in behaving in an entrepreneurial way. As a great emphasis has been put on the emergence of an entrepreneurial dimension characterizing the cultural work (Lindqvist 2011; Enhubner 2014); the same has happened for cultural management's scholars (Gartner et al. 2015). Despite the parallelism between the act of entrepreneurship in business and cultural contexts, business and non-business entrepreneurs (Clark 1972) have been so far studied separately (Nambisan 2015), proposing for the artistic-cultural context a special and different set of tools that refer to a supposed diversity of the context under observation. Accordingly, this chapter is aimed to inquire the following research question: is the academic entrepreneurship a mean to understand deeper the multidisciplinarity nature of cultural entrepreneurship?

Given this starting conditions, the present work aims at observing the academic entrepreneurship and the cultural entrepreneurship, expanding the concept of academic entrepreneurship and trying to frame an academic experience of research that is actually similar to an entrepreneurial one. In order to reach this objective, the chapter is organized as follows: In the first paragraph the linkage between cultural production and entrepreneurship, both in culture and University, is outlined from a conceptual point of view, narrowing how the idea of an entrepreneurial attitude raised both in cultural production and academic activity. Thereafter, the puntOorg International Research Network will be analyzed, taken as an illustrative example of being academic entrepreneurs. In the final section, there will be some conclusive reflection on the role of puntOorg in the general context of academic entrepreneurship.

2 Cultural and Academic Entrepreneurship

We investigate cultural entrepreneurship with specific reference to academic research organizations (Galloway and Dunlop 2007). These organizations are currently included within the realm of "cultural production," as well as products traditionally studied in the literature of art and cultural management: such as museums, theatres, libraries, festivals. But organizations producing these products have one substantial difference: they draw from the past to speak to the present, whereas academic organizations are a mirror of the present scenarios and speak to the future. In addition to carrying out research activities (basic or applied), they are devoted to teaching and dissemination. These three "missions" of the University are present in most of the European countries that are working on the construction of a European harmonization.

For example, Italian legislation distinguishes among I, II and III Mission (Law 240/2010); in France, despite several reforms, the founding principles have remained those of a public system (about 85%), namely accessible, characterized by quality teaching and integrated into national research structures. The United Kingdom (even before Brexit) distinguishes between research and teaching roles.

Until a few years ago, this structure was based on a separation between academic activities (research, teaching, and dissemination) and fundraising activities. The entrepreneurship of researchers and professors (even of those professors who held roles of academic governance) had as its ultimate goal the creation of a "creative-cultural product" (Recaman and Colbert 2016), while business issues (such as risk-taking and people management) were entrusted to other subjects (government, in the case of Italy or France). In short, researchers and faculties were entirely free concerning their research and teaching activities in their "academic model": academic research and teaching activities were (at least formally) separated from governance and budgeting activities. This allowed academics (researchers, lecturers and academics with governance roles) to act, each in their own field, as entrepreneurs.

In the meantime, the new geopolitical arrangements and the need to harmonize budgets between the Member States of the European Union have led to the search for new academic models. In particular, the influence of academic governance and budgetary roles on life (and survival) of academic institutions has also increased in Europe. As a result, academic institutions have become more and more "companies," and less and less "cultural products," stimulating, as a result, new ways of doing academic enterprise. An entrepreneurship that we can imagine (in theory and in a deliberately paradoxical way) with two opposite poles of a *continuum*: on the one hand only fundraising; on the other hand, only cultural research. These two opposite poles are useful because they potentially, and according to the former hypothesis, highlight two aspects:

1. The research activity (in our paradoxical example) becomes a "cultural product" like a museum, a festival, or an opera house.
2. The nowadays entrepreneurial solutions are placed in a "grey zone," intermediate between the two poles of the continuum.

Entrepreneurship in academic context is an intrinsically elusive concept (Wadhwani et al. 2017) and in not so recent times the widespread assumption was that before the 1970s, academic scientists pursued work that "interested them intellectually regardless of whether they had clear economic potential" (Berman 2011: 35). Within the debate around the role of the University, it appears increasingly clear that that role is not education for education's sake (Martin 2012). The entrepreneurship in Academia is usually meant as the researcher's attitude to pursue business drawing on knowledge. Therefore, different ways of being entrepreneurial have been carried out during the year. Patents and licenses are more than easy tools for being academic entrepreneurs, and they are especially employed in the realm of natural sciences.

3 The Italian (Academic) Context on the Threshold of the New Millennium[1]

It was the year 2010 and in Italy a "moralizing government" proposed new rules of the game in the Italian academic world. This was the Berlusconi Government, whose Minister for Research and University was Mrs. Maria Stella Gelmini.

The model of university that came out of that experience of government was the law 240 December 30, 2010 (Gelmini Law). This law has rewritten the rules of university life: the official objective (or, at least, the officially declared one) was to complete the harmonization of the Italian academic system with the one pursued by other European countries (a process that had already begun in previous years with the creation of the European Community).

At the same time, this law changed the professors recruiting process, with the precise objective, declared in all the main national and international press, to crush ancient privileges and "baronies," in the name of concepts such as "meritocracy," "transparency," and "evaluation" of researchers and the quality of published research (Abravanel 2011).

In the climate of those early years of the new millennium, a series of ancient traditions, consolidated in the way of academic research, were discouraged, since they were considered obsolete in the face of the "new" that was advancing (Sicca and Viscardi 2008).

For example, it was told that monographic research work with the publication of books was not be useful to make career advancements. Conversely, it was essential (and still is) to publish in selected and starry top journals, recognized as high quality by the international academic community. In short, the well-known motto "Publish or Perish (PoP)," already in force in the English-speaking countries, was also gaining ground in Italy. "Publish or Perish" also became the rule in the soft underbelly of the Mediterranean, which has its own stories of research, teaching, and commitment by academics in the area.

These activities are what today, about 10 years after Law 240/2010, are called "research," which is the first mission of the university, the second mission of the university is "teaching," and the third mission of the university is the "work on the territory."

In that *Renaissance* climate of Italian academic culture, younger precarious scholars had to follow the new rules of the game, while various possible ways were opened for those who were already part of the academic world and had grown up with old rules, prior to those given by the moralizing government chaired by Silvio Berlusconi. Let's assume two of them:

---

[1]The present paragraph is part of the empirical context in which the idea of this chapter lies. It represents a brief report of some political decisions taken concerning university and researcher's life. It is necessarily incomplete as political analysis per se but rather frames a familiar feeling of how the university role within society has been perceived in a partially concluded political season.

1. Following the new rules of the game in the name of key words such as "meritocracy," "science," assigned to any type of knowledge, even very ancient and pre-scientific, such as music and philosophy, but also more recent such as "economics-and-management" (as a unified—with the dash—theoretical disciplinary corpus)
2. Resisting to the change in the twofold meaning of resistance: to contrast a healthy change by preserving the worst part of the existing system or to initiate "acts of resistance" and "wars of position" (Gramsci 1964), so to not adhere to the rhetoric of the mainstream political and academic language

4 puntOorg International Research Network: An Illustrative Case of Academic Entrepreneurship

Referring to the case of the entrepreneurial experience in academic research of puntOorg International Research Network, we used a self-narrative (Chang 2008) to discuss how the necessity to clamp the research results concerning the relationships between arts and organization has generated several typical entrepreneurial strategies: from diversification to the maintenance of the competitive edge, from taking the risk in terms of resource to managing the exploitation of the research results.

puntOorg begins its first steps in the second half of the nineties of the last century, starting from the following observation, of common sense, but of great interest to some *élite* of academics of organizational studies: business organizations pursue not only visible objectives (efficiency, effectiveness and cost-effectiveness), but also invisible objectives. Business organizations give order to the life of adults and mark the rhythms of dream and wakefulness, the rhythms of food and sexual life and so on. In this sense, organizations are "containers," like boxes that contain the actions at the service of business productivity, but also in the sense of producing limits in the human action.

This observation is reflected in the constructivist epistemology, whereby business organizations are "a social construction of reality" (Berger and Luckman 1967).

In the first decade of the second millennium, puntOorg began to query the reading keys in order to read the micro- and macro-economic systems. That is, glasses through which look over the world, not the only ones able to explain how to design and how to manage the organizational behavior.

The hard sciences, for example, have achieved reassuring successes (thanks to laboratory reductionism) that, from biology and genetics, improve the health of most of the inhabitants of the industrialized part of the world. And to follow, also the humanistic knowledge and even the "arts and culture" have demanded their own label of "sciences."

puntOorg established itself internationally at the end of the first decade of the new millennium, with the proposal to question (not to answer) knowledge and consider it not as "objects," but as partial and imperfect glances on the world.

Each academic discipline gives its intradisciplinary contribution and can make a partial contribution.

On the proposal made by prof. Luigi Maria Sicca, puntOorg International Research Network began officially structuring itself as a sense-making network in 2010, although its informal origins date back to the first half of the 1990s, among academics of critical management and art management (Sicca 1997, 2000; Sicca and Zan 2005). It was a historical moment in which the logic of the "PoP" in the top starry journals became so much violent and aggressive to arrive at a reversal between means and ends, up to the point of generating a pathology, concerning the logic of the ideal-typical Weberian bureaucracies. The prevailing logic in the world of academic research can be expressed in the following sentence, which is even more consolidated today: one publishes to make career and not, otherwise, one makes a good career because it had published in journals with a recognized reputation. Therefore, one has to publish into the "proper" journals, and the "proper" journals are those that use a method based on objective quality. Other types of publication are not required. This point of view has been commented by Professor Luigi Maria Sicca, during an interview to inquire the origins of the puntOorg International research network, concerning the importance of being cultural entrepreneurs proposing cultural products:

> If we want to understand organizational action, we have the opportunity to consider the action itself as a true and proper text, with a grammar and syntax that generate a plot, that we need learning to read it, with the final aim of making an interpretation. Interpretation is the task of the actor, either the dramatic actor or the organizational actor who therefore "plays", in the many meanings of "to play", that is also is spielen (Deutsch), jouer (french) and so on in all the languages of the world, because «to play» is a fundamental part of human character development and we human play all our lives, even in adulthood. Perhaps this is also our distinctive trait compared to other animal species with whom we share the planet.

The roots of puntOorg International Research Network in art management research (Sicca 2016) help to understand this paradox well, thanks to the work carried out closely between musicians and philosophers, holders of knowledge of ancient origin, and "younger" scholars such as management scholars, neuroscientist, or physicist.

4.1 The puntOorg Paradox: The Relationship Between Means and Ends

The question of the relationship between means and ends and its inversion is a typical question of studies in philosophy and music. These disciplinary corpora, even though they have their own internal and intradisciplinary evolution, represent reliable, granitic but alive, and compact theoretical corpora that have passed the test of obsolescence.

Philosophy and music are a few 1000 years old and still represent an essential reference for social knowledge: this is the hypothesis on which the experience of puntOorg International Research Network is built.

This is why the following is an example of how difficult it is to unravel the Gordian Knot in the relationship between means and ends. A knot that puntOorg challenges and tightens further without any claim to untie it. In this regard, we indicate below at least three working tables within puntOorg that operate separately, but then intertwine, as a "plot" (Czarniawska and Rhodes 2006; Czarniawska 2004) starting from a research work that, rather than giving answers (as it happens for "hard sciences" that must produce valid results, "until proven otherwise"), asks questions[2]:

- Critical thinking tables.
- Self-determination of identity and of one's own intra-psychic world, carrying out research that can be labelled as "diversity management" (DM).
- Creative thought table, which studies improvisation (in Jazz, electronic music and contemporary musical composition, but also in decision making that interests entrepreneurs and managers). In the next paragraph we dwell with an exemplum of synthesis on this third working table.

4.1.1 One of Our Gordian Knot

An example of how puntOorg conducts interdisciplinary research is an example of a research process that began in 2012 and is still active today.

It is a research table where researchers and composers of contemporary and electronic music, management scholars, and scientists (accustomed to experimental reductionist work) work together: each one, each with its own epistemological position, compares with the others around a keyword, for example, "gift," "interest," "disinterest," (Mauss 1990; Vv. Aa. 2013; Diana et al. 2017; Mallozzi and Tortora 2017), and "resonance." How does each of these words resound in the disciplinary field of each researcher? This question is accompanied by a series of seminars that have been held regularly for about 10 years. There is often no lack of violent

[2]It is hardly necessary to remember that puntOorg has edited the Italian editions of some classics of organizational literature, bringing in Italian language works present in the official language (English) of the scientific community. For example (Sicca 2018, 2019), the book by Barbara Czarniawska (Narratives in Social Science Research, Sage, 2004) and the book by Peter Senge (The Fifth Discipline, London: Sage. 2006). What is the meaning of these editions? puntOorg, as an Italian player of the internationalization, leverages a dual register: on the one hand, the belief that internationalizing means—even before using a shared language (English, for example, on which the academic debate is based)—a constant commitment to rooting in tradition, which is also translation (and betrayal) of the codes and meanings (Eco 2003) that guide an academic or professional population. On the other hand, the idea that every research is never only the arriving point of a scholar or a research group, but always the starting point for a wide circulation of ideas, multiplying the filter through which to read and write the world. Or those portions that we are led to live and/or narrate.

epistemological clashes, precisely because the idea of puntOorg International Research Network is that research work should be primarily **intra**disciplinary. Only then will it be **inter**disciplinary. In this way, unprecedented doubts and questions can be raised in the disciplinary field of the others. And the latter remains fertilized, enriched, and increased. Learning to see "through the eyes of the other" (von Foerster 1991): this approach taken from radical constructivism makes evident a contradiction of puntOorg. Indeed, on the one hand, each research group carries out its own work as a "subject"; on the other hand, it risks being an "object," just as it happens in the laboratories, where through the reductionist method the complexity of the world is reduced by experimenting, for example, on laboratory mice. Therefore, the researchers of puntOorg are, at the same time, researchers who contest the research on laboratory mice (not for an anti-anthropocentric position, but for a free epistemological choice) and, on the other hand, they are themselves laboratory mice.

In other words, every research table within puntOorg International Research Network is a round table, where scholars from different disciplines sit in a participative way: economists and researchers from management and human resources, artists (musicians, composers, musical performers, visual artists), neuroscientists, physicists, philosophers. Each of them is invited to do work first of all intradisciplinary, questioning their disciplinary field in depth, vertically. Only then each researcher is invited to question horizontally, through an interdisciplinary method that is rigorous, distinct, and distant from straightforward logic, and instead characterized by a healthy, profitable, and prudent crossbreeding.

puntOorg International Research Network is now a research association like many other scientific societies. puntOorg brings together scholars from over 30 research and academic institutions, including universities on three of the five continents, CNR centers and music conservatories. The latter, in Italy, refer to the Higher Artistic and Musical Education (Alta Formazione Artistica e musicale—AFAM), a department of the Ministry of Research and University (MIUR). This has happened since Italy also started the process of harmonization with other European countries in the teaching of musical instruments.

The results of the research work are all traced on a bilingual website, built directly by the group of puntOorg founders (www.puntoorg.net): the latter aspect of self-entrepreneurship (Pettigrew 1979) is an important aspect in the construction of puntOorg's identity.

4.1.2 En Attendant... International Conferences

Another experience of puntOorg, in line with the hard Gordian Knot that binds the means ("how do you", which is technical, from *techne*) for the ends ("what do you do", hence products, result), is the project entitled "En attendant... International Conferences." In a way, this is an initiative that offers a response to the risk of inversion of the relationship between means and ends in the Weberian sense, activated by the "moralizing government" that has introduced also in Italy the dictatorship of the "PoP."

En attendant... designed and developed by Luigi Maria Sicca within puntOorg International Research Network was born from a certain impatience with institutional academic conferences that are often a showcase for researchers to tell their work, without allowing comparison in depth and content. Usually, in those venues, researchers present their work, then open a few minutes of questions and a few minutes of debate in which the speakers must "defend" their paper. The most authentic moments of authentic debate in those conferences are often gala dinners, or beer on the side among a few good friends who can argue with slowness. In short, this cycle of international seminars is a cloistered experience. To be together, with a high level of focus.

The first edition of En attendant was held in November 3, 2014 at the University of Milan (Seminar Room—Department of Political and Social Sciences). Many subsequent editions were held at the University of Bologna, others in Rome at the LUISS, University of Naples, Tor Vergata, Link Campus. On June 7, 2019, the tenth edition of En attendant was held at the University of Naples Federico II, representing one of the first celebration of a puntOorg experience.

The success of this project is linked to the fact that "En attendant" reverses the traditional logic of institutional conferences, giving a few minutes (from 10 to 15) to researchers who present their research in terms of objectives, methodology, and theoretical framework and about 30 min to the community in the room to give suggestions for improvement. Researchers do not have to defend anything at all. They do not have to defend their own paper, but to act as "sponges" that absorb the comments of the community. Conversely, the community has the time to dig deep into the works presented, giving real advice on the work of the others, rather than showcasing the personal sharpness. The structure of En attendant reaffirms thus how delicate and important is the relationship between "being a scientific community" and participating in the life of "scientific societies." En attendant aims to prepare the academic community and its members for competition at the international conferences of leading international scientific societies. A year-round 365 days. Or perhaps, more correctly, 360 days a year, leaving the other five all free to travel to conferences.

Witnessing, in this way, that the internationalization of research is both an object and a method and pays great attention to the processes (how to do), as well as on the results and performances (what to do/get). When participating in the life of "scientific societies" means presenting significant research products that represent the result of an underlying research endeavor effectively.

5 From Process to Product: puntOorg International Research Network as an Italian Player in Internationalization

Drawing on the approach previously outlined, puntOorg International Research Network is an Italian player of internationalization that realizes four output lines (research products) with major international publishers:

1. **Books** (even those monographs poorly seen by the "moralizing government") that follow all the rules of double-blind review. Therefore, puntOorg is a publisher accredited by the Italian Academy of Business Administration (AIDEA).[3]
2. **block #.** While blocks are printed with an ISBN in the form of a book, they are a section of the catalogue of the publishers that wants to provide a less structured space for original insights. Blocks comprise handwritings and drawings, and are composed of incomplete sheets of paper to flick through and sketch ideas on. These blocks start with a brief foreword by a Block mentor and ends with a brief afterword by an expert far from the Block experience. This is the description of the blocks# on puntOorg website[4]: "**block** is a space where people can take notes. **/block** is a site for original insights. **Block** is the concrete social block. / **block** is the well-known block: between parents and their offspring. **/block** is when young generations are less wise than older generations, and the latter are older but not the wiser. **/block** is the absence of social spaces. **/block** is the unexpected roadblock. **/block** is the lack of social time. **/block** is the dread of death. **/block** is the dread of love. **/block** is before and during my being examined. **/block** is lack of genuine unlocking, after the exams. **/block** brings together painters, musicians and photographers. Poets. Stories that are to be written. / **block** is a way of building bridges across and among people, who are learning to progress *en bloc*. Without focusing on details that can place a block.
3. **puntOorg notebooks** are printed books devoted to the technique. The first published notebook, in example, is about a method for using extensions and their application in the orchestral repertoire. Again from the description on PuntOorg website[5]: "**puntOorg notebooks** are the place for technique. / **puntOorg notebooks** are artefacts, artis and factum, the effect of art: a state of mind that meets the means for carrying out actions that generate other actions, even before perceptions. **/puntOorg notebooks** are slim objects: to discuss how critical is the question of how to do things. **/puntOorg notebooks** are an

---

[3]Within which we find the scientific societies related to the different scientific sectors ASSIOA (The Association that aims to contribute to the development and diffusion of knowledge in the field of Organization studies and HR management), that yearly organizes the Workshop of Organization Studies (WOA). With the aim of celebrating the twentieth edition (Palermo, 7–8 February 2019), it has established the first ASSIOA-puntOorg best paper prize.

[4]Retrieved from https://www.puntoorg.net/en/2015-01-25-16-06-31/block

[5]Retrieved from https://www.puntoorg.net/en/2015-01-25-16-06-31/notebooks

opportunity. For civilization and the development of subjectivity: always intertwined and, at the same time, means and scope. **/puntOorg notebooks** collect techniques: instrumental, learning, computing, gaming, listening, and thinking techniques. And not only. **/puntOorg notebooks** contribute to the declination of multiple identities between forms and pluralism of knowledge. / **puntOorg notebooks** invest in moulding the mind, an evolution of both a physical and military exercise."

4. **puntOorg International Journal** is an online, peer-reviewed, international journal, whose primary aim is to publish outstanding research rooted in the tradition of social sciences. The journal features contributions that adopt ethnographic, critical, and interpretive approaches as practiced in such disciplines as organizational, communication, media, and cultural studies. Therefore, PIJ bridges the arts, humanities, and the social sciences, drawing on the disciplinary practices, advanced theory formation, methodologies, and discourses of philosophy, the performing arts, literary and art criticism, and historical analysis (amongst others), and applies them to the organizational and relevant social arenas. pIJ challenges the international scientific community on the sources of managerial knowledge and, thus, particularly welcomes contributions in the field of philosophy and music, since these disciplines both represent pre-capitalist sources of knowledge, which have successfully passed the test of time and obsolescence. In this sense, we welcome contributions embedded in archetypal knowledge structures (i.e., Linguistics, Literature and History, Applied epistemology, etc.) that have much to say and teach to most recent social sciences, such as (critical) management studies.[6]

6 The puntOorg Entrepreneurial Saga

Today puntOorg is an international research association and welcomes researchers from over 30 European and non-European academic institutions, including universities, CNRs, Music conservatories, and some independent researchers, along the entire chain of education. puntOorg has several aims: the first aim is to diagnose business dynamics through the contribution that some forms of ancient knowledge (music, theatre, literature, the philosophy that have overcome the problem of obsolescence) can make to managers and business professionals. Playing with the "hard sciences" on the banks (neuroscience, physics, psychology of psychoanalytic brand) in order to diagnose and find long-term solutions to companies operating in a "new world" that is becoming smaller and smaller, thanks to the impact of information technologies inside and outside the organizations that we live every day. Today puntOorg has numerous research communities where musicians, managers,

[6] Adapted from https://www.puntoorginternationaljournal.org/

entrepreneurs, scientists interact equally each with its own key to understanding the organizational action. On this point Sicca affirms:

> If, we understand organizational action as a text to be interpreted, the aim of puntOorg International Research Network as a "cultural product" is not so much to "convince" others to take a step backwards (and then maybe two steps forward); to carry out a "War of position"[7] Aimed at taking back the lost territories. The objective, once the winners have been announced, is to make a methodological and temporal leap and declare together: "forward to the past".

Affirming in this way an idea of "anthropology of organizations," something very different from the mainstream thinking.

In the perspective of the "anthropology of organizations," puntOorg is an artefact (Gagliardi 1990) that makes tangible the meeting between University and business, so to generate retention of value, valid in the long run, avoiding both the redundancy of abstraction, and the risks associated with an investment in practices without theory.

Accordingly, academic institutions have become more and more business companies and fewer and fewer cultural products, and it becomes essential to understand why. If it is true that Law 240/2010 (namely Gelmini Act in Italy) has captured many subjects in its canvas of "meritocratic" ideas, it is also true that it is necessary to understand that, when a change is received with wide consent, behind it is often hidden a longer and gloomy underground conditioning work.

It is precisely the concept of "resistance to change" evoked above that brings with it responsibilities and often frustrations. But also opportunities, as the puntOorg International Research Network shows, considering that it is part of this renewal context, proposed in a historical moment of Italian and international academic life in which the stimuli for innovation and a sense of personal fulfillment.

The change of orientation experienced by Italian academic organizations in recent years is certainly also the result of a change more subtle and difficult to grasp, but of high social impact, which sees, in a structural and narrative analysis of the facts, what in linguistics we call "actant," moved from one narrative genre to another. When the plot and the genre change, the warp threads that create the plots also change: this is what has happened in modern times and dared "post-modern." Bruner (1990) argued that the attractiveness of a narrative is negotiated from situation to situation; therefore negotiation becomes the first tool that accompanies us in

[7]The concept of "War of position" has long been at the center of a Track "Institutional Resistance, War of Positions and Power Maintenance", which later became Standing Track with the title "Institutional Change, Power, Resistance and Critical Management", within European Academy of Management—EURAM, coordinated by Luigi Maria Sicca (University of naples Federico II) and Edoardo Mollona (University of Bologna), together with Ilaria Boncori (University of Essex); Jean-Francois Chanlat (Paris Dauphine); Stewart Clegg (University of Technology, Sydney); Alessia Contu (UMass Boston); Xavier Deroy (NEOMA Business School Reims Campus); Andrea Fumagalli (Università di Pavia); David Levy (UMass Boston); Mariella Pandolfi (Université du Québec à Montréal); Luca Solari (University of Milan).

choosing the narratives we want to give the power to influence our choices, those narratives in which we decide to become extras or even actors.

From a cultural perspective, the puntOorg research network embodies the interdisciplinarity of cultural ecosystem, as addressed in this edited book. The reason for taking over this particular research network is that interdisciplinarity is not only taken as either a perspective or a method, but rather as a daily practice, demonstrated by the fact that management represents just a focal point of the puntOorg analysis, but it is equally and inclusively inquired with arts and humanities, with physics and neuroscience, in a melted tower of knowledge that is no longer built with ivory, but rather with a tempered glass, thus giving back the sense of entrepreneurship that looks implicit within the project.

puntOorg is committed to re-founding a choice in the subject, without the pretension of screaming louder or dabbling in the attempt to suppress the voice of the other. The attempt is to reconstruct a realm of cultural stories and beliefs, such as to attract back to itself the faith of the subjects concerned. In so doing, generating (not necessarily new) organizational sagas (Clark 1972), within academic organizations, capable of freeing from sleep the many who believe they are speaking an absolute truth. Organizational saga, in short, is understood as the organizational consciousness of an extraordinary enterprise in a group formally constituted and well adherent to its roots (the Italian University) and constantly open to the most qualified part of the academic world everywhere in the world. For this reason puntOorg International Research Network is both a cultural product and an academic enterprise: it has both faith in modern means, but it is always convinced of the importance of building bridges between science and knowledge of which the latter still carry within the former.

References

Vv. Aa. (Ed.). (2013). *I linguaggi dell'organizzare. Musica e testo tra dono e disinteresse*. Napoli: Editoriale Scientifica.

Abravanel, R. (2011). *Meritocrazia. Quattro proposte concrete per rendere il nostro paese più ricco e più giusto*. Milano: Garzanti.

Berger, P. L., & Luckmann, T. (1967). *The social construction of reality*. Harmondsworth: Penguin.

Berman, E. P. (2011). *Creating the market university: How academic science became an economic engine*. Princeton, NJ: Princeton University Press.

Boltanski, L., & Chiapello, E. (2005). *The new spirit of capitalism*. London: Verso.

Bruner, J. (1990). Acts of meaning. Cambridge, MA: Harvard University Press.

Buswick, T., Creamer, A., & Pinard, M. (2004). *(Re)educating for leadership: How the arts can improve business: Arts & business*. Retrieved from http://www.aacorn.net/members_all/buswick_ted/ReEducating_for_Leadership.pdf

Chang, H. (2008). *Autoethnography as method*. Walnut Creek, CA: Left Coast Press.

Clark, B. R. (1972). The organizational saga in higher education. *Administrative Science Quarterly, 17*(2), 178–184.

Czarniawska, B. (2004). *Narratives in social science research*. New York, NY: Sage.

Czarniawska, B., & Rhodes, C. (2006). Strong plots. Popular culture in management practice and theory. In P. Gagliardi & B. Czarniawska (Eds.), *Management education and humanities* (pp. 195–220). London: Edward Elgar.
Diana, R., Sicca, L. M., & Turaccio, G. (2017). *Risonanze. Organizzazione Musica Scienze*. Napoli: Editoriale Scientifica.
Eco, U. (2003). *Dire quasi la stessa cosa. Esperienze di traduzione*. Milano: Bompiani.
Enhuber, M. (2014). How is Damien Hirst a cultural entrepreneur? *Artivate: A Journal of Entrepreneurship in the Arts, 3*(2), 3–20.
Gagliardi, P. (Ed.). (1990). *Symbols and artifacts: Views of the corporate landscape*. Berlin: Walter de Gruyter.
Galloway, S., & Dunlop, S. (2007). A critique of definitions of the cultural and creative industries in public policy. *International Journal of Cultural Policy, 13*(1), 17–31.
Gartner, W. B., Roberts, J., & Rabideau, M. (2015). Perspectives on arts entrepreneurship, part 2. *Artivate: A Journal of Entrepreneurship in the Arts, 4*(2), 3–9.
Gramsci, A. (1964). In G. Ferrata & N. Gallo (Eds.), *2000 pagine di Gramsci: Vol. 1. Nel tempo della lotta (1914–1926)*. Milan: Il Saggiatore.
Lindqvist, K. (2011). Artist entrepreneurs. In I. Zander & M. Scherdin (Eds.), *Art entrepreneurship* (pp. 10–22). Cheltenham: Edward Elgar Publishing.
Mallozzi, C., & Tortora, D. (2017). *La bottega del suono. Mario Bertoncini. Maestri e allievi*. Napoli: Editoriale Scientifica.
Martin, B. R. (2012). Are universities and university research under threat? Towards an evolutionary model of university speciation. *Cambridge Journal of Economics, 36*, 543–565.
Mauss, M. (1990) [1925]. *The gift: Forms and functions of exchange in archaic societies*. New York: Norton.
Nambisan, S. (Ed.). (2015). *Embracing entrepreneurship across disciplines: Ideas and insights from engineering, science, medicine and arts*. Cheltenam: Edward Elgar.
Pettigrew, J. D. (1979). Binocular visual processing in the owl's telencephalon. *Proceedings of the Royal Society of London. Series B. Biological Sciences, 204*(1157), 435–454.
Recaman, A. L., & Colbert, F. (2016), Tania Pérez-Salas, dancer and choreographer: From artistic creation to management of a cultural enterprise. *International Journal of Arts Management*: 232–245 (Special Edition Latin America).
Sicca, L. M. (1997). The management of opera houses. The Italian experience of Enti Autonomi. *International Journal of Cultural Policy, 4*(1), 201–224.
Sicca, L. M. (2000). Chamber music and organization theory: Some typical organizational phenomena seen under the microscope. *Culture and Organization, 6*(2), 145–169.
Sicca, L. M. (2016). Diversity management, Inclusione, Analisi organizzativa. Spunti, punti e appunti. *Prospettive in Organizzazione, 4*.
Sicca, L. M. (2018). Nota editoriale. In *La narrazione nelle scienze sociali* (first Italian edition of Czarniawska 2004. Narratives in social science research. London: Sage). Napoli: Editoriale Scientifica.
Sicca, L. M. (2019). Ieri, oggi e domani. Ma dopodomani? Learning organization, prospettiva e percezione del cambiamento. In P. Senge (Ed.), *La Quinta Disciplina* (Italian edition of Senge, P. 2006 The fifth discipline. London: Penguin random House) (pp. 17–31). Napoli: Editoriale scientifica.
Sicca, L. M., & Viscardi, R. (2008, November 26–28). Neophilia and managerial rhetoric: How can managers be constructors of organizational realities? Conference proceedings of the 3rd Australasian caucus of the standing conference on organizational symbolism neophilia and organization, University of Technology, Sydney.
Sicca, L. M., & Zan, L. (2005). Much ado about management: Managerial rhetoric in the transformation of Italian opera houses. *International Journal of Arts Management, 7*(3), 46–64.
von Foerster, H. (1991). Through the eyes of the other. In F. Steier (Ed.), *Research and reflectivity* (pp. 63–75). London: Sage.

Wadhwani, R. D., Galvez-Behar, G., Mercelis, J., & Guagnini, A. (2017). Academic entrepreneurship and institutional change in historical perspective. *Management and Organizational History, 12*(3), 175–198.

Davide Bizjak, PhD, is Research Fellow in Organization Studies at the University of Naples Federico II, Italy, where he teaches Human Resource Management. His primary research interests lie in the areas of Arts Organisations, Organisational Identity, and Arts Entrepreneurship. He has been awarded several prizes and research grants, EURAM Grant Scheme 2019, and Best Reviewer at the EURAM Conference—SIG 01 (Paris 2016). He actively serves as a reviewer for several international journals and conferences, and has published extensively on the abovementioned areas both in refereed international journals and edited volumes. He is part of the puntOorg International Research Network.

Luca Pareschi is an Assistant professor researcher in business administration organization and Human Resource Management at the University of Rome 'Tor Vergata'. Previously, he worked at University of Venice 'Ca' Foscari', and at the University of Bologna, where he got his PhD in management. His main research interests regard the performative power of words and dictionaries, the management of arts and culture, and the neo-institutional approaches to organizational fields. He conducts research through qualitative and semi-automatic text analysis techniques to extract meanings from texts and reconstruct narratives.

Silvio Ripetta is graduated in Business administration at the University of Naples Federico II. His main interests lie in narrative methodologies and learning organizations.

Part II
The Meso Level of Analysis

Promoting Collaboration Through Creative Networks: The Puglia Music Industry

Manuela Barreca and Marco Meneguzzo

Abstract This research aims to investigate collaborative processes through the development of networks in a creative context. The theoretical framework lay on network governance and management for the enhancement of collaboration settings in the cultural and creative sectors. This chapter presents the early development of this kind of networks in the music industry in the Italian Region of Puglia. We have analysed the results from a 2-year project that monitored activities launched by the regional programme Puglia Sounds in 2010–2011. The programme, financial support from the EU and the regional government's wise vision to enhance an innovative and creative environment based on participatory approaches, today, have made Puglia a best practice in the field of cultural and creative management. Data were collected from both primary (semi-structured interviews and focus groups) and secondary (e.g., official documents and institutional reports, press, and media channels) sources. Results indicate that public networks in creative environments are considered an effective solution for dealing with the cultural and creative sector complexity and an influential tool to foster strategies and collaborations for regional development. Therefore, to guarantee a long-lasting network, a clear vision, procedures and rules are essential to coordinate and manage the network's complexity over time.

Keywords Cultural and creative sector · Public networks · Network governance · Participation · Music industry

M. Barreca (✉)
Faculty of Economics, Università della Svizzera italiana (USI), Lugano, Switzerland
e-mail: manuela.barreca@usi.ch

M. Meneguzzo
Department of Management and Law, University of Rome "Tor Vergata", Rome, Italy
e-mail: meneguzzo@economia.uniroma2.it

P. Demartini et al. (eds.), *Cultural Initiatives for Sustainable Development*,
Contributions to Management Science,
https://doi.org/10.1007/978-3-030-65687-4_9

1 Introduction

Nowadays, cultural and creative sectors (hereafter CCS) have internationally proven themselves to be both resilient and progressive in driving local economies, supporting employment, social inclusiveness, citizen participation and urban attractiveness (KEA 2006; UNCTAD, 2008, 2018; EENC 2013; OMC 2014a, b). Cultural and creative industries (hereafter CCI) are considered the core of the so-called creative economy (Cunningham 2002; O'Connor 2010). CCI can unlock potential, help us overcome the old-fashioned idea that culture does not generate economic value but just a symbolic and aesthetic ones (Foord 2008; EU 2010; Moore 2014) and, as stated by the UNCTAD Creative Economy Outlook (2018), make a valuable contribution to the achievement of sustainable development goals. Considering how national economies were affected by the financial global crisis in 2008, cultural and creative sectors are now strongly characterised by two main elements: **enduring budgetary constraints and poor handling capacity and lack of knowledge**. Concerning the enduring budgetary constraints, due to the stress of the spending review process, many western European countries have faced drastic cuts in public funding for cultural investments (Council of Europe 2011; Bonet and Donato 2011). Furthermore, activities and projects generated by the CCS, despite their considerable potential, remain underestimated, especially in terms of financial access. Regarding poor handling capacity and lack of knowledge, nowadays, organisations, in general, operate in dynamic environments characterised by fast changes and transformations. Therefore, there is a demand for a high skill set, know-how and creativity from all the heterogeneous actors involved to fuel the sector's growth in the long run. In the last few years, the effects generated by the economic crisis has challenged many European states and local institutions to rethink how the public sector and the constellation of actors characterising the CCS should be involved in the delivery of both cultural policies and services (Bonet and Donato 2011). The theoretical discussions going on in the public sector in part reflect the debate of cultural academics and professionals that support the idea of promoting a collaborative approach to stimulate the cultural and creative sectors' potential even more (Cogliandro Beyens and Ortega 2014).

Under the umbrella concepts of New Public Management first, and then with the Public Governance, a new model emerged, based on collaborative relationships and processes among public and private actors, non-profit organisations and social enterprises to solve complex societal problems (Cristofoli et al. 2017). Since the 1990s, scholars have suggested that, in a complex and changing time, using participative approaches supports the accomplishment of outcomes more effectively (Agranoff 1992; Mandell 1999; Milward and Provan 2003; Agranoff and McGuire 2003). Precisely, demanding collaborations, especially for the public administrations, means understanding their role in the 'crowded' system of institutional, social, creative and economic agents and strategically managing administrative processes cooperatively (Agranoff 2006; Klijn 2008; Isett et al. 2011). For academics (Agranoff and McGuire 2001; Milward and Provan 2003; Milward and Provan

2000a, b, 2006; Meneguzzo and Cepiku 2008, 2010), practitioners and even policy-makers, public networks are considered the multi-organisational arrangement 'par excellence' to achieve solutions that are difficult to obtain individually. Network advantages are noteworthy, especially nowadays in the complex and shifting time that is characterising contemporary societies. As cultural activities belong to the wider set of public services, public networks help to reduce fragmentation and improve coordination among stakeholders to raise the efficiency and effectiveness of delivering public services (Cristofoli and Macciò 2012; Cristofoli et al. 2014a, b). In the cultural sector, public administrations have to strengthen the connections among the actors involved in the governance through shared values, competencies and resources but also have to improve inter-institutional cohesion to overcome the traditional interpretation of private and non-profit involvement as mere sponsors (Badia et al. 2015).

This work aims to highlight good practices in supporting collaborative settings through the development of cultural and creative networks in the music industry in the Italian Region of Puglia. The case study is based on a 2-year project carried out by the Università della Svizzera italiana and aimed to enhance the cultural and economic values of the music industry in Puglia. Precisely, its objective was to evaluate the main actions of the project Puglia Sounds, a regional program funded by the European Union (ERDF—European Regional Development Fund 2007–2013)—which was designed as a facilitator for the regional creative network driven by the regional music industry. The study also aimed to understand how the network is formed, governed and managed, identifying possible mechanisms and strategies applied by the public sector to guide and build stakeholders' involvement and commitment to achieving both the network's goals and a sustainable long-term vision effectively.

This chapter consists of four sections, in addition to the introduction. Section 2 presents the theoretical framework of network governance and management for the enhancement of collaboration settings in the cultural and creative sectors. Section 3 outlines the applied methodology and its phases. Section 4 introduces the case study analysis, and the last part defines an overall conclusion as well as some implications for academics and cultural managers.

2 Theoretical Framework

2.1 *From Government to Governance Perspective*

Since the mid-1980s, the public sector has been exposed to global changes and increasing complexity due to the growth of our information and knowledge-intensive society (Osborne and Gaebler 1992). The public sector has managed to confront these so-called 'wicked'[1] and twist problems, addressing a more effective delivery of public services, spreading information, introducing learning and innovative practices, and lastly improving the more efficient use of limited resources. It has uncovered the inefficiency of traditional bureaucratic, hierarchical structures and shown the lack of skills and capabilities of many administrations to cope with the fast changes and new demands that characterise contemporary societies, thus generating a need to re-think new models. Accordingly, national governments and public administrations have started to experiment with innovative forms of horizontal governance (Klijn 2008), such us public–private partnerships (Osborne 2000), stakeholder involvement (Edelenbos and Klijn 2006) and other forms of citizen involvement (Lowndes et al. 2001). Consequently, with the spread of the public governance paradigm (Kooiman 1993; Rhodes 1997; Meneguzzo 1997; Osborne 2010), both scholars and practitioners have extensively recognised networks as an important form of multi-organisational governance (Provan and Kenis 2008). Kettl, in 2005, called out the 'global public management revolution', observing how government and governance have deeply moved towards a 'networked government' (Kamarck 2002). Isett et al. (2011) highlighted how networks in the public sector are used *as mechanisms to encourage collaborations* and defined a network as *a group of goal-oriented interdependent but autonomous actors that come together to achieve a goal that no one actor could produce efficiently on its own*. Recent recognition of the current studies on public networks was given during the tenth EGPA-TAD Conference 'From Public Administration to XXI Century Collaborative Administration. The role of public networks' in 2014 and focused on how to make public networks succeed over time. Some authors have emphasised the importance of network governance in structures and contexts; others have shed light on network management and coordination tools and mechanisms. Recently, some authors have enlightened the importance of "soft factors" such as trust (Klijn et al. 2010), culture (Vangen 2017) and language (Mandell et al. 2017).

2.2 Network Governance and Management

This study refers to conceptual models and interpretative schemes that characterise both network governance and management. Network governance[2] emphasises the procedures for the coordination and governing of networks at both strategic and operational levels of decision-making. This kind of network is considered the essence of public administration studies, which both aim for policy formation and

---

[1]Clarke and Stewart (1997) stated: "*wicked problems cannot be dealt with as management has traditionally dealt with public policy problems. They challenge existing patterns of organisation and management*"

[2]Keith G. Provan, Patrick Kenis, Modes of Network Governance: Structure, Management, and Effectiveness, Journal of Public Administration Research and Theory, Volume 18, Issue 2, April 2008, Pages 229–252.

Table 1 Example of network governance parameters used in cultural and creative networks

Structure	What kind of structure has been chosen? Is it democratic (enough)? What kinds of management tools are applied?
Resources	Which human, financial and infrastructural resources are available? How are they generated and managed?
Communication	How is communication organised—Internally and externally? What kind of information is communicated? To whom and how?
Knowledge	How is the flow of knowledge organised—Inside out and vice versa? How is knowledge generated and made available? What kind of management systems support the flow, transfer and safeguarding of knowledge?
Social capital	Who are the members? Are they heterogeneous enough? How can people access the network? How often do face-to-face meetings happen? How are relationships strengthened? How are conflicts handled? How is leadership and participation organised? Are members committed (enough)? How do you maintain their commitment?
Skills and capabilities	What kinds of skills are available? Are they used? Is training available to improve skills? What kind of specific capabilities does the network have? How is a learning environment assured?
Diversity	What kinds of activities are realised? Are they relevant? How are they implemented? By whom? Are results communicated and evaluated?
Innovation	Is the network diverse with regard to all parameters?
Legitimacy	Is the network (still) legitimate? How is social and political relevance monitored?

Source: Anna Steinkamp (2013) and IFACCA (2016)

implementation through coordinating organisations and actors to achieve a common goal (Macciò 2013). Network management (Kickert et al. 1997) is a process used to investigate how networks are formed, governed and managed. It identifies the mechanisms and strategies applied by the network manager to guide and build stakeholders' commitment to achieving the network's goals effectively. The network governance and management perspective enrich and integrate the consolidated approach of cultural management, which has always been directed towards a micro-economic perspective of cultural organisations and institutions (Baumol and Bowen 1966; Blaug 1976; Adler 1985). This vision focuses on internal governance, based on relationships between administrators and professionals (art manager), and the management of practices and techniques, as drivers for creative network creation. In the cultural sector, as in the public sector (Bouckaert and Halligan 2008), the meso and macro dimensions are linked to the understanding of the wider system of partnerships, formulation and application of policies and the assessment of performance (Barreca et al. 2017). The governance in cultural and creative networks is extremely complex, given the interdependence of different parameters to achieve sustainability in networks' activities (see Table 1). As stated by Steinkamp in the D'Art Report 49-International Culture Networks (IFACCA 2016), *Networks are like a phoenix—they appear suddenly and disappear just as quickly, sometimes without making a significant impact. Their success and failure are both a result of their characteristics, and their effectiveness depends mostly on their governance.*

2.3 Cultural and Creative Networks

Network experiences occurring in the cultural and creative sectors are considered *the art of collaboration* (DeVlieg 2001). Cultural networks are recognised as *the organisational form of collaboration in a complex and global world* (Steinkamp 2013). According to IFACCA (2016), the literature on cultural networks has tried *mapping them, analysing their role, work, structure and models of function*. The popularity of cultural networks reflects their ability to facilitate flexible cooperation in solving complex problems, bring together various actors to pursue a common interest and provide a positive environment to create a common and stable platform to work on shared projects (Uzelac 2006). The cultural and creative sectors can be seen as networks themselves, with various actors that support the creative atmosphere. According to Bagdadli (2001, 2003), three elements are considered during the establishment of cultural networks. First, *the complexity of relations,* which could generate two paths. **Formal networks** can converge into a single central organisation, or rather an inter-institutional network (formulation of agreements between public institutions), and **informal networks**, the dominant model today, support collaborations between organisations who are free to join or leave the network at any time. The second element is *network activation.* Networks can be **public** (managed by public institutions, special agencies and public enterprises). It is also possible to activate networks based on a **public–private partnership**, or **collaborative networks** by encouraging the engagement of for-profit, non-profit organisations, social enterprises and citizens. In this case, the public actor could play a different role (Agranoff 2005). The third element is *synergies developed.* **Horizontal networks** are made up of the same public organisations and institutions in terms of cultural services offered (e.g., museums, inter-institutional museums and theatre networks). **Vertical or integrated networks** (e.g., library network, museum networks and theatre networks) produce synergies for local territories (e.g., synergies and economies of scale, cross-fertilisation, sharing of expertise and innovative practices). In Italy, several attempts have been made to categorise existing experiences into cultural networks (Trimarchi 1992; Salvemini and Soda 2011), in which single units of cultural and creative industries are investigated, including theatres and museums (Bagdadli 2001, 2003), television production (Soda et al. 2004), the music recording industry (Cillo and Soda 2001; Montanari and Mizzau 2007), libraries (Fiorani and Meneguzzo 2012) and archaeological sites (Meneguzzo 2008), until the more recent discussion on the integrated cross-sectorial perspective (Meneguzzo and Trimarchi 2009; Borin 2015; Patuelli and Donato 2018). Best practices promoted by both the public sector and national representative associations, such as Federculture, have supported bench-learning in the field. Federculture has promoted, since 2020, some significant knowledge-based innovative practices in the Italian cultural and creative sectors thanks to the establishment of *Premio Cultura di Gestione*—Management Culture Award.

Over the last few years, research and projects on the cultural and creative networks have been investigated from multiple perspectives (DiMaggio 2011) in many international research networks in arts and cultural management (e.g., ENCATC, AIMAC and ASEF). The main purpose is to, on the one hand, figure out the conditions that make networks successful and sustainable in cultural and creative sectors, and, on the other hand, to comprehensively understand the future trajectories that existing cultural networks will have to face in the following years.

3 Methodology

A qualitative approach (Denzin and Lincoln 2011) has been used to emphasise the intention to carry out an exploratory and descriptive case study on Puglia's music industry and its main stakeholders to achieve a comprehensive understanding of a *contemporary phenomenon within its real-life context* (Yin 2014). As a result, this work highlights good practices in supporting a collaborative approach and building cultural and creative networks in Italy. Specifically, it focuses on how music industry professionals and regional artists, supported by their regional institutional setting, can form, manage and enhance cultural participation by building a form of cultural and creative networks in the field of music. We have analysed the results from a 2-year project that monitored activities launched by the regional programme Puglia Sounds in 2010–2011. Data were collected from both primary (semi-structured interviews and focus groups) and secondary (e.g., official documents and institutional reports, press, and media) sources. This chapter presents findings from an articulated interview research process conducted with music industry professionals, associations, music foundations, as well as regional artists and musicians. Table 2 summarises the methodological protocol used for this study. The first part mapped out and analysed the state of the art of the music industry, in which a two-step procedure was required. First, an online survey was structured based on the CAWI (computer-assisted WEB interviewing) method. CAWI is based on an internet surveying technique in which the interviewees follow a structured script provided on a website, and a software application then speeds up the collection and editing of data. The census had several sections and aimed to define the characteristics of the universe of the businesses involved in the Puglia music industry. The firms were selected based on their enrolment in the regional Chamber of Commerce by AtEco (code system used to classify the economic activities by Italian National Institute of Statistics—ISTAT), the regional register of suppliers working in live performance and concert production, and some trade associations in the sector were also considered (e.g., FIMI—Federazione Industria Musicale Italiana, the organisation that keeps track of the music recording industry in Italy). There were 1372 firms in the first sample. All the firms received an official letter with the guidelines for filling in the online survey. The survey ran for 2 months, from November 2011 to January

Table 2 Methodology applied

Stage of the research	Method applied
1—Context	**1. Literature reviews** Public governance, network theory (network governance and management) and cultural management studies
	1.2 Document analysis Official documents and institutional reports, policy agenda, strategic plan, press and media statements
2—Investigation	**2.1 Online survey—CAWI** 1372 firms (first sample) 579 firm out (first screening) 793 firms (final sample) – 344 did not answer the survey – 157 refused to fill in the survey 292 responses—37% response rate
	2.2 Semi-structured interviews 16 interviews (music industry professionals, artists, music associations and foundations)
	2.3 Focus group Four festival networks
3—Data analysis	**Narrative approach**

Source: Authors' elaboration

2012, to provide a state of the art of the regional music industry according to four main dimensions: localisation, macro-area of activities, and level of specialisation, size and revenue. Some further recall by telephone was required. After this screening, we found that 579 firms are not working anymore or have changed their sector. Therefore, the final poll was composed of 793 firms; however, 344 did not answer the survey (despite many calls), and 157 refused to fill in the survey. In the end, 292 responses were collected with a response rate of 37%. Subsequently, 16 semi-structured interviews with music industry professionals (business owners), artists, music associations and foundations were conducted between November and December 2011 to provide enlightenment on particular issues or distinctive traits of the industry, as well their perception of future scenarios within the public institution in supporting the music sector. At this point, semi-structured interviews were selected as one of the sources of data collection. First, they are well suited for the exploration of the perceptions and opinions of respondents. Secondly, the broad professional profiles in terms of the core business, skills and expertise excluded the use of a standardised interview format (Barriball and While 1994). A total of nine categories were identified (record label, recording studio, music publisher, music distributors, musical education, booking agents, music service, production and musical instrument rentals, entertainment and festivals, live music and artists). The interview was divided into three sections and aimed to understand: the state of the art of their business (from 2006 to 2011), challenges and limits to new strategies and collaborations expected shortly after that (from 2012 to 2016) and, finally, to test their knowledge about the Puglia Sounds program and their perception and idea of the role of the public sector in supporting the music industry. Based on the McCracken model (1988), each interview took about 40–60 min. The second part of the study focused on evaluating the effectiveness of the first public calls launched during the first few years of activity. For this part, the participants' perspective was central to the analysis. Focus groups can provide information about a range of ideas and feelings that participants have about certain issues; in addition, it reveals the differences of perspective between groups of individuals. This part of the study focuses on one of the three focus groups performed with the winners of the public calls. Specifically, the one designed to build a 'network of festivals' was selected. There were four different experiences of music network management in the focus group out of the 10 network winners. The main point of analysis was to show the potential of Puglia Sounds as a mediator in supporting collaboration among professionals, artists and public institutions. In both phases of data collection, a team of researchers carried out both the interviews and the focus groups. The interviews and focus groups were recorded and fully transcribed in the following 24 h.

4 Case Study Analysis

The narrative of the analysis will be structure as follows: First, the regional institutional settings that made Puglia an interesting case for the development of cultural and creative networks will be briefly described. Then, based on the results of the qualitative analysis, the numbers, features and issues of the regional music industry will be underlined. Finally, the functioning and management of the Puglia Sound network will be defined, and one of its programs will be presented Richeri et al. (2012).

4.1 The Regional Context

Puglia is a region in the southeast of Italy, established in 1970 as part of a new system of decentralised regional government (Putnam et al. 1981). Since 2001, the regional administration has acquired extensive regulatory power, especially in the area of economic development and innovation policies *to sustain the conditions for the growth, competitiveness, innovation and international development of regional businesses, while enhancing and promoting the distinctive strengths and factors of attraction of the local economy* (EDRF ROP 2007–2013). Financial support from the EU and the regional government's wise vision to enhance an innovative and creative environment based on participatory approaches, today, have made Puglia a best practice in the field of cultural and creative management. All this was possible thanks to the great efforts made to gain resources and set the policy agenda to develop the proper background to support an ideal integrated cultural creative network (Montalbano et al. 2014). At the beginning of the programming period for the Regional Operational Programme EDRF 2007–2013, previous analysis of the cultural sector, as well the music field, pointed out both major and consistent potentials and failures between the two sectors. The strong presence of different kinds of events (e.g., cultural, art, live performance, festival, etc.) in the region and a rich, articulated cultural and natural heritage was not enough to properly move towards renovating the region alongside the creative industries. Moreover, the lack of organisation and collaboration among the local actors, the relevant low level of specialisation and, lastly, limited resources led the music industry to realisation that it was necessary to bring the main stakeholders together into an organised form, with the purpose of optimising resources and competencies. There are many reasons why the Puglia region, its music industry and the Puglia Sound programme make this such an interesting case. First of all, we should point out the long-term outlook that the regional administration and the Regional Minister for Mediterranean Culture and Tourism have had since 2000. They have highlighted concepts such as creativity, innovation, participation and collaboration as keystones for their future sustainable development agenda. Next, they adopted a comprehensive policy framework that supports cultural and creative industries and implemented several regional programs and initiatives, including financial support from the European funds. As stated by several authors (Montalbano and Valentino 2015; Calaizzo et al. 2018), Puglia is also considered a best practice in shifting its cultural policies from the preservation to the enhancement of its wide cultural heritage, which has been integrated into the creative industries, operating as an 'entrepreneurial regional authority' (Mazzucato 2013, 2015). As part of the regional planning strategy, the music sector was identified as the cornerstone for the local economic renovation based on the concept of 'Puglia Creativa',[3] which aimed to develop an innovative hub for cultural and creative industries. The institutional setting and the intangible music heritage developed so far has supported the creation of the Puglia Sound program. The program is an interesting alternative (for the first time in Italy, especially in the southern part) for the use of the ERDF convergence fund, to stimulate the cultural sector, in particular the music industry, and build and implement a regional music system supported by participatory forms of collaboration.

4.2 The Music Industry

According to the results of the online survey (292 responses collected with a response rate of 37%), in 2012, the music industry in Puglia was composed of

[3]"Puglia Creativa" is considered the creative district of Puglia. It was institutionally recognised as an association in 2012 according to Regional law 23\2007. Today, it represents more than 100 regional cultural and creative industries (Performing arts: theatre, dance, music, festival—Cultural industries: cinema, media, audiovisual, gaming, software, publishing—Creative industries: design, Made in Italy, architecture, communication agencies—Heritage: visual arts, cultural hub, cultural heritage, entertainment places—Creative driven: social innovation, services, training).

793 businesses, which were geographically located in the two main provinces of the region, the capital city Bari 52% and Lecce 39%, and employed about 3000 workers. Based on a previous analysis carried out by TPP in 2004, the industry has shown an increase of 24% in the number of businesses operating in the music sector. Regarding the macro-area (core business), over 50% were involved in the record label and recording, 25% in the production and organisation of music events (festivals, concerts, etc.) and 15% made up the service supporting the live performance. The remaining percentage stated that they did other activities (clubs, music publishers, distribution, musical education, music hire shops, media, and press agency, etc.). Most of the businesses remained small, with a staff of no more than 3–4 workers and revenue less than 50,000 euros. The tendency of the actors to engage with the international markets was limited. The analysis confirmed the need for specialisation: only 25% were highly specialised in producing and recording activities. The percentage drastically decreased in the service, where only 5% was considered specialised. During the interviews, most of the interviewees shared a common point of view about the regional music industry. Regarding the narrative, it is evident that the music industry in Puglia is quite stuck on the dichotomy between potentials and unfavourable situations that prevent suitable development. All the interviewees shared the same awareness about the intangible value that music has for the local communities: *Our music is strongly linked with our land; it is something that allows us to be unique on stage…to offer a different kind of entertainment that other musicians cannot offer to their public*. The unexpressed potential of traditional Apulian music and the role it could play in relaunching the regional economy also emerged. An interviewee cited some best practices in which traditional music has become a characterising element of the country itself: *Our strategy is to turn Puglia into a Region characterised by the music, like in Portugal, Spain, Brazil or the Balkan countries….* Among the interviewees, the need to create a 'creativity district' has been hypothesised. Accordingly, there are high expectations in the pilot role that the Puglia Sounds network will play in setting various dynamics for creating collaborations: *I hope that the Puglia Sounds program will build a real network among the actors, to help support us in the projects and initiatives that could promote and valorise both our traditional music and our sector*. During the interviews, some participants shed light on some issues that represent the Apulian music industry and that could affect the collaborative environment. The identified issues can be clustered into four topics: (1) **Low level of specialisation**: *Unfortunately, our business is full of unprofessional people, not specialised and, most of the time, improvised*. (2) **Labour market**: the main problems here referred to the lack of transparency and regulations comparable with the general European standards: *We are talking about a sector in which everything is vague, and without any appropriate regulations* concerning the first issue, here the black market is considered *a real plague for our sector*. In this case, one of the main values of the Puglia Sounds network is expressed as *a platform that will be able to change and innovate a sector, based on the culture of the rules, (trust, transparency and honesty), and highly specialised professionals*. (3) **Lack of entrepreneurial approach and strategic planning**: *There is a lack of managerial skills. Few businesses can work according*

to a structured plan and *we need a new entrepreneurial approach to respond to the challenges that are affecting our industry globally, and this can only be possible through continued education based on a multidisciplinary perspective*. The last issue is crucial for the interviews. (4) **Access to financing**: *We did not even try because we do not have a balance sheet with income to justify our intention to ask for a loan.* One interviewee said: *Timing is not adequate; in our sector, all the decisions must be made quickly, but when you talk about music production with a bank, the project is considered unclear, vague and not economically convenient [...] the communication process is complex.*

4.3 The Puglia Sounds Network

Puglia Sounds (hereafter PS) is a government-sponsored program that was developed in 2010 with the main *goals to boost Puglia's musicians and music industry and help establish the region as a major player in the Italian—and general European—music scene* (www.pugliasounds.it). It was established to address the region's strategy in the Region's strategic development and plan for culture. It was promoted by the Department of Tourism, Economy of Culture and Valorisation of the Region of Puglia and settled as a public office under the umbrella activities of *Teatro Pubblico Pugliese* (hereafter, TPP). TPP is a public agency supported by the regional government, responsible for spreading culture throughout Puglia (e.g., theatre, performing art, dance, music). The European Union financed most of the project with the ERDF funds,[4] and the project was also supported by national and regional contributions[5] from 2010 to 2016. It operates in collaboration with the Apulia Film Commission (cinema and audio-visual production) and *Puglia Promozione* (regional tourism agency) to optimise resources and have more successful results. The main goal of PS is to support the first regional music network in Italy. *It is the first of its kind in Italy: a government-sponsored program that focuses on music projects and recognises the important role that music continues to play in the economic growth and cultural life of its home region and how important it is to connect Puglia with major national and international music markets* (www.pugliasounds.it). Despite the main issues already presented, which characterise the regional music industry, Puglia is one of the most important and recognised music capitals in all of Southern Italy. Music plays a fundamental role in the native communities, and it is perceived as a vital part of the intangible heritage of the

[4]The program is funded by the EDFR ROP contributions based on Axis IV—activity 4.3 in "Development of cultural activities and entertainment", sub-lines 4.3.2 concerning cultural activities and entertainment, and 4.1.2 for marketing destination on national and international markets.

[5]According to the ex-post evaluation of Cohesion Policy programs 2007–2013, the PS project was funded with a total of 6,800,000 million euros (ERDF 4,243,200 million euros), national contribution 1,367,480 million euros and regional one for 1,189,320 million euros).

region. The folk Apulian music, 'Pizzica',[6] is recognised worldwide in the field of world music. The value of PS is to place music at the core of planned and organised initiatives and to enhance the role of the public sector to support the music industry as a noteworthy part of the creation of a regional music network. The Puglia Sounds network supports the music system through systemic actions, building strategic partnerships and promoting high-profile public events. The network achieves its vision based on public grants, managed by public calls, in order to finance various and promising projects, in which the main beneficiaries are musicians, music professionals, entrepreneurs, as well as companies and institutions involved in music production, distribution and promotion.

The strategic plan run by Puglia Sounds is developed around three main programs: **Puglia Sounds EXPORT** is dedicated to promoting the international mobility of musicians abroad. Funding tours abroad has increased the image and presence of popular music and artists around the world. **Puglia Sounds RECORD** is based on the idea of supporting and promoting the existing regional music productions and attracting new recording productions (e.g., release new albums, music compilations, or web compilation for free download), which allow the local artists to produce their music and increase their reputation in the national and international music markets. **Puglia Sounds LIVE** is the program that best describes the tendency of public cultural policies to support network governance and the management approach to foster cross-sectoral collaborations to organise live events in Puglia and Italy.

For the first year, **Puglia Sounds LIVE** financed 47 festivals and supported the creation of 10 festival networks. There were four experiences of music network management in the focus group out of the 10 winners from the first public call—'Development of festival networks-Summer 2011'. As shown in Table 3, the participants had different public cultural networks experiences according to the level of the public sector's involvement and synergies developed with other sectors (e.g., performing art, museums and theatres). Some of the main points discussed during the focus group highlight the advantages of collaborating and the synergies created in terms of economies of scale: *We also used the public funding, with all the partners, to create a common fund to invest in new equipment* (e.g., *stage, equipment, new musical instrument, different public cultural network experiences) and communication (marketing and advertising)*. For coordinate processes: *We shared a common vision and strategy, and the network helped us to reach new artists and manage the schedule and activities better*. To improve the impact of their communication strategy on the festival spectators: *We used territorial marketing in our communication strategy to get more attention outside of Puglia* and *working together facilitated the purchase of more advertising spaces in specialised*

[6]*La notte della Taranta* is one of the biggest music festivals in Italy and one of Europe's most important events dedicated to traditional culture. Established in 1998, in 15 years, the festival grew, including its audience and international prestige, attracting nearly 200,000 participants (Source: https://www.lanottedellataranta.it/en/the-festival/festival).

Table 3 General information–focus group–network composition

Networks	Role in the network: Leader	Role in the network: Partners	Cities involved	Projects
A	Cultural association	Cooperative Cultural association, booking agency and entertainment and festival	4	Jazz festival network
B	Municipality	Cultural associations	4	Early music festival
C	Consortium	Cultural associations	5	Integrated music festival
D	Record label	Cultural associations Health protection agency association of social promotion	5	Integrated music festival

Source: Authors' elaboration based on Monitoraggio Puglia Sounds, 2012

magazines and local and national newspapers; moreover, *thanks to the strong investment in communication, our festival reached a new audience.*

5 Conclusions

The latest economic crisis has globally exposed the structural weakness of the worldwide economy. Many national governments in Europe have been facing massive cuts in public funding in most of the public sectors, including the cultural one. Some scholars have suggested that turbulent times should be seen as an opportunity to review consolidated models of cultural management, which are characterised by self-referential attitudes, financial sustainability, lack of know-how and a micro-economics perspective towards meso and macro ones to address the potential synergies of an articulated system of relationships among public and private actors, non-profit organisations and social and cultural enterprises. The narrative presented in this chapter aims to contribute to the study of cultural management by emphasizing how network governance and management perspective can produce interesting insights into the investigation of collaborative processes through the development of networks in creative contexts as an effective way to deliver public cultural services. Public networks are considered an effective solution for dealing with the many and complex problems that characterise our society and tools to foster strategies for local development. The rise of the creative network 'Puglia Sounds' in the Puglia Region is considered a best practice for several reasons. Puglia is the first example of regional government that frames a collaborative setting based on the implementation of a developmental model that focuses on investing in the cultural and creative sectors. At the European level, it has pointed out a different way to generate regional growth, especially for countries that are financially supported by the ERDF (European Regional Development Fund). The research shows the importance of a common cultural background, a significant pre-existing distinctive cultural heritage; in this case, traditional music is necessary to enhance a strong commitment. Puglia Sounds could be considered the first example of a public creative network in Italy, in which the central policy authority (the Puglia region) plays a key role in the governance and management of the network over the years. Since the beginning, through the use of public calls, the music network has effectively been designed to finance collaborative projects (e.g., creation, promotion, and distribution of new music recordings) and events (e.g., live shows and concerts in Puglia and abroad, International Festival & Music Conference MEDIMEX), rather than an individual ones, and to support the exchange of ideas and innovative practices for the development of new forms of integrated networks. From its establishment to 2020, Puglia Sounds has improved the presence of Apulian music and musicians around the world by helping with the production of more than 1532 concerts in 89 countries (Puglia Sounds EXPORT); supporting the creation, promotion and distribution of new music recordings with 264 new albums on the market (Puglia Sounds RECORD); increasing the number of live music performances in Puglia and Italy with a special focus on festivals and one-off concerts, involving more than 70 Apulian municipalities; and funding over 1600 live shows, including 65 new music shows of local, Italian and international artists performing in Puglia. In addition, 106 music festivals and more than 300 concerts were performed throughout Italy (Puglia Sounds LIVE). Moreover, in 2017, two new programs were launched. Puglia Sounds YOUNG is dedicated to younger musicians and aims to help organise events focusing on younger audiences and educational programs designed to increase the purchase of music by younger users. Puglia Sounds GREEN supports musical growth by way of a green delivery system that falls in line with the European Community's 2020 Principles of Sustainable Growth.

The study carried out in 2011 is not free of limitations. The lack of systematic monitoring and a clear evaluation system to assess mid-term impacts have been perceived as a major deficiency during the design of the project. On the other hand, one of the challenges that were identified early on is the strong dependency on public funding; thus, the future of the program's financial sustainability remains a key point for the music network members. In fact, it seems that the Puglia Region will overcome these issues in the extended and integrated PS program as one of the main pillars of the PiiiL[7] Cultura in Puglia 2017–2026—Strategic Culture Plan for the Puglia Region.

From a managerial perspective, this study suggests some possible toolkit of guidelines for cultural managers. In light of all the initial considerations and analysis produced, collaborations are not an option for the cultural and creative sector, and it is no longer just considered a symbolic sector; thus, integration with other sectors is not an exception to the rule. Therefore, to guarantee a long-lasting network, a clear vision, procedures and rules are essential to coordinate and manage the network's complexity over time. From an academic perspective, the topic of cultural and

[7] https://www.piiilculturapuglia.it/piano-strategico-della-cultura-della-regione-puglia/

creative networks has strong potential for future research, especially concerning the public network evolution and possible comparisons of the specificities of public networks in the cultural and creative sector among clusters of countries, involving Northern European countries, Continental European countries and Southern European countries (e.g., Italy, Spain, Portugal and Greece).

References

Adler, M. (1985). Stardom and talent. *American Economic Review, 75*, 208–2012.

Agranoff, R. (1992). *Intergovernmental policy-making: Transitions and new paradigms.* School of Public and Environmental Affairs, Indiana University, Bloomington and Fundación Ortega, y Gasset, Madrid. Unpublished manuscript.

Agranoff, R. (2005). Managing collaborative performance: Changing the boundaries of the state. *Public Performance and Management Review, 29*(1), 18–45.

Agranoff, R. (2006). Inside collaborative networks: Ten lessons for public managers. *Public Administration Review, 66*(s1), 56–56.

Agranoff, R., & McGuire, M. (2001). Big questions in public network management research. *Journal of Public Administration Research and Theory, 11*(3), 295–326.

Agranoff, R., & McGuire, M. (2003). *Collaborative public management: New strategies for local governments.* Washington, DC: Georgetown University Press.

Badia, F., Borin, E., & Donato, F. (2015). *How the financial crisis affected models of public-private partnership in the cultural sector: Empirical evidence from France, Germany and Italy.* https://doi.org/10.13140/RG.2.1.3041.9929

Bagdadli, S. (2001). *Reti di musei in Italia e all'estero. L'organizzazione a rete per i beni culturali.* Milano: Egea.

Bagdadli, S. (2003). Museum and theatre networks in Italy: Determinants and typology. *International Journal of Art Management, 6*(1), 19–29.

Barreca, M., Meneguzzo, M., Fiorani, G., & Colasanti, N. (2017). *L'impact des évènements culturels dans le soutien des réseaux culturels et créatifs. Le cas de Puglia Sounds in Evènements et territoires—Aspects managériaux et étudas de case.* Administration Publique aujourd'hui, 1re édition. Ed. Bruylant, pp. 311–329.

Barriball, K. L., & While, A. (1994). Collecting data using a semi-structured interview: A discussion paper. *Journal of Advanced Nursing, 19*(2), 328–335.

Baumol, W. J., & Bowen, W. G. (1966). *Performing arts—The economic dilemma. A study of problems common to theatre, opera, music and dance.* New York: Twentieth Century Fund.

Blaug, M. (1976). *The economics of the arts.* London: Boulder Westview Press.

Bonet, L., & Donato, F. (2011). The financial crisis and its impact on the current models of governance and management of the cultural sector in Europe. *ENCATC Journal of Cultural Management and Policy, 1*(1), 4–11.

Borin, E. (2015). Fostering the creation of cross-sectorial networks: Key drivers for culture-related collaborations in Italy's Po Delta region. *ENCATC Journal of Cultural Management and Policy, 5*(1), 27.

Bouckaert, G., & Halligan, J. (2008). *Managing performance: International comparisons.* New York: Routledge.

Cillo, P., & Soda, G. (2001). La Social Network Governance nella musica popolare. In S. Salvemini & G. Soda (Eds.), *Artwork & network—Reti organizzative e alleanze per lo sviluppo dell'industria culturale* (pp. 181–201). Milan: EGEA.

Clarke, M., & Stewart, C. (1997). *Handling the wicked issues: A challenge for government.* Discussion Paper, University of Birmingham: Institute of Local Government Studies.

Cogliandro Beyens, G., & Ortega Nuere, C. (2014). Exporting culture in a global world: War economy or Warhol economy? In *Exporting culture* (pp. 109–123). Wiesbaden: Springer VS.
Colaizzo, R., Letta, M., & Montalbano, P. (2018). The impact of cultural policies in Apulia. Factual and counterfactual assessment on touristic attractiveness. *Economia della Cultura, 4*, 491–508.
Cristofoli, D., & Macciò, L. (2012). *Governance, management e performance dei network pubblici*. Management Pubblico in Ticino. Teorie, esperienze e prospettive. Salvioni Edizioni, pp. 43–59.
Cristofoli, D., Macciò, L., Meneguzzo, M., & Markovic, J. (2014a). Managing strategically in collaborative networks. In P. Joyce, J. Bryson, & M. Holzer (Eds.), *Developments in strategic and public management* (pp. 242–253). Houndmills: Palgrave Macmillan.
Cristofoli, D., Markovic, J., & Meneguzzo, M. (2014b). Governance, management and performance in public networks. How to be successful in shared governance networks. *Journal of Management and Governance, 18*, 77–93.
Cristofoli, D., Meneguzzo, M., & Riccucci, N. (2017). Collaborative administration: The management of successful networks. *Public Management Review, 19*(3), 275–283.
Cunningham, S. (2002). From cultural to creative industries: Theory, industry and policy implications. *Media International Australia incorporating Culture and Policy, 102*(1), 54–65.
Denzin, N. K., & Lincoln, Y. S. (2011). *The SAGE handbook of qualitative research*. London: Sage.
DeVlieg, M. A. (2001). *Evaluation criteria for cultural networks in Europe*. Brussels: EFAH.
DiMaggio, P. (2011). Cultural network. In J. Scott & P. J. Carrington (Eds.), *SAGE handbook of social network analysis*. London: Sage.
Edelenbos, J., & Klijn, E. H. (2006). Managing stakeholder involvement in decision-making: A comparative analysis of six interactive processes in the Netherlands. *Journal of Public Administration Research and Theory, 16*(3), 417–446.
EDFR-CF, Work Package 9: Culture and Tourism—Mini Case Study Puglia Sounds.
ERDF ROP 2007–2013 of the Puglia Region (POR FESR 2007–2013 Regione Puglia).
ERDF ROP 2007–2013 of the Puglia Region Annual Implementation Report, 2012, 2013, 2014.
ERDF ROP 2007–2013 of the Puglia Region Axis IV Implementation Programme 2012–2014 (Programma Pluriennale di Attuazione Asse IV 2012–2014).
European Council. (2011). Council conclusion on the contribution of culture to the implementation of the Europe 2020 strategy. *Official Journal of the European Union, 54* (2011/C 175/01).
European Expert Network on Culture. (2013). In J. Staines & C. Mercer (Eds.), *Mapping of cultural and creative industry. Export and internationalization strategies in EU members States*. https://doi.org/10.2139/ssrn.2261595
European Union. (2010). *Green paper. Unlocking the potential of cultural and creative industries*.
Fiorani, G., & Meneguzzo, M. (2012). I livelli di governance nel distretto culturale dei Castelli Romani. *Azienda Pubblica, 25*(2), 187–204.
Foord, J. (2008). Strategies for creative industries: An international review. *Creative Industries Journal, 1*(2), 91–113.
IFACCA. (2016). *D'Art 49: International culture networks*. https://ifacca.org/media/filer_public/53/e2/53e2a765-42b5-40a0-9c06-c7e18ef5c44e/dart_49_international_culture_networks.pdf
Isett, K. R., Mergel, I. A., Leroux, K., Mischen, P. A., & Rethemeyer, R. K. (2011). Networks in public administration research: Understanding where we are and where we need to go. *Journal of Public Administration Research and Theory, 21*(S1), 157–173.
Kamarck, E. C. (2002). The end of government as we know it. In B. D. Donahue & J. S. Nye Jr. (Eds.), *Market-based governance* (pp. 227–263). Washington, DC: Brookings Institution.
KEA European Affairs. (2006). The Economy of Culture in Europe: A Study Prepared for the European Commission (Directorate-General for Education and Culture). EC.
Kickert, W. J. M., Klijn, E. H., & Koppenjan, J. F. M. (1997). *Managing complex networks*. London: Sage.
Klijn, E. (2008). Governance and governance networks in Europe. *Public Management Review, 10* (4), 505–525.

Klijn, E., Edelenbos, J., & Steijn, B. (2010). Trust in governance networks: Its impacts on outcomes. *Administration and Society, 42*(2), 193–221.

Kooiman, J. (Ed.). (1993). *Modern governance. New government-society interactions*. Newbury Park: Sage.

Lowndes, V., Pratchett, L., & Stoker, G. (2001). Trends in public participation: Part1—Local government perspectives. *Public Administration, 79*(1), 205–222.

Macciò, L. (2013). *Do public network really work? An essay on public network performance and its determinants*. PhD Dissertation. Lugano: Università della Svizzera italiana.

Mandell, M. P. (1999). Community collaborations: Working through network structures. *Policy Studies Review, 16*(1), 42–64.

Mandell, M., Keast, R., & Chamberlain, D. (2017). Collaborative networks and the need for a new management language, public management review. *Public Management Review, 19*(3), 326–341.

Mazzucato, M. (2013). *The entrepreneurial state: Debunking the public vs. private myth in risk and innovation*. London: Anthem.

Mazzucato, M. (2015). *Building the entrepreneurial state: A new framework for envisioning and evaluating a mission-oriented public sector*. Levy Economics Institute of Bard College Working Paper (824).

McCracken, G. (1988). *The long interview*. London: Sage. Business & Economics.

Meneguzzo, M. (1997). Ripensare la modernizzazione amministrativa e il New Public Management. L'esperienza Italiana: Innovazione dal basso e sviluppo della governance locale. *Azienda Pubblica, 10*(6), 587–606.

Meneguzzo, M. (2008). Partnership locali per finanziare la cultura: dai bond culturali al social venture capital in Trimarchi M., Barbieri P., L'azione pubblica a sostegno dell'offerta culturale Quaderni Formez n. 62

Meneguzzo, M., & Cepiku, D. (2008). *Network pubblici. Strategia, struttura e governance*. Milan: McGraw-Hill.

Meneguzzo, M., & Cepiku, D. (2010). Coordinamento e gestione strategica delle reti: profili e competenze dei network manager. *Sviluppo & Organizzazione, 238*, 75–84.

Meneguzzo, M., & Trimarchi, M. (2009). Creatività e territorio: l'azione pubblica regionale. In *Economia della Cultura, Rivista trimestrale dell'Associazione per l'Economia della Cultura* (Vol. 1/2009, pp. 11–18). https://doi.org/10.1446/29677.

Milward, H. B., & Provan, K. G. (2000a). Governing the hollow state. *Journal of Public Administration Research and Theory, 10*(2), 359–379.

Milward, H. B., & Provan, K. G. (2000b). How networks are governed. In C. J. Heinrich & L. E. Lynn Jr. (Eds.), *Governance and performance: New perspectives* (pp. 238–262). Washington, DC: Georgetown University Press.

Milward, H. B., & Provan, K. G. (2003). Managing the hollow state: Collaboration and contracting. *Public Management Review, 5*(1), 1–18.

Milward, H. B., & Provan, K. G. (2006). *A managers guide to choosing and using collaborative networks*. Washington, DC: IBM Center for the Business of Government.

Montalbano, P., & Valentino, P. A. (2015). Cultural policies and local development: Theory and practices. *Economia della Cultura, 25*(3–4), 377–386.

Montalbano, P., Palumbo, F., & Valentino, P. (2014). *Culture and creativity as drivers for local sustainable economic development*. Project Wealth, Local Sustainable Economic Development Research Group Working Papers.

Montanari, F., & Mizzau, L. (2007). The influence of embeddedness and social mechanisms on organizational performance in the music industry: The case of mescal music. *International Journal of Arts Management, 10*(1), 32–44.

Moore, I. (2014). Cultural and creative industries concept—A historical perspective. In I. Meidutė-Kavaliauskienė & J. Stankevičienėn (Eds.), *Procedia—social and Behavioural sciences, volume 110, in the 2-nd international scientific conference contemporary issues in business, management and education 2013*.

O'Connor, J. (2010). *The cultural and creative industries: A literature review* (2nd ed.). Creativity, culture and education series. London: Creativity, Culture and Education.
OMC. (2014a). *Good practice report on the cultural and creative sectors' export and internationalization support strategies*. European agenda for culture—Work plan for culture 2011–2014.
OMC. (2014b). *Policy handbook. How to strategically use the EU support programmes, including structural funds to foster the potential of culture for local, regional and national development and the spill-over effects on the wider economy*. European agenda for culture—Work plan for culture 2011–2014.
Osborne, S. B. (Ed.). (2000). *Public-private partnerships; Theory and practice in international perspective*. London: Routledge.
Osborne, S. B. (2010). *The new public governance. Emerging perspectives on the theory and practice of public governance*. London: Routledge.
Osborne, D., & Gaebler, T. (1992). *Reinventing government: How the entrepreneurial Spirit is transforming the public sector*. New York: Addison-Wesley.
Patuelli, A., & Donato, F. (2018). Developing local cultural networks: The case of Dante 2021 in Ravenna. *ENCATC Journal of Cultural Management and Policy, 8*(1), 44–55.
Provan, K. G., & Kenis, P. (2008). Modes of network governance: Structure, management, and effectiveness. *Journal of Public Administration Research and Theory, 18*, 229–252.
Putnam, R. D., Leonardi, R., & Nanetti, R. (1981). Devolution as a political process: The case of Italy. *Publius: The Journal of Federalism, 11*(1), 95–118.
Rhodes, R. A. W. (1997). *Understanding governance*. Buckingham: Open University Press.
Richeri, G., Doglio, D., Meneguzzo, M., Barreca, M., Giordano, F., & Fiorani, G. (2012). *Monitoraggio Puglia sounds*. Lugano: Università della Svizzera Italiana.
Salvemini, S., & Soda, G. (2011). *Artwork & network. Reti organizzative e alleanze per lo sviluppo dell'industria culturale*. Milano: Egea.
Soda, G., Usai, A., & Zaheer, A. (2004). Network memory: The influence of past and current networks on performance. *The Academy of Management Journal, 47*(6), 893–906.
Steinkamp, A. (2013). *Network governance—Governance models of international networks of cultural cooperation*. Master thesis—HUMBOLDT-VIADRINA School of Governance, Berlin.
Trimarchi, M. (1992). *The cultural sector: Empty box or institutional network?* Quaderni per la discussione dell'Istituto di Economia e Finanza, 1992, n. 2, Messina.
UNCTAD. (2008). *Creative economy report 2008. The challenge of assessing the creative economy towards informed policy-making*. https://unctad.org/system/files/officialdocument/ditc20082cer_en.pdf
UNCTAD. (2018). *Creative economy outlook: Trends in international trade in creative industries*. https://unctad.org/en/PublicationsLibrary/ditcted2018d3_en.pdf
Uzelac, A. (2006). Cultural networks in virtual sphere—Between infrastructure and communities. In B. Cvjetičanin (Ed.), *Dynamics of communication: New ways and new actors* (pp. 305–317). Zagreb: Institute for International Relations.
Vangen, S. (2017). Culturally diverse collaborations: A focus on communication and shared understanding. *Public Management Review, 19*(3), 305–325.
Yin, R. K. (2014). *Case study research: Design and methods* (5th ed.). London: Sage.

Manuela Barreca is a lecturer and researcher at the Università della Svizzera italiana (USI, Lugano) in Switzerland, where she received her Ph.D. in economics in 2016. Her teaching experience is in the fields of public and strategic management in the public and nonprofit sectors. Her main research interest is in the area of public art and cultural management, creative industries, network governance, public and private partnership, and civic crowdfunding. Since 2013, she has been the coordinator of the XIX Permanent Study Group on Public Network Policy and Management for the European Group of Public Administration, and member of the IIAS—International Institute of Administrative Science in which she is co-chair for the social innovation, commons, and administration panel. She is part of the CEPS Research Fellows—Center for Philanthropy Studies at the University of Basel.

Marco Meneguzzo is a professor of strategic management at the Università degli Studi di Roma "Tor Vergata" in which he is the coordinator of the Government and Civil Society research group, which includes the MIMAP Master (Innovation and Management of Public Administration) and the MEMIS Master (Social Innovation Management), of which he is the scientific director. In Switzerland, he is a professor of public management and nonprofit at the Università della Svizzera Italiana, where he is the co-director of the Master PMP (Public Management and Policy), a joint degree programme organized by the Swiss Universities of Lugano, Lausanne, and Bern. Internationally, he has coordinated various projects funded by the Italian Ministry for the University and Scientific Research and Swiss National Foundation on public networks and governance (social-health care, leisure, and cultural sector); governance and performance evaluation of public and non-profit networks in the multimedia and cultural sector; governing, managing, and developing networks of public administration and public/private partnership in labour and knowledge-intensive sectors.

Change in Perspectives in Cultural Tourism: A Sustainable Managerial Model for Cultural Thematic Routes Creating Territorial Value

Pamela Palmi, Marco Esposito, and M. Irene Prete

Abstract The negative externalities due to mass tourism and its unprofitable concentration in some artistic city centers or in some periods of the year can be overcome by both a diversification of the cultural heritage offerings and the use of new technologies. Firstly, cultural thematic routes (CTRs) allow for the promotion of new itineraries through the rich cultural heritage of peripheral and landscape areas, supporting them with a system of information services aiming to communicate their value. Secondly, ICT approaches such as business process management (BPM) and product lifecycle management (PLM), already existing in many industrial contexts, could be extended to the cultural tourism sector. To intercept and satisfy a wider and more complex demand for knowledge, culture, and heritage, this chapter is aimed at proposing a sustainable managerial model applied to a specific case study: the "Francigena Route" in the Apulia Region. It uses information and communication technologies to develop new cultural services, improving management effectiveness and efficiency, going beyond the traditional approach of mere conservation of cultural heritage. The result is a model capable of managing organizational variables and highlighting critical issues in the evolution of cultural routes, providing strategic guidelines to both cultural and touristic operators and policymakers.

Keywords Cultural thematic routes · Business process management · Product lifecycle management · Territorial value · Cultural tourism

P. Palmi (✉) · M. I. Prete
Department of Scienze dell'Economia, University of Salento, Lecce, Italy
e-mail: pamela.palmi@unisalento.it; mariairene.prete@unisalento.it

M. Esposito
CORE Lab, University of Salento, Lecce, Italy
e-mail: m.esposito@unisalento.it

P. Demartini et al. (eds.), *Cultural Initiatives for Sustainable Development*,
Contributions to Management Science,
https://doi.org/10.1007/978-3-030-65687-4_10

1 Introduction

Sustainable development is widely recognized as one the most appropriate approaches to achieving environmental, social, and economic benefits for all types of tourism (Lazano-Oyola et al. 2012). Already in 2005, the World Tourism Organization defined sustainable tourism as *tourism that takes full account of its current and future economic, social and environmental impacts, addressing the needs of visitors, the industry, the environment and host communities* (UNEP and UNWTO 2005; World Tourism Organization 2017, 2019). Managing tourism based on sustainable principles ensures long-term life quality for local communities as well as a high-quality visitor experience (Bramwell and Lane 1993; Choi and Murray 2010; Choi and Sirakaya 2006). Therefore, tourist destinations and activities should be planned, prepared, and experienced respecting the local community and their culture, environment, and ecosystem while still being profitable. Several studies have shown that multiple benefits arise from the social, cultural, and environmental aspects of tourism—including a better preservation of historical buildings, and other cultural assets—adding value for better quality of life in general (Choi and Murray 2010; Gursoy et al. 2002; Madrigal 1995). Notwithstanding the dissemination of guidelines for the development of sustainable tourism by international organizations (UNEP and UNWTO 2005) and its effective application by policy makers and operators is still challenging.

Cultural tourism is a relevant component of the tourism industry (Sharareh and Badaruddin 2013), which combines "classic" tourist products—such as accommodation, restaurants, and entertainment—with cultural heritage products—such as museums, performances, art exhibitions, and so on. Cultural and heritage assets are a particularly delicate issue, for many reasons. On one hand, their management requires obtaining the consensus of tourism planners/managers and the local community. On the other, they are often physically vulnerable and must be handled with care, as they can be threatened when tourism development damages the maintenance of local traditional values (Coccossis 2008). The search for a balance between the consumption and safeguarding of cultural products is, therefore, essential. The negative externalities due to the mass tourism and its uneconomic concentration in some artistic city centers or in some periods of the year can be overcome by both an appropriate diversification of the cultural heritage offer and the use of new technologies. Not only is the need to increase cultural tourism becoming urgent, but also the need to spread it over the year and in areas different than the traditional circuits of art.

To this aim, a decisive role can be assumed by the cultural thematic routes (CTRs), which allow for the promotion of new itineraries through the rich cultural heritage of peripheral and landscape areas, supporting it with a system of information services aiming to communicate the value of these historical places. In an era characterized by the support of digital technologies, ICT approaches, such as business process management (BPM) and product lifecycle management (PLM), already existing in many industrial contexts, could be extended to the tourism and cultural sectors. BPM is a set of activities and methods focused on improving corporate performance by managing a company's business process (Harmon 2010); a PLM system is a cooperative backbone which allows people in large companies to act jointly in an efficient way (Saaksvuori and Immonen 2008), thus guiding the integration of IT solutions across the different phases of product and service lifecycles.

The objective of this study is to propose a managerial model for the sustainable development of CTRs through the employment of the ICT approaches, which could be generalized for the management of the these kinds of cultural routes, such as complex cultural products with a touristic and cultural nature for their interactions with the territory and local stakeholders. The research question concerns how we can manage CTRs in the different phases regarding project management and data integration throughout the itinerary lifecycle. To answer this question, we have used a case study methodology, starting from the reconstruction of processes, and its application presentation of a specific case study: the "Francigena Route" in the Apulia Region.

Focusing on the new concepts of cultural heritage (Cerquetti 2007, 2010)—defined as "any material evidence of civilization" (Dragoni 2005; Montella 2003, 2006; Valentino 2003)—we study the strategic value of CTRs, with the aim to propose a managerial model for the adequate redevelopment of the cultural and tourist offer on a territorial scale. ICTs are used to create new cultural services, improving management effectiveness and efficiency and going beyond the traditional approach of mere conservation of cultural heritage. The chapter is structured as follows. Starting from the literature on the subject, we analyze the decisive role assumed by the CTRs for both the production and the organization of the territorial cultural and tourist offer in relation to the changed demand for culture. Then, the proposition of a sustainable model is described, including methodology, results, and discussion. Finally, conclusions are summarized.

2 The Need for New Policies of Cultural Heritage: Cultural Thematic Routes

The trend of tourist flows demonstrates that in recent years (WTO 2019), cultural tourists, aware of the importance of their free time and increasingly selective in their choice of holiday, are rejecting standardized models and mass tourism and are choosing destinations that offer attractive, experiential, and educative situations (Brown 2013; Pigram 1993). They prefer special events (i.e., festivals) and, particularly, thematic itineraries, which balance the resources of cultural landscape with the heritage of living landscape (Berti 2013; CUEIM 2006; Pencarelli 2005; Primicerio 1993; Trono et al. 2015), in a sort of reconstruction and regeneration of body and soul (Krippendorf 2013).

In the next few years, the tourism and travel industry could be disrupted and suffer the short-, medium- and long-term economic effects caused by the massive contagious and infectious disease COVID-19, as a consequence of travel restriction and social distancing. According to an evaluation of its impact in China, in the tourism sector demand will decrease by 75% (Ruiz Estrada et al. 2020); therefore, it is likely that there will be a reduction in the international demand for the mass and overly visited touristic places in favor of the more peripheral and landscape areas.

Turning from the side of the demand to the side of the offer, the policies for the enhancement and management of European cultural heritage continue to leverage the values of monumental excellence of nineteenth-century matrix and aesthetic-idealistic nature, aiming at an entertainment tourism driven by the attractiveness of the masterpiece (Cerquetti 2007). In this way, the places that have a cultural heritage of reduced quantity, limited fame, and minor monumental value discount the effects of competitively weak strategies by acting on the same markets where large museums and cities of art operate with an already consolidated brand. These strategic choices derive mostly from the uncritical assumption of models offered by the great European and American museum structures. The management of CTRs that collect heterogeneous landscape beauties and local history cannot be compared to the refinement, the result of a careful selection that the museums flaunt, preserving worldwide objects, often disconnected from the surrounding environment of the museum that hosts them (Causi 1998; Settis 1998).

This difference, normally treated as a competitive weakness, contains a potential significant advantage. Indeed, European—and especially Italian—privilege lies in the capillary territorial distribution of a cultural and landscape heritage, whose quality 'consists in ubiquity and diffusion' (Cerquetti 2007; Toscano 1998). Since the 'museum Italy' (Chastel 1980) stands out because of the intensity and capillarity of the presence of cultural and landscape heritage in the territory, the presence of minor touristic places—such as a historical route, an archaeological park, or a forest—should be preserved and valued. These areas should be protected not only for their individual value, but above all because they are the systemic components of a single very dense network of phenomena and relationships, and the product of a secular accumulation of civilization, in which the whole far exceeds the union of the parts (Settis 2005).

For effective policies of cultural heritage enhancement, it is crucial to recognize the extraordinary potential of the relationship between the CTRs and the territory so as to understand the multifaceted function of the art products, as well as the values rooted in the cultural continuum between agriculture and eno-gastronomy, urban planning, and landscape. In doing so, it is necessary to avoid falling on the territory of values extraneous to it or forcing it into a simple asset and with a marked and restrictive connotation. In order to develop deep-rooted and long-lasting paths, we need to focus on the systemic enhancement of the historical-cultural heritage and the multiplicity of productive resources of a territory, as components of an integrated managerial model (Casini and Zucconi 2003).

Starting from these concepts, the intervention methodologies for the enhancement of the territories are also increasingly based on a value-driven or value-based approach to the planning of sites, which is oriented to the identification and evaluation of the set of values which form the significance of cultural sites, as a condition that determines the choice of policy-makers (Greffe 2003; Valentino and Misiani 2004).

3 Management of CTRs

3.1 How New Technologies Can Help Culture

The success of CTRs depends on their management and the promotion techniques used, especially those based on new technologies (Trono et al. 2015). To adapt the systemic management plan of the activities envisaged in the cultural projects to the tourist itinerary, the resources must be organized according to a process of feedback (planning and control), taking into account cost constraints, time, and quality (Archibald 1994). The problems that may arise during the implementation of the program must be foreseen, analyzed, and resolved (Amato and Chiappi 1997; Gaio 2010). As with any cultural products, project management of CTRs will have to foresee five different phases of planning (Argano 2012). In particular: (1) ideation and defining the general characteristics of the project in terms of concept, activities, objectives, and contents; (2) activation and study of the feasibility of the project, identification of the organizational components, and the fundamental structural aspects; (3) planning, based on the available information, the execution of the event, the definition of the tasks to be performed, the necessary resources, the economic and financial path to follow and the activity program; (4) implementation of the project including control mechanisms; and (5) project completion, ex post evaluation and process feedback.

In the process complexity (Fig. 1), the predicted stages can be seen as macro-processes, subdivided into sub-activities (sub-processes), which, in turn, contain numerous interdependent variables, whose dynamics and impacts influence the

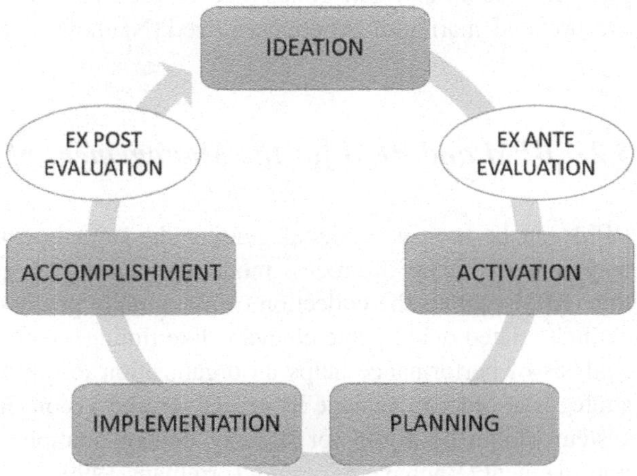

Fig. 1 Phases of the development of a cultural project. Source: Adapted from Argano (2012)

whole process of development (Argano 2004). The evolution of this complex cultural project, which may depend on the demand in the reference market, is an integral part of the project itself, according to a timetable that starts from the study, planning, and implementation of the creative product to the development phase, maturity, decline, and, finally—when the product has become obsolete—abandonment.

According to this perspective, it should be possible to develop a set of variables useful for evaluating the feasibility of the cultural itinerary (potentially to be included in a guided tour) and assessing its sustainability. In order to obtain an effective analysis, it must be assumed that a tourist itinerary activates a complex network of relations among places, people, and activities. Every city, country, place, or cultural site affected by the thematic route represents a node on which to plan, organize, and develop a series of events and activities (e.g., festivals). Taking into account the specific characteristics and functional relations activated for each stopping place on the itinerary, in turn, each node is connected to the other nodes of the cultural network. It is therefore essential to establish how to link the individual nodes in accordance with the distinctive characteristics of each territory, its identity, sounds, and the interests of the parties involved in the cultural project.

ICTs have become a critical factor in the management of company's innovation, both for processes reconfiguration and organizational changes. Performance enhancement is one of the main objectives to achieve to gain market competitiveness and, since the early 1990s, many firms have revised their business models in the light of this goal. The introduction of new ICTs allows for the achievement of a more efficient management and exploitation of cultural heritage, as well as a control of the logic of the protection and conservation of cultural heritage. To innovate the cultural tourism offer, the challenge is to test innovative models already successfully applied to other market sectors subject to changing business dynamics. In this work, the proposed models are BPM and PLM, here applied in order to enhance a resource management procedure, which facilitates the organization and production of cultural content concerning awareness, the search for preliminary information, comparison, planning, the purchase of goods and services, assistance, and, finally, the communication and sharing of experiences lived (Neuhofer et al. 2012).

3.2 BPM and PLM for the Management of CTRs

BPM can be seen as a signal guiding the organization towards a process-based approach, by means of process modeling, automation, management, and optimization. BPM enables the collection of measurable process performance data, making explicit related quantifiable elements like timing, costs, and quality indicators. The analysis of performance helps an organization to build and to implement its own strategy, in order to achieve its objectives and vision. In this view, a measurement system should be a tool for strategy implementation, organizational change, and improvement (Vanhaverbeke and Torremans 1999).

Organizations redesign themselves grouping concurrently all those activities that logically belong together in terms of creating value for the customer, in the view of a process-based organization. At first, companies have to redefine their key processes, and then, they can start to reconfigure the organization around such processes. The following step is to identify new kinds of processes that enable the units to cooperate in an efficient way (Vanhaverbeke and Torremans 1999). In this regard, Majchrzak and Wang (1996) have shown that collective responsibility and collaborative culture are basic attributes to guarantee improved performance when a company redesigns its organizational structure from functional units into process-centered departments; moreover, the lack of such attributes invalidates the success of innovation. The processes represent the way to achieve objectives and implement strategies, the core of business. However, companies' culture is still far from this concept. Processes are a set of coordinated activities that contribute to value creation, and they need adequate resources to be effectively implemented, after being properly planned and scheduled.

Directly linked to the management of processes, the PLM approach is strictly related to the strategic vision of a process-oriented organization. It is an "integrated, information-driven approach comprised of people, processes/practices, and technology, to all aspects of a product's life, from its design through manufacture, deployment and maintenance, culminating in the product's removal from service and final disposal. By trading product information for wasted time, energy and material across the entire organization and into the supply chain, PLM drives the next generation of lean thinking" (Grieves 2006, p. 39). In the industrial sector, this approach enables the tracking of product and parts constituting it throughout manufacturing processes in real time, helping management to resolve the critical issues, make the right decision at the right time, based on updated data, therefore saving time and costs and ensuring better quality in the final product.

In the cultural heritage sector, PLM allows for an appropriate system of communication among different actors through an effective information management (Ding et al. 2007). Recently, it has been applied to the management of historical and artistic knowledge in museums, enabling the collection of the available knowledge and the creation of virtual links between the data, thus enhancing the knowledge database with several contents (such as geographic or historic information, semantics) (Hervy et al. 2013). The analysis of the processes in the cultural lifecycle promotes the standardization of both procedures and activities, which produce the data output, and allows for the identifying of the actors and the criticalities related to them (Hervy et al. 2013).

From this perspective, it is possible to improve the level of effectiveness and efficiency of a cultural tourism project management by adopting the "Design for X" company philosophy (Kuo et al. 2001), already used in the industrial sector to overcome the limits of the design of the serial product. In the details, the design methodology is strictly correlated with the product life cycle, which adopts an integrated approach structured in phases involving actions aimed at improving product quality and optimizing times and costs of the project implementation. Managing a cultural project in terms of a life cycle allows for the possibility to change the feature of the tourism product and, hence, to reconfigure the itinerary based on customers' and stakeholders' needs or feedback and outcomes of previous experiences. To create this opportunity, the involved processes must be identified with the related multiple variables. This will allow operators and policy makers to design and develop guided tours with a more scientific approach based on real data rather than subjective evaluations. Furthermore, following this approach it is possible to modify and reshape each path obtained on the basis of the availability of resources or eventual new services to be provided.

The BPM and PLM engineering approaches can be powerful instruments for the integrated governance of tourism-cultural heritage belonging to a local system. Indeed, through the application of a meta-model based on a logic of processes, it will be possible to trace the replicable success factors of a cultural project, and to exploit the know-how emerging from other events held in different territorial contexts. Recently, these approaches have been implemented to improve the sustainable valorization of an archaeological site, by improving the flow of data, collecting information, and sharing knowledge generated during the archeological process (Mangialardi et al. 2016). The management of digital data is crucial. Digitization, intended as a copy of a physical original, is commonly used in different types of cultural institutions, such as historic archives, libraries, and picture galleries (Guccio et al. 2016). The digitization of information (such as size, date, origin, title, description, context) resulting from earlier documentation or from personal knowledge generates a huge set of metadata which are useful to identify, describe, understand, and value heritage material objects. For example, such amounts of data can be stored, categorized, and managed in order to create engaging initiatives, like interactive narrations through the use of augmented reality to guide tourists through a path toward a point of interest. The digitization of cultural data enables new opportunities to generate and disseminate content. In this sense, the Transmedia Storytelling concept is used to create contents, exploiting available data, which can be delivered through different media platforms. Transmedia designers and cultural operators have common interests, starting from the shared need to attract different types of people and communities by spreading stories across different media in the view of catching different audience needs (Mateos-Rusillo and Gifreu-Castells 2018). At the same time, the CTRs can be a sort of puzzle of Transmedia experiences, which has the task of engaging visitors and stimulating them to explore the different fragments of contents across multiple media, along the path's points of interest, like chapters of a story told with different and innovative technological tools (Corallo et al. 2019).

4 Methodology

4.1 Case Selection Strategy

The need to bridge the literature and managerial gap on how ICTs can foster the sustainable management of a CTR, and the early stages of research on this issue, has led us to conduct an inductive study (Easton 1995; Eisenhardt 1989). The in-depth case analysis (Easton 1995; Eisenhardt 1989; Yin 1994) seemed to be consistent with the exploratory nature of this research. The case study can be considered a well-suited research strategy as it offers "depth and comprehensiveness" (Easton 1995, p. 475) to understand the organizational complexity of the topic. Building on Berends and Deken's (2019) methodology on the composition of qualitative process research, our narrative follows the "conceptualized composition" format.

4.2 Data Collection

Following Yin (1994), the phenomenon was analyzed by relying on several sources of evidence to exploit the synergistic effects of combining them via triangulation (Jick 1979). We have collected data and documents from the European Institute of Cultural Routes (EICR) (2004), a technical agency of the Council of Europe, the European Association of the Vie Francigene (AEVF), the Department of Tourism, Economy of Culture, and Valorization of the Territory of the Apulia Region and the

Table 1 Description of data source

Primary sources			
Semi-structured interviews	Francesco Palumbo, Head of Policies Area for the Prevention of the Territory, Knowledge and Talents	March 20th 2015	1 h
	Giancarlo Piccirillo, Director of Regional Agency Puglia Promotion	June 7th 2015	2 h
	Silvia Godelli, Regional councilor for the Mediterranean Cultural Activities and Tourism	June 8th 2015	1 h
Secondary sources			
Reports on the Francigena Route	Newspapers: La Repubblica; Il Corriere del Mezzogiorno; La Gazzetta del Mezzogiorno	2015–2020	Several pages and minutes
	TVs/Radios, Rai 3, Rai news, Rai 1	2015–2020	
Official communications	Vie Francigene website European Institute of Cultural Routes (EICR) website European Association of the Vie Francigene (AEVF) website "Puglia Promozione" Agency and Apulia Region website	2015–2020	About 300 pages

Regional Agency of Tourism, "Puglia Promozione," an in-house public body of the Apulia Region. We conducted semi-structured interviews with the Head of the Policies Area for the Prevention of the Territory, Knowledge and Talents that started in 2010 the establishment of the Environmental and Cultural Systems (*Sistemi Ambientali Locali*, SAL) of the Apulian Region, with the Regional Councilor for the Mediterranean, Cultural activities and Tourism, and with the Director of Regional Agency "Puglia Promozione" (Table 1). On average, interviews last from 1 to 2 h each; then, they were transcribed to avoid misinterpretations. Following Spradley (1979), we used a semi-structured approach, which consisted of two main kinds of questions, as "grand tour" questions allowing interviewees to deal with the arguments they believe important, and follow-up questions, which guide the interview to analyze the arguments of the present research. We collected data and information on the main purposes, processes, activities, and criticalities concerning the implementation of the CTR "Via Francigena" in the Apulia Region. The following themes were considered: the role of the thematic route in the creation of a regional sustainable tourism; valorization and promotion of the tangible and intangible regional assets; deseasonalizing of the cultural tourism; territorial development and social cohesion; enhancement of the local identity; relationship with the trans boundary regions; and the use of technologies in the management of the CTR.

We also included secondary sources, such as archival records and documentary information. We believe that the combination of the above different sources helped overcome the limitations of an analysis based on data of separate sources (Yin 1994, p. 92).

5 Case Study: The Francigena Route in the Apulia Region

The Francigena Route, formerly called "Via Francesca" or "Romea," is part of a series of routes that lead from northern Europe to Rome, and then continue in the direction of Jerusalem: an itinerary of history, a main road traveled by thousands of pilgrims, merchants, armies, and fairs (Fig. 2). It is a real crossroad of cultures that has contributed substantially to the construction of European culture. For this reason, in 2004 the Francigena Route—along with the Camino de Santiago de Compostela and other routes—became part of the "Council of Europe Cultural Itineraries Program" which promotes a multiplicity of connections between large cities of art and small medieval villages, monasteries and abbeys, cultural heritage, and landscapes.

The itineraries constitute a concrete affirmation of the fundamental values of the Council of Europe, namely: human rights, cultural democracy, diversity and cultural identity, dialogue, exchange, and mutual enrichment, beyond borders and centuries. On December 18th 2008, the first Accreditation Agreement for the Francigena Route was signed in Rome between the Directorate General Libraries, Cultural Institutes and Copyright of Italian Ministry of Cultural Heritage and Tourism (MIBACT) and

Fig. 2 The Francigena cultural thematic route. Source: Own reproduction (The Via Francigena—Gaetano Dini)

AEVF, which opened up new perspectives for the development of the Francigena project as far as regards usability of the itinerary, reception, and communication.

The main objectives are: (1) promoting awareness of the common cultural identity and of a European citizenship, based on shared values, which become tangible around itineraries that trace the history of the influences, exchanges, and evolution of cultures in Europe; (2) promoting dialogue between cultures and religions through a better understanding of European history; (3) protecting and enhancing the cultural and natural heritage understood as a factor in improving the quality of life and a source of social, economic, and cultural development; and (4) assigning a high position to cultural tourism, aimed at achieving sustainable development.

The criteria for the recognition of the CTRs are contained in the European Resolution n. 66/2013 and the annexed Regulations that establish a formal framework of collaboration among the 47 States' signatories of the European Convention. Countries can formulate proposals for the creation of new itineraries which, in order to be included in the program, must meet the following criteria: (1) be centered on a theme representative of European values and common to several countries; (2) be developed around a historical route or, in the case of cultural tourism, creating a new one; (3) develop long-term multilateral cooperation projects in some priority sectors (scientific research, conservation and enhancement of heritage, cultural and educational exchanges between young Europeans, contemporary practice of culture and the arts, cultural tourism, and sustainable development); and (4) be managed by one or more independent organizations and structured in the form of associations or federations of associations.

In December 2010, the Committee of Ministers of the Council of Europe established—with the Resolution 22/2010 (updated with Resolution 67/2013)—an Enlarged Partial Agreement on Cultural Routes to strengthen cultural cooperation, sustainable territorial development, social cohesion, attributed to the certification functions of the "Cultural Itinerary of the Council of Europe" and periodic evaluation of certified itineraries.

The Council of Europe defines the political orientations of the program, adopts the new themes, empowers the networks, and attributes the recognition of "cultural itinerary." It is a network of international cooperation that supports and promotes the cultural routes project. It has the aim of increasing the feeling of belonging to a European culture by enhancing these historically "common" itineraries and fostering the relationship between different itineraries and the relationships in the interested rural and urban areas.

For each itinerary, the Council of Europe recognizes a *reseau porteur*. This qualification allows a single organization to activate the processes necessary to develop routes and paths according to shared parameters at all levels, from the Council of Europe to national governments to the Regions. This complex governance has led to the establishment of a coordination center, namely, the European Institute of Cultural Routes based in Luxembourg.

5.1 European Association of the Vie Francigena Route

On April 7th 2001, 34 local authorities (30 municipalities and 4 provinces) located along the Italian Sigerico route signed the constitutive charter of the current European Association of the Via Francigena Ways (AEVF) AEVF, to which more than 100 local and regional governments adhere. The Association plays an impulse role by connecting all the interlocutors of the itinerary enhancement project at local, regional, national, and European levels. AEVF has consolidated an effective model of governance, which earned it the authorization to *réseau porteur* of the Francigena Route in 2007, confirmed in March 2013. The prestigious recognition makes AEVF a European reference model for the development, protection, and promotion of the Francigene Routes. Hundreds of European municipalities are crossed by the imaginary spinal column that outlines the main way to achieve the peaceful Europe of peoples, where national identities give strength and value to the comparison of cultures and the roots of European identity. In the light of the Valorization Agreement signed with the Italian Ministry of Cultural Heritage and Tourism (on May 31st 2014 by the Director General for Libraries and Cultural Institutes), AEVF is engaged in an intense communication activity with each of the 140 Italian municipalities of the Sigerico route.

5.2 Developments of Routes and Francigena Networks

Along the Francigene Routes, the idea of the network finds different variations, both as regards the physical path in its historical-cultural values and as regards the associative project. From the point of view of the route, the Francigena Route represents: (1) A network of paths, a physical path, a road-area consisting of a network of paths; the main route—defined on the basis of the 79 sub-mansiones of the Sigerico diary—presents historical variations that, merged with the main track, contribute to creating a network of paths; and (2) a network of itineraries that—beyond the multiple readings that are suitable for tourism and culture—was born as a journey of faith which, together with the Route, leads to the three "peregrinationes maiores": Rome, Jerusalem, and Santiago.

From an organizational point of view, the Francigena Route represents:

- A network of institutions found in different administrative territories: In modern geography, the Streets touch five States (England, France, Switzerland, Italy, and the Vatican State), 18 regions, 25 provinces/departments, and over 400 municipalities.
- An authoritative part of the network of routes recognized by the Council of Europe. The Francigena Route is one of thirty-three cultural routes certified by the Council Program of Europe.
- A transnational cooperation network. AEVF has been part of the European Group of Economic Interest "Culture-Routes Europe," born from the initiative of the European Institute of the Cultural Routes of Luxembourg.
- A European interregional coordination network, promoted by the AEVF, with the aim of putting the potential of the routes to the system by launching lobbying actions towards the European institutions and promoting Europe as the world's leading tourist destination.
- A network of networks for the integrated enhancement of territories. Given that the Francigena Route supports the development and growth of the territories crossed, the project aims to develop a common action strategy among the countless actors (profit and non-profit) engaged in integrated tourism development, consortia, and associations.

5.3 The Role of the Apulia Region

In the process of recognition of the Francigene Routes directed to Jerusalem, the Apulia Region has assumed a decisive role, in particular through the impulse of the local agency "Mediterranean, Culture and Tourism." Briefly, the actions taken to enhance and promote the Francigena Route are listed below:

- Deliberation of the Regional Council, April 5th 2011, n. 643: participation of the Apulia Region in intervention programs for the realization of projects of excellence for the development and promotion of the national tourism system ("Monti Dauni Project: Integrated enhancement of cultural, religious, landscape and food and wine excellences along the Francigena Routes")
- Deliberation of the Regional Council, June 15th 2011, n. 1333 and deliberation of the Regional Council August 7th 2012, n. 1675: adhesion to the European Association of the Francigene Routes
- Deliberation of the Regional Council, December 12nd 2011, n. 2807: approval of the Program Agreement for the realization of projects of excellence for the development and promotion of the national tourism system (Monti Dauni project)
- Deliberation of the Regional Council, June 1st 2013, n. 1174: approval of the route of the Apulian itineraries of the "Francigene Routes"
- Approval of the 2007–2013 Greece-Italy Cross-Border Cooperation program: "CULT.ROUTES Cross Border Religious and Cultural Heritage"

These strategic acts have made it possible to achieve important goals: (1) membership of the European Francigena Routes Association; (2) entry into the Council of Europe Cultural Itineraries Program; (3) establishment of internal coordination between the strategic sectors for the development of slow mobility and the management of cultural heritage; (4) approval of the official route to be included in the Regional Landscape Plan by providing the necessary protection restrictions; and (5) approval of projects aimed at infrastructure and enhancement of the routes, as well as activating a branch dedicated to the Francigena Route at the offices of the Regional Agency of Tourism, "Puglia Promozione," an in-house public body of the Apulia Region.

5.4 The Tourist Product "Francigena Route"

The recognition of the route as a "Council of Europe Cultural Itinerary" allows territories and tourist destinations to take advantage of the added value of an international brand of great importance and authority. Like other network brands (e.g., "Borghi più Belli d'Italia," "Borghi Autentici d'Italia," "Città Arancioni" of the Italian Touring Club, and "Città Slow"), the certification does not exhaust its own task. More similar to the entry in the UNESCO World Heritage List, it requires the drafting of "management plans" that allow for the enhancement of the tangible and intangible assets and the correct planning of the routes in order to meet the international travel demands and to realize the "microeconomics of the path," for local development in economic, occupational, landscape, environmental, and quality of life terms. In other words, procedures, requirements, and disciplinary necessary to build the destination and the resulting network of services must be identified within strategic territorial planning constraints. Moreover, the tools of international promotion to meet the demand of the markets with the territorial offer of information, reception, hospitality, and services must be considered.

The Apulia Region has dedicated a "Project of Tourist Excellence" to the first part of the Francigena Route in Apulia, in the Dauni Mountains, the Apennine area crossed by ancient roads from Rome, and in particular the Via Appia Traiana—as well as an extraordinary concentration of morphological, scenic, and anthropological uniqueness. The valorization of the Francigena Route and of the network of the Routes contributes transversally to at least three fundamental objectives for the whole region: (1) introducing new models of infrastructure for the soft mobility system, determining the Francigena routes as unavoidable references in the plot of the regional greenways for the qualified use of the landscape; (2) introducing new territorial systems for the fruition of cultural, tangible, and intangible heritage; and (3) creating a tourism center dedicated to the landscape and the environment and, therefore, to the segment of market demand that, in addition to hiking and devotional motivations, develops active and slow tourism flows, particularly interested in discovering the destination in all its identity aspects. The specific objective of redeveloping and recovering the reuse of historical infrastructure—roads, railways, itineraries, and *tratturi*—is part of the general objective of enhancing the cultural identity and settlement, contributes to supporting the redevelopment of historic rural landscapes, and strengthens the slow use of landscapes.

However, the Apulian Francigena Route includes many other territories and reaches the deep Salento, extending over 400 km of itineraries surrounded by 800 km of coasts and unraveling on the ancient roads from the Appia Traiana to the Appia Antica down to the Salento town of Santa Maria di Leuca, the Italian *finisterre*. The routes intercept all the Apulian territories of the Cultural Environmental Systems (*Sistemi Ambientali Culturali*, SAC), launched by the Apulia Region in 2010. They arise from territorial aggregations aimed at the enhancement and integrated management of already existing and usable environmental and cultural assets (protected areas, monumental and archaeological assets, museums, historical theaters, libraries, archives, etc.). They are characterized by a strong and original idea that specifies the possible paths of development, attractiveness, and quality of the territorial system, to be pursued through innovative use, integration, and increase of available environmental and cultural resources.

6 Findings

With reference to the evolutionary phases of a cultural project, the "Francigena Route" in the Apulia Region is still in an embryonic phase, with many developed territories and others still in the activation phase.

Having conceptually established the guidelines for its creation—by selecting the phases, the events, and the actors involved based on the reference context—the project is currently in the phase of defining the structural and organizational aspects.

The implementation of the Francigena Route has involved a plurality of subjects organized in sub-groups, which, in pursuit of a common goal, interact in the various operational contexts, associated with four main areas. Specifically, the following spheres are interested: the scientific area, including the conceptual phase, with the preliminary draft of the project; the technical-organizational area, in which the logistic aspects are defined; the administrative area, including the economic and financial management of resources and bureaucracy; and the marketing area, including the related information system (Acerboni 2009).

From the analyzed documentation and the interviews with the regional policy-makers, the analysis showed that the main strengths are related to the scientific dimension, where the scientific and cultural itinerary, identified in the conceptual phase, is clear and well defined. Moreover, critical issues emerged in the technical-organizational area (logistics, services, etc.), in the administrative area (budgeting, reporting, fund raising, etc.), and, although to a lesser extent, in marketing (management information, communication, forms of promotion, and other activities).

For the benefit of all the stakeholders and, above all, for the policy-makers, we have identified the problems and potentials that have emerged in the early stages of implementation and the strategic elements and require more attention and interventions, together with the variables to be managed through the chosen model. The SWOT analysis (Table 2) highlights the main opportunities, threats, strengths, and weaknesses that have emerged so far. This step lays the foundation for the subsequent phases of planning and implementation and for the creation of an adequate marketing and communication plan.

Another relevant finding is that the relationship with the other trans-boundary regions, which is needed in order to create an effective European interregional network, appears to be less developed. Currently, the "Via Francigena" in Apulia is a small part of a long route that crosses several Italian and European regions from Canterbury to Jerusalem. This represents an important criticality, for the reason that the success and efficacy of CTRs require investment in co-operative networks among a variety of different stakeholders. The implementation of these networks has a crucial role for social and cultural growth and regional development, as it involves cooperation among the embedded local communities (Zabbini 2012), cultural tourism and development for sustainable tourism, enhancement of memory, history and European heritage, cultural exchanges, and contemporary practice of culture and arts (Fig. 3).

Table 2 SWOT analysis of the Francigena Route in the Apulia region

Strengths	**Weaknesses**
• Strong historical-cultural identity of the territories and populations involved • Significant and high cultural and historical value • Fascinating geographical route, easy to travel, low cost • Enhancement of transnational partnership • Promotion and fostering of research and education • Scientific and cultural itinerary, identified in the conceptual phase, clear and well defined	• Shortage of a «local» perception of the itinerary • Poor cooperation among the countries involved in the CTRs • Low participation of local community • Low user-territory interaction • Inadequate planning and management rules • Insufficient monitoring and coordination among partners • Lack of sharing and archiving of documentation • Inadequate cooperative models able to identify communication, promotion, and compliance strategies in order to connect public and private partnership • Difficulties on the part of the local communities in recognizing and enhancing resources, both cultural (highly localized and ready for use for social purposes) and human • Inadequacy of skills, competences and management, organizational resources
Opportunities	**Threats**
• Growing tourist flows in the Apulia Region, especially in the territories involved in the itinerary, an excellent driver for advertising the CTR • Great attention of the Council of Europe and the international community to CTRs • Large demand-side attention among tourists in thematic routes as an opportunity to learn the history and better know the common heritage • Involvement in international cultural networks for the construction of a distinctive and high-quality tourism product	• Broad negative economic outlook of tourist flows, and an economic global crisis due to the "COVID-19" health emergency • Competition in the cultural tourism sector by operators better able to intercept the tourist flows in the regions involved in the route • Possible geopolitical tensions in the countries involved • Bureaucracy issues

Source: Own elaboration

7 Discussion

The aim of this study was to propose a managerial model for the sustainable development of CTRs through the employment of the ICT approaches, specifically, BPM and PLM. The suggested managerial model, based on the case study concerning the implementation of the "Francigena Route" in the Apulia Region, can be a valid pattern for the management of similar complex cultural products.

CTRs require adequate analytical methods and interpretative models to place them in their correct cultural and historical contexts: they can be seen as a single product composed of various routes, each one associated with single events. In the

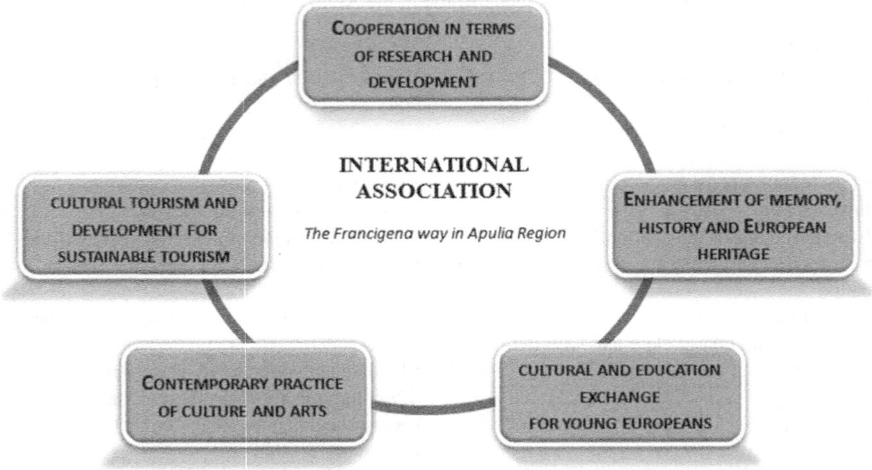

Fig. 3 The Areas of interest of the Francigena CTR in the Apulia region. Source: Adapted from Trono et al. (2015)

considered case study, the application of BPM and PLM approaches has required the preliminary modeling of the all processes and the related sub-processes constituting the different phases of the lifecycle (AS-IS), with the explication of documents and technologies involved.

Before the application of a single digital management model, it is necessary to detect and gather the requirements, in terms of both quality and functionalities. The digital management model involves events, tangible and intangible cultural heritage all together. The digital management model has to be scalable and modular, so that it can be adapted to any kind of cultural product, such as a tourism itinerary and its associated regional resources, and it represents a critical instrument for policy and decision-makers, similarly to the manufacturing context.

Considering the CTR as a product with different phases, it is possible to list the main processes to be mapped and managed, leading to a number of benefits, deleting redundant activities and bottle-necks, highlighting the interactions among partners and the related issues (Fig. 4). In tourist itineraries, the stream is mainly composed of data and documents, as these are the most important instruments in decision-making. Hence, activities regarding project management and data integration assume great importance throughout the itinerary lifecycle. More specifically, the project management activities focus on the integration and management of all tasks within the different phases and the assessment of data and associated feedbacks. The management of each active project is performed by involving:

- Staff management, regarding the coordination and management of different professionals collaborating across the lifecycle phases and processes, in the view of generating value for the territories

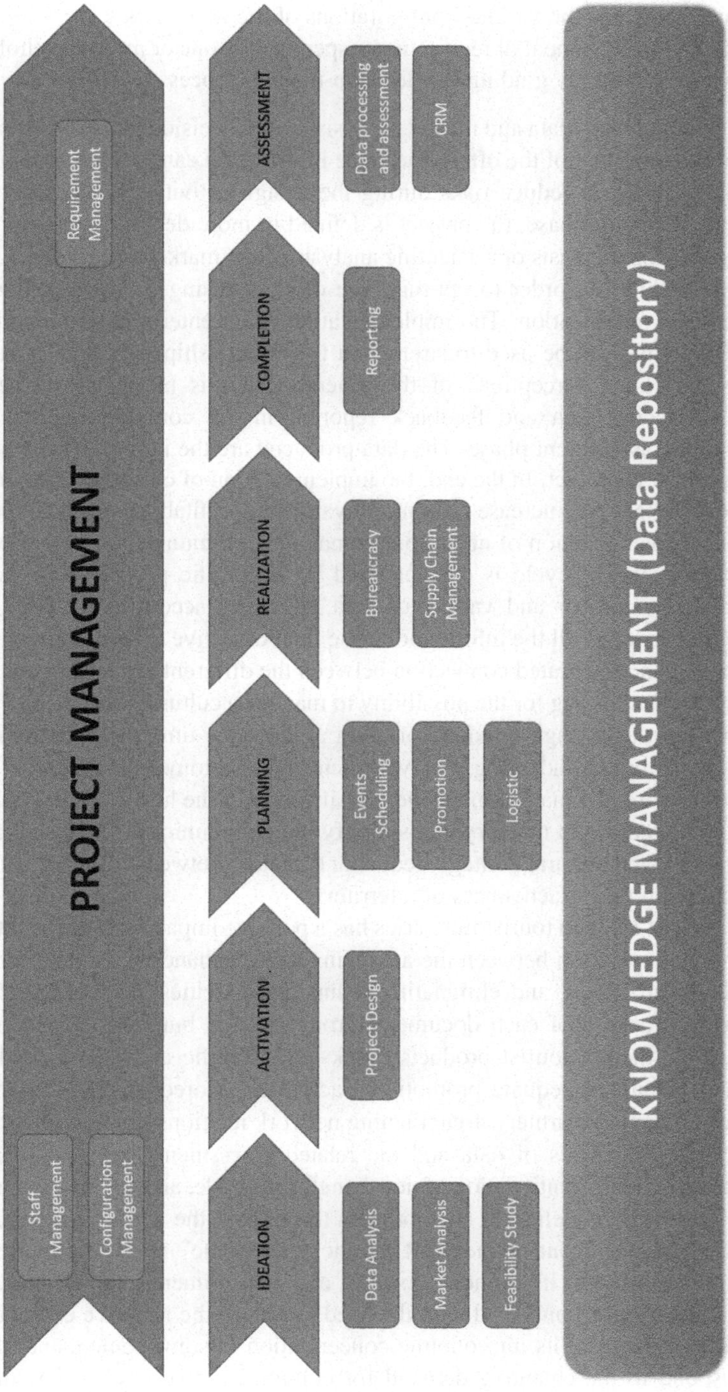

Fig. 4 PLM-based framework for the management of a CTR. Source: Own Elaboration

- Configuration management (including supply chain management), regarding the differences between the various configurations of the itineraries
- Analysis and management of requirements specified by one or more stakeholders, in the view of properly guiding the decision-making process

In the ideation phase, data and market analysis guides decision and policy-makers toward the enhancement of the offer. The same information can be used as input for the feasibility study, to reduce risks during the design activities of the activation phase. In the planning phase, the project is defined in more detail, the promotional plan is designed on the basis of a scientific analysis of the market and feedback. The events are scheduled in order to optimize the offer, focusing on time saving and logistic services optimization. The implementation of an enterprise resource planning (ERP) module can be used to strengthen the relationship with local partners, enhancing the "local perception" of the itinerary. In this phase, the project is executed, producing data and feedback reported in the completion phase and processed in the assessment phase. The data produced are the input for the conceptual phase of a new project. In the end, the implementation of customer relationship management (CRM) can increase customer loyalty and profitability by enabling the ideation and implementation of new promotional and communication strategies.

Each step of the lifecycle is characterized by a specific process flow, which produces a large number and variety of data, requiring access to a shared data repository that contains all the information. The final objective of such a framework is the complete and integrated connection between the different organizational units and stakeholders, allowing for the possibility to manage a cultural product, such as a tourism itinerary, as a single product, offering at the same time the opportunity to change its configuration adapting to new policies and customer needs (Prete et al. 2018). Furthermore, the management and classification of the historical and cultural information related to the itinerary can simplify the communication and sharing of common history and cultural heritage, reducing distances between different cultures and strengthening the attractiveness of a territory.

Application of PLM to tourist itineraries has a positive impact as it contributes to strengthening cooperation between the actors involved, enhancing the decision and policy-making processes, and eliminating redundant activities and bottlenecks. It allows for the tracking of each document throughout the bureaucratic processes, generating high-quality tourist products thanks to scientific analysis of feedback (CRM), and planning adequate promotional activities. Moreover, PLM enhances communication among partners strengthening network relationships and integrating logistics (SCM). Analysis of data and the related assessments can make further improvements and the identification of new benefits possible, and the following step is the definition and modeling of each process throughout the whole lifecycle. The main areas of management of the CTR "Francigena Route" in the Apulia region (Fig. 4) can be related to its economic, social, and environmental dimensions. The economic dimension is connected with the need to reduce the negative externalities due to mass tourism and its uneconomic concentration in some centers and, therefore, to respond to the changing demand for culture, by diversifying the cultural offer. CTRs allow for the tracing and promoting of new itineraries through the rich cultural heritage of the outlying areas, supporting it with a system of information services aimed at communicating the value of historical significance incorporated in the heritage as a territorially circumstantiated system.

The social dimension is related to the fact that CTRs embrace new values and specificities, including the intangible manifestations of human activity, understood, like material ones, as responses to human needs and factors of identity and social cohesion. The reference dimension becomes the landscape as a complete manifestation of cultural heritage, "an all-encompassing context whose form refers to the highest possible degree the spatial and temporal relations among individual things and measures the individual and overall values in the light of the overall meaning" (Montella 2003, p. 38). CTRs are able to re-establish the link with the territory, becoming the keystone of wider itineraries, a hub for new and more capillary and significant tourist routes. In comparison with the hurried trajectories of mass tourism, they allow for the balancing of the distribution of flows in different places and throughout the whole year (Montella 2006), with particular attention to "geographically close endogenous tourism, a harbinger of relevant and immediate social and economic benefits" (Montella 2006, p. 359). Moreover, if designed in accordance with effective cooperation processes and well managed at the political level, CTRs can become an instrument of integration and social inclusion (Trono et al. 2015).

Finally, the environmental dimension is connected to the fact that CTRs generate flows associated with the new needs of society, increasingly anxious to combine physical well-being with the environmental context and attentive to regional assets. They encourage new and alternative types of tourism, such as religious tourism, eco-tourism, slow tourism, food-and-wine tourism, wellness tourism, and health tourism.

8 Conclusions

When properly managed, CTRs can be advantageous and strongly contribute to innovation, economic growth, and sustainable development of tourist destinations (Macbeth et al. 2004). By fostering visitors to spend more time at destinations (Vengesayi 2003), they can generate a chain of economic opportunities, such as developing employment, producing revenue, and offering different and superior hospitality facilities, in addition to offering education and training. Thus, cultural capital has an essential role in the tourism industry (Grangsjo 2003; Vengesayi 2003), and becomes a critical resource for positioning, strengthening, forming, developing, and creating the identity of a specific territory (Li et al. 2008).

The cultural and tourist products, and above all CTRs, tend to be configured by its nature as a complex, varied and not easily standardizing product. For this reason, we propose that the proper use of ICTs to culture, in order to obtain effective and efficient results also in terms of the organization of tangible and intangible resources, can be a precious benefit for both tourist operators and policy-makers. The introduction of ICTs in the cultural and tourism sector, with the experimental approach of BPM and PLM, tools traditionally used in the industrial sector could actually represent a turning point even in the management of the intangible.

Acknowledgments The authors would like to thank Francesco Palumbo, Head of Policies Area for the Prevention of the Territory, Knowledge and Talents, Silvia Godelli, Regional Councilor for the Mediterranean, Cultural activities and Tourism, and Giancarlo Piccirillo, Director of the Regional Agency "Puglia Promozione," for providing kind collaboration at the time in which the research was carried out.

References

Acerboni, F. (2009). La gestione dei progetti nelle organizzazioni culturali. In M. Rispoli & G. Brunetti (Eds.), *Economia e management delle aziende di produzione culturale*. Bologna: Il Mulino.

Amato, R., & Chiappi, R. (1997). *Tecniche di Project Management*. Milano: FrancoAngeli.

Archibald, R. D. (1994). *Project Management: La gestione di progetti e programmi complessi*. Milano: FrancoAngeli.

Argano, L. (2004). *La gestione dei progetti di spettacolo. Elementi di Project Management culturale*. Milano: FrancoAngeli.

Argano, L. (2012). *Manuale di progettazione della cultura. Filosofia progettuale, design e project management in campo culturale e artistico*. Milano: FrancoAngeli.

Berends, H., & Deken, F. (2019). Composing qualitative process research. *Strategic Organization*. https://doi.org/10.1177/1476127018824838.

Berti, E. (2013). Cultural routes of the Council of Europe: New paradigms for the territorial project and landscape. *Almatourism, Journal of Tourism, Culture and Territorial Development, 4*(7), 1–10.

Bramwell, B., & Lane, B. (1993). Sustainable tourism: An evolving global approach. *Journal of Sustainable Tourism, 1*(1), 1–5.

Brown, L. (2013). Tourism: A catalyst for existential authenticity. *Annals of Tourism Research, 40* (1), 176–190.

Casini, A., & Zucconi, M. (Eds.). (2003). *Un'impresa per sei parchi. Come gestire in modo imprenditoriale e innovativo il patrimonio culturale e ambientale pubblico*. Milano: Il Sole 24ORE.

Causi, M. (1998). Stato e mercato nella gestione dei beni culturali. In A. Mattiacci (Ed.), *La gestione Dei Beni artistici e culturali nell'ottica del mercato*. Milano: Guerini e Associati.

Cerquetti, M. (2007). La componente culturale del prodotto turistico integrato: La creazione di valore per il territorio attraverso i musei locali. *Sinergie, 73–74*(07), 421–428.

Cerquetti, M. (2010). Strategie di branding del cultural heritage nella prospettiva esperienziale. *Sinergie, 82*, 123–142.

Chastel, A. (1980). L'Italia museo dei musei. In A. Emiliani (Ed.), *Capire l'Italia. I musei*. Milano: TCI.

Choi, H. C., & Murray, I. (2010). Resident's attitudes toward sustainable community tourism. *Journal of Sustainable Tourism, 18*(4), 575–594.

Choi, H. C., & Sirakaya, E. (2006). Sustainability indicators for managing community based tourism. *Tourism Management, 27*(6), 1274–1289.

Coccossis, H. (2008). Cultural heritage, local resources and sustainable tourism. *International Journal of Service Technology and Management, 10*(1), 8–14.

Corallo, A., Esposito, M., Marra, M., & Pascarelli, C. (2019). Transmedia digital storytelling for cultural heritage visiting enhanced experience. In L. De Paolis & P. Bourdot (Eds.), *Augmented reality, virtual reality, and computer graphics. AVR Lecture Notes in Computer Science* (Vol. 11614). Cham: Springer.

CUEIM. (Ed.). (2006). La valorizzazione del patrimonio culturale periferico per lo sviluppo del territorio. *BIC Notes, 2*(5).

Ding, L., Ball, A., Matthews, J., McMahon, C., & Patel, M. (2007, September 19–21). Product representation in lightweight formats for Product Lifecycle Management (PLM). In *The Fourth International Conference on Digital Enterprise Technology*, Bath.

Dragoni, P. (2005). Antimarketing dei musei? *Sinergie, 68*(Settembre–Dicembre), 55–74.

Easton, G. (1995). Case research as a methodology for industrial networks: A realist apologia. In *Proceedings from the 11th IMP Conference* (pp. 368–391). Manchester: Manchester Federal School of Business and Management.

Eisenhardt, K. M. (1989). Building theories from case study research. *Academy of Management Review, 14*, 532–550.

European Institute of Cultural Routes (EICR). (2004). *Cultural tourism or tourism and culture?* Luxembourg: EICR.

Gaio, L. (2010). *Project management: Elementi teorici e applicazioni. Metodi di evidenze empiriche per il turismo*. Milano: FrancoAngeli.

Grangsjo, Y. F. (2003). Destination networking: Competition in peripheral surrounding. *International Journal of Physical Distribution and Logistics Management, 33*(5), 427–448.

Greffe, X. (2003). *La gestione del patrimonio culturale*. Milano: FrancoAngeli.

Grieves, M. (2006). *Product lifecycle management: Driving the next generation of lean thinking*. New York: McGrawHill.

Guccio, C., Martorana, M. F., Mazza, I., & Rizzo, I. (2016). Technology and public access to cultural heritage: The Italian experience on ICT for public historical archives. In *Cultural heritage in a changing world* (pp. 55–75). Cham: Springer.

Gursoy, D., Jurowski, C., & Uysal, M. (2002). Resident attitudes: A structural modeling approach. *Annals of Tourism Research, 29*(1), 79–105.

Harmon, P. (2010). The scope and evolution of business process management. In J. vom Brocke & M. Rosemann (Eds.), *Handbook on business process management 1* (International handbooks on information systems). Berlin: Springer.

Hervy, B., Laroche, F., Bernard, A., & Kerouanton, J. L. (2013). Co-working for knowledge management in cultural heritage: Towards a PLM for museum. In A. Bernard, L. Rivest, & D. Dutta (Eds.), *Product lifecycle management for society* (IFIP advances in information and communication technology) (Vol. 409). Berlin: Springer.

Jick, T. D. (1979). Mixing qualitative and quantitative methods: Triangulation in action. *Administrative Science Quarterly, 24*, 602–611.

Krippendorf, J. (2013). *Le vacanze e dopo? Capire l'impatto del tempo libero e dei viaggi* (Italian edition by G. Montemagno). Milano: Cesano Boscone, Egea.

Kuo, T. C., Huang, S. H., & Zhang, H. C. (2001). Design for manufacture and design for "X": Concepts, applications, and perspectives. *Computer Industrial Engineering, 41*, 241–260.

Lazano-Oyola, M., Blancas, J. F., González, M., & Caballero, R. (2012). Sustainable tourism indicators as planning tools in cultural destinations. *Ecological Indicators, 18*, 659–675.

Li, M., Wu, B., & Cai, L. (2008). Tourism development of world heritage sites in China: A geographic perspective. *Tourism Management, 29*, 308–319.

Macbeth, J., Carson, D., & Northcote, J. (2004). Social capital, tourism and regional development: SPCC as a basis for innovation and sustainability. *Current Issues in Tourism, 7*(6), 502–522.

Madrigal, R. (1995). Residents' perception and the role of government. *Annals of Tourism Research, 22*(1), 86–102.

Majchrzak, A., & Wang, Q. (1996, September–October). *Breaking the functional mind-set in process organizations* (pp. 93–99). *Harvard Business Review*.

Mangialardi, G., Corallo, A., Esposito, M., Fortunato, L., Monastero, A., & Schina, L. (2016). An integrated and networked approach for the cultural heritage lifecycle management. *ENCATC Journal of Cultural Management and Policy, 6*(1), 80–95.

Mateos-Rusillo, S. M., & Gifreu-Castells, A. (2018). Transmedia storytelling and its natural application in museums. The case of the Bosch project at the Museo Nacional del Prado. *Curator: The Museum Journal, 61*(2), 301–313.
Montella, M. (2003). *Musei e beni culturali. Verso un modello di governance*. Milano: Electa.
Montella, M. (2006). Approccio manageriale ai musei italiani. In R. Fiocca & L. Battaglia (Eds.), *Il museo tra cultura e management* (pp. 51–70). Milano: McGraw-Hill.
Neuhofer, B., Buhalis, D., & Ladkin, A. (2012). Conceptualising technology enhanced destination experiences. *Journal of Destination Marketing and Management, 1*(1–2), 36–46.
Pencarelli, T. (Ed.). (2005). *Letture di economia e management delle organizzazioni turistiche*. Trieste: Edizioni Goliardiche.
Pigram, J. (1993). Planning for tourism in rural areas: Bridging the policy implementation gap. In D. G. Pearce & R. W. Butler (Eds.), *Tourism research: Critique and challenge* (pp. 156–174). London: Routledge.
Prete, M. I., Palmi, P., Piper, L., & Guido, G., (2018, July). *Creating satisfying visitors' experience in cultural organizations*, IFKAD, International Forum on Knowledge Assets Dynamics, University of Delft.
Primicerio, D. (1993). In giro per il paese dei musei: realtà e potenzialità. In P. A. Valentino (Ed.), *L'immagine e la memoria. Indagine sulla struttura del museo in Italia e nel mondo* (pp. 109–115). Roma: Leonardo Periodici.
Ruiz Estrada, M. A., Park, D., & Lee, M. (2020, March 9). *The evaluation of the final impact of Wuhan COVID-19 on trade, tourism, transport, and electricity consumption of China*. Available at SSRN: https://ssrn.com/abstract=3551093 or https://doi.org/10.2139/ssrn.3551093
Saaksvuori, A., & Immonen, A. (2008). *Product lifecycle management*. Berlin: Springer.
Settis, S. (1998). Gli Stati Uniti come modello per la gestione dei beni artistici e culturali in Italia? In A. Mattiacci (Ed.), *La gestione Dei Beni artistici e culturali nell'ottica del mercato*. Milano: Guerini e Associati.
Settis, S., (2005, November 23). Pietre dell'identità. Noi e le città: Perché gli italiani sono così legati al proprio patrimonio storico-culturale, *Il Sole 24 Ore—Domenica*, n. 308.
Sharareh, A. D., & Badaruddin, M. (2013). Local perception of tourism development: A conceptual framework for the sustainable cultural tourism. *Journal of Management and Sustainability, 3*(2), 31–39.
Spradley, J. P. (1979). The ethnographic interview. In *Re-published in 2016*. Long Grove, IL: Waveland Press.
Toscano, B. (1998). Il problema della tutela: La peculiarità italiana come chiave organizzativa e come fattore di sviluppo. In A. Mattiacci (Ed.), *La gestione Dei Beni artistici e culturali nell'ottica del mercato*. Milano: Guerini e Associati.
Trono, A., Corallo, A., Vestito, D., & Esposito, M. (2015, March 18–21). Cultural events and tourist services. A management model for guided tours. In *Proceedings 4th. International Research Forum on Guided Tours*, Estoril, pp. 130–148.
UNEP, & UNWTO. (2005). *Making tourism more sustainable—A guide for policy makers*, pp. 11–12.
Valentino, P. A. (2003). *Le trame del territorio. Politiche di sviluppo Dei sistemi territoriali e distretti culturali*. Milano: Sperling and Kupfer.
Valentino, P. A., & Misiani, A. (2004). *Gestione del patrimonio culturale e del territorio. La programmazione integrata nei siti archeologici nell'area euro-mediterranea*. Roma: Carocci.
Vanhaverbeke, W., & Torremans, H. (1999). Organizational structure in process-based organization. *Knowledge and Process Management Review, 6*(1), 41–52.
Vengesayi, S. (2003, December 1–3). A conceptual model of tourism destination competitiveness and attractiveness. In *ANZM 2003 Conferences Proceedings*, Adelaide.
World Tourism Organization. (2017). *UNWTO annual report 2016*. Madrid: UNWTO.
World Tourism Organization. (2019). *International tourism highlights*. Madrid: UNWTO.
Yin, R. K. (1994). *Case study research design and methods* (2nd ed.). Thousand Oaks, CA: Sage.

Zabbini, E. (2012). Cultural routes and intangible heritage. *Almatourism-Journal of Tourism, Culture and Territorial Development, 3*(5), 59–80.

Pamela Palmi (Ph.D.) is an associate professor of business organization at University of Salento, Lecce (Italy), where she teaches business organization and knowledge management and organization. Her research interests focus on creativity, cultural industries, innovation, and organizational change. She has published two research monographs and several articles in international peer-reviewed journals, e.g., *European Management Journal, Kybernetes, Journal of Public Affairs, International Journal of Business and Management, Law and Economics Yearly Review*. Currently, she is member of the Network UNITWIN-UNESCO "Culture, Tourism, Development," at the Paris 1 Panthéon-Sorbonne.

Marco Esposito is a research fellow at the CORE Lab of the University of Salento, Lecce (Italy). His research is based on conception, design, and implementation of technological and organizational solutions aimed at optimizing business processes and business models, both in the manufacturing sector and in radically different and heterogeneous industrial environments, such as cultural and creative industries. His studies focus on the management of change in complex industries. The author of national and international papers, he writes and manages research and development projects, business modeling and planning, fundraising, and technology consulting.

M. Irene Prete (PhD, MSc) is an assistant professor of business management at the University of Salento, Lecce (Italy), where she teaches digital marketing and digital management. She also teaches at LUMSA (Libera Università Maria Santissima Assunta), Rome (Italy), and has taught heritage marketing at the University of Salento. She has held seminars at many international universities, has produced about 80 publications and participated in numerous national and international conferences on the topics of consumer behavior and marketing. In the heritage field, she has studied visitors' behavior and satisfaction, and the creation of visitors' experience in cultural and touristic organizations. Currently, she is member of the Network UNITWIN-UNESCO "Culture, Tourism, Development," at the Paris 1 Panthéon-Sorbonne.

Detecting the Social and Economic Impact of Cultural Initiatives: A Case Study of the Taormina Film Fest

Pierluigi Catalfo, Martina Giustra, and Agata Cardillo

Abstract The aim of this research is to investigate, evaluate, and give representation to some of the characteristics of the social, cultural, and economic values generated by the activities of cultural institutions, such as the *Taormina Arte Foundation*, which was established in Taormina in 1983, with the aim of promoting and organizing prestigious cultural initiatives. Every year, the *Taormina Arte Foundation* organizes the internationally recognized *Taormina Film Fest* that takes place at the Greek Ancient Theater of Taormina, which has a variety of dramatic impacts as a rich cultural event in the northeast Sicilian territory.

The impact of cultural and artistic value that is generated within a territory cannot be expressed purely by economic value, as there are others, such as the social, aesthetic, educational, and symbolic, each of which contributes to the overall value that is given by the institution or cultural experience.

Therefore, cultural organizations need tools to measure their performances in terms of social impact.

From a methodological point of view, this work is based on a case study approach and focuses on the *Taormina Film Fest 2017*, and, according to some epistemological assumption coming from the relevant work of Edmund Husserl, we apply the SROI methodology in order to evaluate and represent the impact on the territory.

Keywords Heritage · Cultural events · Public choice · Territorial development · Social impact

---

P. Catalfo (✉) · A. Cardillo
Department of Economics and Management, University of Catania, Catania, Italy
e-mail: pcatalfo@unict.it

M. Giustra
Faculty of Architecture, Marcell Breuer Doctoral School, University of Pécs, Pécs, Hungary

P. Demartini et al. (eds.), *Cultural Initiatives for Sustainable Development*, Contributions to Management Science,
https://doi.org/10.1007/978-3-030-65687-4_11

1 The Generation of Cultural Initiative as Expression of Territorial Identity

In most cases, creating cultural initiatives in locations that are significant to a community is a way to enhance cultural paths. Through them we are able to appreciate historical heritage and places of importance from the past that have a feature of cultural identity, creating a relationship between material and cultural resources and intangible ones.

Making cultural initiatives, therefore, tends to define and generate new values from other pre-existing ones, such as the architectural-archaeological endowment in combination with the realization of an activity that has the value of cultural performance, and which tends to develop, promote, or propose art forms capable of attracting and involving subjects sensitive to the cultural proposal.

First of all, it is relevant and determinative to consider the concept of culture as a value defined in the socio-anthropological sense. The actual shape of the concept of culture came to its current meaning through an evolutionary path that started at the beginning of the twentieth century. The first great result on this path was the perception of culture as the result of a system of relationships, the product of a relational capital whose qualitative-quantitative dimension was defined by components of value and flows in relation to actions generated by individuals or institutions. Culture, as a social fact, therefore, is a collective product capable of generating cohesion and social appreciation with its precise endowment (Daum 2001).

Culture protects values and the relevance of historical heritage.

In this sense, culture is based on control systems linked to both social and remuneration-incentive mechanisms which drive production and innovation processes, and, at the same time, inside this systema there are mechanisms which select preferences and define the system of purposes to aim for (Durkheim 1912). So, culture in a certain sense, is a system in which quality and relevance of production are strictly related to interaction of several factors and actors. The economic relevance of culture is an effect of the interaction inside this system that is able to select, promote and inspiring specific innovation processes and specific—cultural—products and obviously a specific economic process based as already said mainly on remuneration and incentive of artists. Inside this system the cultural interaction generates cultural evolution and knowledge for "client" enhancing the evolution of culture in itself (Colbert 1994; Frey and Pommerehne 1989; Peacock and Rizzo 2008; Biondi et al. 2020).

Therefore, the role of public institutions in this context could be to promote or to actively contribute to the generation of cultural initiatives with resources that, as public, must be justified in their use, through accountability and disclosure of the self-realized effects of the impacts generated. In fact, for public institutions, the main reason for intervening in the implementation of cultural initiatives remains to be the political consideration of the impact.

For this reason, measurement of impact and accountability become essential; therefore the question of the precision of the measure is combined with the need for reasonable significance of the measurement made. Thus, the focus of the present work is the relation between measurement and value representation; it is not the methodology of measurement itself under a narrow methodological angle, but it takes into consideration the value of the rhetoric and managerial value of measurement, and, therefore, the work focuses on the characteristics that a measurement process must also have in reaction to some epistemological cornerstones. In this direction, the case study that will be proposed empirically supports this research, and wants to propose a methodological solution not in relation to its precision but in relation to its capability to represent and highlight the value of the different quality of performance, and the framework of the political choices made through the reconstruction and weighing up of the benefits acquired by the various stakeholders. Definitively, the problem is not to look for the perfect precision of an accounting measurement or a methodology of measurement, but the reasonableness and meaningfulness of the measurement obtained. If we deeply analyze accounting measurement processes or the representation of accounting results, most of the time, we have to accept that, at the end, perfection is something theoretical to obtain because we measure sociological-anthropological effects of wide interactions through our cognition-perception skills, so we have to admit discretionality and imperfection by definition.

Furthermore, the aim of this research is to investigate, evaluate, and represent the complexity of the social, cultural, and economic values generated by public institutions, using a case study approach and adopting the SROI methodology in order to represent the complexity of value generated, like in the case of Taormina Arte Foundation, which was established in Taormina in 1983, aiming to promote and organize prestigious cultural initiatives.

2 Choosing the Epistemological Approach and the Methodology to Understand the Social and Economic Impact of Cultural Initiatives

The measurement and understanding of the impact related to the creation of cultural initiatives in an architectural-archaeological heritage contest is a topic that cannot be treated without taking into consideration both the methodological choice and the definition of a reporting and representation approach, which can contextualize this choice. The process of cultural initiative creation involves information of a different nature that needs to be integrated. It must be taken into account that traditional methods like cost/benefit analysis are based on the need to look for some dimension of objectivity through a strong effort in transforming every result observed in quantitative data, most of them financial. This approach is clearly very difficult to follow, especially when it is applied on results and performance of cultural initiatives in which, by definition, value created cannot be always translated into financial/quantitative measures. As a consequence, the methodology chosen to carry out the research has to avoid the loss of value in the representation of the phenomena observed, due to reductions and overt simplifications of the complexity of the reality. Simplifications are generally accepted, for example, as a necessity in order to implement a statistical and mathematic methodology to make precise measurements and applications of quantitative models. To define this process of choice, we have taken into account the two traditional ways of operating: (a) aiming at achieving precision in measurement using incontrovertible data and avoiding estimations; (b) accepting reasonable discretional assumptions, based on estimations, to represent the complexity of the elements involved, including elements that cannot be represented in an incontrovertible way only with financial data coming from cashflow.

In this research, we have decided to use the dialectic approach in order to reach, as a result of this process of choice, an "apodictic" method, with the intention of enhancing the qualified perception of reality in its entirety without reductions.

According to the needs of defying a choice, in order to represent the complexity of the impacts of cultural initiatives, the eidetic intuitions (Husserl 1936) on which the evidence of the elements is assumed are not the result of arbitrary assumptions (meaning "not scientifically trustable"), but the precise use of a methodology that, during the evolution of the modern epistemology, has already largely demonstrated its qualities.

To underline the richness of the followed research path, it should be noted that the apodictic method guarantees the scientific nature of the research path. The orientation to the Husserlian approach, for understanding the essence and relevance of the non-financial statements, is motivated by the need to overcome what has been achieved by the technical vision of the financial statements, unable to coordinate and integrate the different information from the current financial reporting system.

From these considerations comes the conscious adoption of a methodology for structuring the case study, such as social return on investment (SROI), which integrates trustable balance sheet data, derived from the recorded financial data, with data resulting from estimates and conjectures. Thereof, the methodology takes into account both discretionary choices and data derived from conditions of incontrovertibility.

In particular, this approach stands out from those pure financial-quantitative ones that are largely adopted, because it tries to look at the balance sheet not just as a tool of representation based only on measurement operations, which has long since been demonstrated to be a partial vision.

This approach, in a certain sense, follows the path drawn by Husserl with the indication of looking, when carrying out research or an evaluation activity, at "Lebenswelt's"; that is to say, in the "living environment." Within this philosophy, the research/evaluation becomes an experience and it identifies its essential meaning in the "plenum" (Husserl 1912).

As a consequence, balance sheet data has to be used in an evaluation process, like the one that will be presented in the case study, as part of a plenum that has to be reconstructed and examined, taking into consideration both the internal and external context of the company and all its functions. The evaluation process, therefore, becomes a social communication tool (Hopwood 1983; Hopwood and Miller 1994; Potter 2005) for accountability.

More deeply, this methodology can be interpreted as a new theoretical practice, because it transfers the focus of the evaluation process and the methodological choice from the technical precision to the richness of the conscious experience.

Science and research, in this sense, are part of the philosophical thought that overcomes the attitude of specialized technical sciences and contemporary research of Galilean origin, which have largely lost the link of unitary meaning, caring too much for the technique as a simplification and categorization of reality.

Contemporary science research has taken refuge, to a large extent, in the professional technical dimension and, with respect to unitary phenomena, has disarticulated the systems, delimiting partial scientific skills. The recovery of the dimension of meaning and the search for the Husserlian plenum, therefore, depend on the loosening of the "pre-categorization" that generates the fragmentation of research and the recourse, according to Husserl, to the "not unaware doxa," qualified as a scientific basis for the reconstruction of meaning. Instead, Galilean science, as an objective science, counts on its effectiveness. The application of the above approach to cultural initiatives, set in archaeological territorial heritage, leads to an observation of the phenomenon in their imperfect completeness and complexity, accepting the research based on the "unconscious doxa" and rejecting the prevailing categorization (Bourdieu and Passeron 1979; Bourdieu 1991; Myles 2004).

Why should managerial and stakeholders needs of information be satisfied, uncritically, by a flat representation of reality based on precise measurements when evaluating the impact of an initiative? Research without the "pre-categorized" reality would drop the meaning of many research questions that are satisfied through precise measurements, but that often have neither a reasonable sense nor a contextualization in a significant plenum.

Contemporary science takes refuge in technical efficiency in specialist methodological approaches, losing, most of the time, the ambition of understanding the whole sense of things. Galilean relativism has pushed us to a new conception of knowledge that takes into account the accuracy of the method and measurement, by reducing the understanding of the facts to measurement and the aspiration of predictability.

Instead, complexity requires an approach that can recover the unity of meaning; therefore the Husserl "plenums" path is more appropriate. The methodological model to be preferred is therefore intended as a model and tool of the language of "corporate entity life" in its different dimensions, and in its different directions of relationship, recomposed in unity, reconsidering reality as not fully comprehensible with an objective key, as well as being known since its origins, in a sense, through a system of "doxa not unaware."

The limit in the choice of precision is that of not following a path of research of this kind; it is that of not facing the incompleteness of the sense of the balance sheet, which is derived from the use of specific methodological approaches. Management science cannot lose the sense of the system in the generation of measurement and representation models, in the face of the only technical efficiency of specific studies, and at the cost of a lack of specific value.

The methodological alternative, and the path proposed by the Husserlian approach, allows the business economy to recover the sense of more significant ontological positions, by recontextualizing more analytical efforts based on reducing reality to objectivity.

A countertrend approach, already perceived in the evolutionary processes, can be observed, for example, in the balance sheet of companies and institutions which, for practical reasons, integrate information of different quality and reliability in a single structure, in which performance, different final results, and development projections reconstruct a single meaning.

3 The Exigency of Evaluating the Impact Related to the Organization Cultural Initiative, Using Historical and Architectural Resources

Cultural heritage, whether internationally recognized or of proximity to the territories and local communities, through managerial choices, can constitute a strategic driver for the development of the territory and generate an economic and social impact of which the value can be measured (Catalfo and Giustra 2020).

Governments, cultural heritage offices, and local institutions transpose the guidelines set out by the main European Conventions and international agreements on cultural heritage (some of the more relevant EU Conventions)—such as 1954, European Cultural Convention, Paris; 1985, Convention for the Protection of the Architectural Heritage of Europe, Granada; 1992, Convention for the Protection of the Archeological Heritage of Europe, Valletta; 2000, European Landscape Convention, Florence; 2005, Convention on the Value of Cultural Heritage for Society, Faro; 2017, Council of Europe, Convention on Offences relating to Cultural Property;—and implement them by putting adequate policies and measures in place for the protection, conservation, and enhancement of Cultural, Archaeological, Architectural, and Landscape Heritage, also promoting their use for cultural initiatives and events.

In fact, if on one hand the continuous work of maintenance and conservation of the cultural heritage generates an enormous cost for local governments, yet on the other hand, a management that foresees its use for cultural and artistic purposes (as in the case of events, shows and live performances), then it can therefore constitute revenue in economic terms and determine a direct and indirect socio-economic impact of broader scope in favor of the local industry.

This *modus operandi* certainly guarantees greater prestige in terms of the cultural offer, and a more effective involvement of users in scheduled events, and must also necessarily take into account the specific features, often of the fragility and intrinsic rules, that the use of the cultural asset imposes (Lorusso et al. 2016, 2018). Only in full respect of the complex balance between economic, managerial, and protection and enhancement interests can the combination of cultural heritage and cultural business be sustainable, and constitute a real resource for the territory and the population (Longstaff et al. 2010).

The impact generated and its measurable value constitute valuable information for the continuity in strategic planning and the governance of the territories and local communities (Clark 2000; Cristofoli et al. 2014).

Measuring this impact guarantees transparency in respect to the work of the administrations, bodies, and institutions involved, and returns the measure of change, facilitating the understanding and communication of strategic management choices at local, intra-territorial, and transnational levels.

The Taormina Film Fest case study represents a virtuous example, and a model in which relevant issues converge, such as management complexity, strategic choices for the use for artistic, cultural and recreational purposes of an archaeological and monumental UNESCO World Heritage site, and virtuous work of a foundation.

The result of these measures, balances, and alliances has set up a cultural offer of international scope over the years, capable of generating a great impact on the territory, at different levels.

4 The Case of Taormina Festival

In recent years, the business world has shown growing interest in the issue of measuring social impact. Both social and "for profit" companies have understood the importance of considering and measuring the social value created, not only in order to "report" their results to stakeholders, but also to review their actions and strategies (Emerson et al. 2000). The definition of impact is very complex, as is its measurement. An impact is generally defined as the long-term sustainable change—positive or negative; primary or secondary—generated in people or in the environment that the intervention has partially contributed to achieve, since it is also influenced by other exogenous variables (UN-ECOSOC 2008). It is important to underline that impact is determined taking into consideration the results of what is referred to in the scientific field as "counterfactual analysis," that is, the evaluation that allows verifying what would have happened in the absence of the activity implemented by the organization and, consequently, of the causality between the work of the organization and the impact generated. Through the "impact indicators," it is possible to measure the quality and quantity of long-term effects generated by the project/intervention considered. These indicators describe changes in people's lives and development globally, taking into account the exogenous variables that influence those changes. One of the most tested and widespread methods internationally, especially for assessing the social impact in social enterprises, is the "social return on investment" (SROI). This method aims to calculate the economic return of the funding received from the organization, in terms of the relationship between funding and result, monetizing the different elements of social impact generated by the organization's activity.

With the case of the Taormina Film Fest, the return on investments made in 2017 in favor of the aforementioned event, carried out by the Taormina Arte Foundation, was calculated in the form of an SROI ASSESSMENT. The Taormina Arte foundation was founded in 1983 in Taormina, with the production of prestigious musical, theatrical, and dance events, with the Taormina Film Fest constituting its first event. In addition to being the cornerstone of the organization of events including the Taormina Film Fest and the Nastri d'Argento, Taormina Arte also provides logistical support for concerts, performances, and "external" events carried out by private individuals. It consists of a professional team that, for more than a decade, has been committed to keeping up the name of Taormina, famous worldwide for its extraordinary landscape setting, but also for tourism and international stardom. In the activity of tourism and cultural promotion, with a view to enhance the territory, Taormina Arte has always played a decisive role. Therefore, the choice of this work was based on the awareness that the Taormina Arte Foundation is a cultural institution with a strong propensity for the creation of social impact, and on the belief that SROI, among the various existing measurement methods, was the most suitable to verify the investments made in favor of the Taormina Film Fest.

In order to develop the case of the Taormina Film Fest, the necessary steps to carry out an SROI analysis were followed:

1. Definition of the field of analysis and stakeholder involvement.
2. Understanding of the change, identifying inputs (with the relative value), outputs, and outcomes (with the relative indicators, quantity and duration).
3. Definition, measurement, and quantification of the outcomes: The phase that allows us to identify the different steps through which the Taormina Film Fest generates a certain change in people's lives (social impact), proceeding, consequently, with the measurement and quantification of the aforementioned change in terms of outcomes, defining the indicators for each outcome.
4. Proxies: The phase in which the perceived change is quantified and a monetary value is attributed through the identification of financial proxies for each indicator.
5. Calculation of the impact (counterfactual analysis): the phase that aims to identify and measure what has really changed thanks to the realization of the Taormina Film Fest, and what would have happened otherwise.
6. The SROI ratio: the phase that summarizes all of the financial information previously examined. The basic idea is to calculate the financial value of the investment, and the financial value of the costs and social benefits. It may seem like a mere mathematical calculation, when, in fact, it helps to reduce the risk of overestimating the results and strengthens the credibility of the event.
7. Analysis of the results.

Through the SROI analysis of the Taormina Film Fest 2017, it was possible to provide an answer to the following: Does the Taormina Film Fest represent a resource or a cost for the territory? What is the return generated by the Taormina

Film Fest for stakeholders? Is the social impact in the territory generated by the Taormina Film Fest 2017 quantifiable? In general, does the Ancient Greek-Roman Theater of Taormina generate social benefits for the territory?

The quantitative approach of the data required the use of descriptive techniques, averages, correlations, and probability calculation. This made it possible to form a picture of the main stakeholders involved, as well as to measure the changes and the impact generated by the Taormina Film Fest 2017.

5 Some Methodological References and Features of SROI

There is increasing recognition of the need for better ways to account for the social, economic, and environmental value that results from our activities. The language varies—"impact," "returns," "benefit," "value"—but the questions around what sort of difference and how much of a difference we are making are the same. Understanding and managing this broader value is becoming increasingly important for the public and private sectors alike. This is true whether it is a civil society organization working to create value, governments commissioning and investing in activities to create social value, investors seeking to ensure that their investments will make a difference, or private businesses recognizing both risk and opportunities in the wider effects of operations (Zamagni et al. 2015).

All of this means that it is also more important that we have some consistency and a shared language when we talk about value. SROI is the application of a set of principles within a framework that is designed to help bring about that consistency, whilst at the same time recognizing that what is of value will be very different for different people in different situations and cultures (Meynhardt 2009).

Every day our actions and activities create and destroy value; they change the world around us. Although the value we create goes far beyond what can be captured in financial terms, this is, for the most part, the only type of value that is measured and accounted for. As a result, things that can be bought and sold take on a greater significance, and many important things get left out. Decisions made like this may not be as good as they could be, as they are based on incomplete information concerning the full impact. Social return on investment (SROI) is a framework for measuring and accounting for this much broader concept of value. SROI measures change in ways that are relevant to the people or organizations that experience or contribute to it. It tells the story of how change is being created by measuring social, environmental, and economic outcomes, using monetary values to represent them (Nicholls et al. 2009).

The social return on investment (SROI) method is designed to measure the outcome of an intervention, rather than merely tracking outputs, and its monetization technique facilitates the comparison of otherwise incommensurable benefits across different activities, producing a transferable evidence base that can be communicated to a wide range of audiences. The results are distributed using a "return-on-investment" language that is familiar to investors and commissioners and is based on real data collected through qualitative stakeholders' engagement, to ensure that what matters to users is what is being measured. This is realized through a comprehensive method that is robust and replicable due to recent standardization work (Arvidson et al. 2010, 2013; Perrini and Vurro 2013). As a result, social return on investment has achieved a significant deal of traction within the social enterprise sector and, increasingly, public policy and commercial industry also. It therefore has potential as a novel post-occupancy tool to capture the impact of design for building users, and disseminate the findings in a more powerful way across the variety of actors in the design and construction sector (Watson and Whitley 2016).

The intangible impact of design on building users cannot be understood without consideration of the social context that mediates user-experience, yet existing post-occupancy methods measure predetermined criteria about building performance.

A shift in evaluative focus is required, away from measuring building performance from a user perspective toward measuring the outcomes experienced by building users as a result of the dynamic interactions between buildings, users, and the social context that mediates them.

The need to capture post-occupancy feedback from building users in a more meaningful way shares a considerable overlap with the concept of social value and the impact-evidencing activities of mission-led organizations and programs. Recognizing the subjective, malleable, and variable nature of social value is key to the development of metrics suited to its capture and measurement (Watson and Whitley 2016).

In the same way that a business plan contains much more information than the financial projection, SROI is much more than just numbers. It is a story about change, on which to base decisions, that includes case studies and qualitative, quantitative, and financial information (Clark et al. 2004). An SROI analysis can take many different forms. It can encompass the social value generated by an entire organization, or focus on just one specific aspect of the organization's work. There are also a number of ways to organize the "doing" of an SROI. It can be carried out largely as an in-house exercise or, alternatively, can be led by an external researcher (The SROI Network 2012).

There are two types of SROI:

- Evaluative: conducted retrospectively and based on actual outcomes that have already taken place
- Forecast: predicting how much social value will be created if the activities meet their intended outcomes

Carrying out an SROI analysis involves six stages:

1. Establishing scope and identifying key stakeholders. It is important to have clear boundaries about what the SROI analysis will cover, who will be involved in the process, and how.
2. Mapping outcomes. Through engaging with your stakeholders, you will develop an impact map, or theory of change, which shows the relationship between inputs, outputs, and outcomes.

3. Evidencing outcomes and giving them a value. This stage involves finding data to show whether outcomes have happened, and then valuing them.
4. Establishing impact. Having collected evidence on outcomes and monetized them, those aspects of changes that would have happened anyway, or are a result of other factors, are eliminated from consideration.
5. Calculating the SROI. This stage involves adding up all the benefits, subtracting any negatives, and comparing the result to the investment. This is also where the sensitivity of the results can be tested.
6. Reporting, using, and embedding. Easily forgotten, this vital last step involves sharing findings with stakeholders and responding to them, embedding good outcomes processes and verification of the report.

5.1 SROI Methodology: Criticality and Strengths

SROI is a social impact methodology that allows not-for-profit organizations to evidence the wider value of their work. It is based on a traditional cost-benefit analysis and assigns a monetary value to social returns using financial proxies, which are compared against the level of investment to produce an SROI ratio of costs to social outcomes. It was originally developed by the Roberts Enterprise Development Fund (REDF) in the United States in the mid-1990s. More recently, an emphasis on stakeholder engagement, using a standardized methodology, has evolved through the work of the New Economics Foundations (NEF) in the United Kingdom.

SROI has been critiqued in academic and applied literature, as it has been considered to be a reductionist approach to monetizing qualitative social outcomes. It is often argued that reducing social outcomes to monetary measures is neither possible nor desirable, and the methodology is considered to underestimate the true value created.

Supporters of SROI argue that its ability to draw attention to otherwise intangible outcomes, by presenting them in a commonly recognized unit of value, promotes a broader discussion about what is important. In the case of the intangible benefits of good design, SROI has the potential to ensure that user perspectives are taken into account in design and commissioning decisions [. . .]. For the built environment, the variety of data produced by the SROI approach is key for its effective communication to a variety of design professionals, end user clients, and organizations (Watson et al. 2016).

6 Data Processing for Social Impact Under the SROI Methodology

The processing of the data (Tables 1 and 2), necessary for the assessments of the social impact generated by the Taormina Film Fest, is based on assumptions and variables verified and measured through primary sources, interviews, questionnaires, and the analysis of documents. It should be noted that the choice of the type of outcomes to be detected and their method of collection strongly influence the results obtained. In this case, it was decided to adapt a scientific evaluation system (SROI) on an information system not originally set up for this type of analysis; therefore, it is easy to foresee that many aspects and changes, although understandable, have not been collected and measured, and so consequently, their value has been lost. In order to calculate the social return, it was necessary to know the initial investment, i.e., the amount in Euros intended for the realization of the film festival. For this reason, each investment component necessary for the assessment was extrapolated from the 2016 and 2017 financial statements and, where necessary, from that of past years.

For the quantitative result of the data it was necessary to resort to descriptive techniques, averages, correlations, and calculation of probability. The need to explain how it is hypothesized to obtain change and evaluate it through the data collected requires a specific analysis for each stakeholder, as a result of different types of change (expected or unexpected—positive or negative), and also requires an enhancement operation of quantities that often do not approach an economic value because they are abstract. So, it becomes necessary to present useful evidence to demonstrate that the change has occurred, and to contextualize it over time. In the context of the theory of change, the following have been identified: input (with its value), output, and outcome (with related indicators, quantity, and duration).

However, in regard to the determination of the outcomes, and subsequently of the impact, financial proxies were used for the monetization of non-market assets, for which it is not possible to obtain an exact measure and, therefore, not always objectively determinable. Of course, each organization chooses the method that best suits its sector of intervention and its procedures. Hence the extreme fragmentation of standards and indicators used, which penalizes the social impact measurement phase. This stimulates the use of continuous dialogue with stakeholders to reach a shared evaluation, thus taking over a (subjective) interpretative key.

Table 1 Data processing and analysis

	Number of meals	Unit price	Number of nights	Unit price
International tourism	10	€ 60.00	3	300
Italian tourists	6	€ 60.00	2	250
Italian presence	1	€ 30.00		
Students	1	€ 20.00		

Source: data obtained from assumptions and variables verified and measured through primary sources, interviews, questionnaires, and the analysis of documents

Table 2 Data processing and analysis

	Value									
	Information relating to customer segments									
Description	Tickets sold	Ticket value	Tourist services	Travel expenses	Restaurants	Overnight stay	Food and wine shopping	Gadgets	Books and dvds	
International tourism	2244	€ 16	€ 120	€ 450	€ 600	€ 900	€ 350	€ 20.00	€ 75	
Italian tourists	2543	€ 16		€ 200	€ 360	€ 500	€ 200	€ 20.00	€ 51	
Italian presence	10,893	€ 16		€ 50	€ 30		€ 20	€ 20.00	€ 33	
Students	1560	€ 16		€ 20	€ 20		€ 10		€ 12	
Total	17,240	€ 275,840	€ 269,280	€ 2,094,250	€ 2,975,890	€ 3,291,100	€ 1,527,460	€ 313,600	€ 676,182	
Value added margin	Customer satisfaction value		60%	30%	20%	50%	20%	10%	10%	
Incremental added value			€ 161,568	€ 628,275	€ 595,178	€1,645,550	€ 305,492	€ 31,360	€ 67,618.2	€ 3,435,041.2

Source: data obtained from assumptions and variables verified and measured through primary sources, interviews, questionnaires, and the analysis of documents

Then, a counter-factual analysis (dead-weight) was performed for each stakeholder, i.e., the calculation of the impact or outcome with the film event and the value of the outcome without the film event. The weight of other factors or actors in determining the outcome (attribution) and the reduction of the impact, or impact on the impact over time (drop-off), was also considered. This was done even if, for reasons of simplification, it was necessary to consider an evaluation period of only 1 year (12 months), even when changes were foreseen for longer periods, but which would be difficult to evaluate objectively. The assessment was very prudent, which can be seen from the fact that significant deadweight and attribution have been included in the calculation.

Finally, for the calculation of the SROI ratio, all previously analyzed financial information was summarized. The financial value of the investment was calculated, and the financial value of costs and social benefits. Subsequently, the value of the total impact at the time of the realization of the event (money discounting process) was contextualized, applying the inflation rate for the year 2017 (+1.2%) to the amount published on the website of the Ministry of Economy and Finance.

As can be seen, the main operations, envisaged by the SROI analysis, have the need for monetary valuation subject to estimates, consequently meaning that in-depth analyses and investigations are required. However, in the case of the Taormina Film Fest, it has been shown that this additional effort, if rational and well planned, allows achieving important advantages. But this significant effort can be made only if the advantages obtained are considered to be of a relevant level; otherwise it is appropriate to remain with the traditional quantifications. In any case, the ever more frequent use of monetary values facilitates the valuation of operations over time. Through the process of generating and processing the data, it has been possible to demonstrate that the attractiveness of the Taormina Arte Foundation for a large and quality audience, albeit in a very limited period, has a very strong impact on the local economy, and that the Taormina Film Fest event shows the possibility of continuous growth over time, both in importance and territorial impact.

7 Some Outputs of the Data Processing for the Definition of a Final Result

Once the field of analysis has been defined, the stakeholders identified, and the expected or planned changes defined (Table 3), it becomes appropriate to formulate a logical framework that allows adequately understanding how the change has been made by the organization. This logical model, which in the SROI approach takes the name of "impact value chain" (input, output, outcome, impact), allows us to identify the different quality dimensions by which the Taormina Film Fest generates a certain change in people"s lives (impact social), or specific benefits. In a certain sense, this map gives us the opportunity to identify an outcome analytic.

Table 3 Logical framework of impact changes and benefits

	Stakeholder	Expected/planned changes-benefit
1	Event organizers	Strengthening of the Foundation Brand
2	Economic operators	Increase in the sale of goods/services
3	City community	Strengthening social cohesion
4	National and international cinephile culture	Strengthening cinephile culture and Theatre Brand
5	Youth volunteering	Creation of job opportunities
6	Donors/sponsor	Increase in visibility

Source: Elaboration on, TaoArte foundation data and analysis

Subsequently we proceed with the measurement and quantification of the change in terms of outcome, defining the indicators for each outcome (Table 4).

As previously explained, the SROI methodology aims to identify and measure what has really changed thanks to the intervention carried out, and what would have happened anyway. In that sense we can consider a proxy determinate as a measurement of this general value.

Not all changes detected, however, could be the result of the festival. By analyzing this problem it is possible to consider a proxy estimated according to the idea of "plenum"—in order to have a more complete result we have to clean the total impact from some interferences. So, to measure the impact of the Taormina Film Fest, we have to multiply the value of the attributed outcome by the value of the financial proxy, before finally subtracting the size of the deadweight, attribution, displacement, and drop-off.

Analytically, we consider VAO as the value of the attributed outcome, Fpx the value of the financial proxy, SDW the size of the deadweight (indicates to what extent the results would have been achieved without the Taormina Film Fest), ATT the attribution (indicates to what extent the results are attributable to the Taormina Film Fest and not to any other event); DDF the displacement (DSP) (indicates to what extent the results of the Taormina Film Fest damages other events or create costs elsewhere) and DRO the Drop-off, and the Impact IMP per Stakeholder as the amount of outcomes per proxy (less deadweight, attribution, displacement and drop-off).

We can generalize, taking in account that:

$$\mathrm{K}_0 = (\mathrm{VAO} * \mathrm{Fpx})$$

assuming that T is expressed as a percentage and in general $T_1 \neq T_2 \neq T_3 \ldots \neq T_n$ and in this case $T = (T_1 = SDW;\ T_2 = ATT;\ T_3 = DDF)$ and the Impace per stakeholder is IS_{J+1} and J from 0 to n–1.

$$\mathrm{IS}_{J+1} = \mathrm{IS}_J * (1 - \mathrm{T}_{J+1})$$

Table 4 Measurement and quantification of the change in terms of outcome, defining the indicators for each outcome

Stakeholder	Input	Output		Indicator	Value indicator	Quantity	Duration
Who do we affect? Who affects us?	What do they invest?	Summary of activities in numbers	Description				
Event organizer: Foundation Taormina Arte	Know how Qualified personnel Economic resources	Organization, active participation in the event	Foundation brand	Variation in the number of evenings	55,168€ (average revenue value of tickets per evening)	5 (number of evenings for Taormina Film Fest)	1 years
Economic operators	Tourist movements	Nights	Nights overnight	Marginal added value	3,435,041.2€ (tot. Sale of goods/services) *	7 (number of main goods and services offered to participate in the festival)	1 years
		Travel	Transport				
		People	Tourist services				
		Meals eaten	Restaurants				
		Shopping	gadget				
		Shopping	Books and DVDs				
		Shopping	Food and wine				
City community	Inhabitants	Historical cultural heritage	Brand city Social capital and pride of citizens	Variation in the number of inhabitants	130,160€ (total estimated value of the communication campaign)	5000 (estimated number of flyers and printed flyers)	1 years
National film culture E international	Event budget	Value of the show produced	Strengthening cinephile culture Theatre brand	Variation in the number of tickets sold	257,840€ (total value of tickets sold) *	17,240 (total tickets sold)	1 years
Youth voluntary service	Time taken	Contribution to the success of the event	More active volunteers	Variation in the number of dedicated volunteer hours	–	–	1 years
Donors/sponsor	Economic resources	Offer extension	Expansion of networking between partners	Variation of sums for co-planning	796,520€ (estimated budget for the realization of the Taormina Film Fest)	505,680 (average of the contributions received)	1 years

Source: Elaboration on, TaoArte foundation data and analysis. Specifically (*) = data from the 2016 and 2017 financial statements and, where necessary, from previous years

$$IS_{j+1} = IS_j(1 - T_{j+1}), j = 0, \ldots, n-1$$

$$IS^s_{j+1} = IS^s_j\left(1 - T^s_{j+1}\right), j = 0, \ldots n-1; s = 1, \ldots, q$$

$$IS_{j+1} = \sum_{s=1}^{q} IS^s_{j+1} = \sum_{s=1}^{q} IS^s_j\left(1 - T^s_{j+1}\right), j = 0, \ldots, n-1$$

$$IS^1_3 = IS^1_2(1 - T^1_2)$$

$$IS^2_3 = IS^2_2(1 - T^2_2)$$

To run the process, it is important to define and evaluate a proxy for each stakeholder in order to understand in which way the value was produced (Table 5). Specifically, in the following cases.

Having noted the above, the total impact value of the event amounts to 1,648,651.97€ (sum of the impact relating to each individual outcome) (Table 6). Subsequently, the value of the total impact at the time of realization of the event (process of discounting the money) was contextualized by applying the inflation rate for the year 2017 (+1.2%), published on the website of the Ministry of Economy and Finance.

$$\text{Total present value} = \text{total impact value} \div 1.2\% = 1,628,868.14 \text{ EUR}.$$

Table 5 Stakeholder value estimation by proxy

Stakeholder	Financial proxy description	Proxy	Value estimated or recognized in €
Event organizer	It is assumed that the foundation's brand will increase in proportion to the number of events held	Average estimated cost for the realization of an event	200,000.00
Economic operators	It is assumed that the sale of goods and services will increase in proportion to the tourist presence during the festival	Incremental added value	3,435,041.2
City community	It is assumed that the brand of the city increases in proportion to the number of tourist services offered	Indicative cost for the supply of tourist services	120.00
National film culture E international	It is assumed that the value of cinephile will increase in relation to the number of tickets sold (average between the maximum value of the TaoFilmFest ticket and the maximum value of the ticket for the "Nastri d'Argento" award ceremony)	Cost of a ticket for the Taormina Film Fest	16.00
Donors/sponsor	It is assumed that the theater brand increases in proportion to the number of visitors	Cost of a ticket to visit the monument	5.00

Table 6 Estimation of total value impact of the event

Stakeholder	Value by stakeholder €	Total outcome per stakeholder €	Deadw. %	Attribut. %	Displace. %	drop-off %	Impact €
Event organizer	200,000.00	1,000,000.00	20%	80%	0%	0%	160,000.00
Economic operators	3,435,041.20	24,045,288.00	80%	80%	10%	0%	865,630.37
City community	120.00	600,000.00	40%	80%	10%	0%	64,000.80
National film culture E international	16.00	275,840.00	0%	100%	0%	0%	275,840.00
Donors/sponsor	5.00€	2,528,400.00	30%	80%	20%	0%	283,180.80
Total value impact of the event							1,648,651.97

Lastly, the RATIO SROI was calculated, in the ratio between the total present value and total investments made (corresponding to the gross cost of the festival).

RATIO SROI = total present value 1,628,868.14€ (+ ticket revenues 275,840€)/ value of investments (corresponding to the estimated gross cost for the realization of the festival 796,840€) = 2.4.

It has been shown that for every euro invested in the event there was a social return of 2.4€. So for every 10€ invested in the Taormina Film Fest, gains or benefits were induced for all stakeholders, estimated at 24€.

Furthermore, we could have used the official table that is available on the socialvalueuk.org website, but instead we have chosen to report the data with the tables shown for the type of disclosure and the detail of the data collected.

However, it is clear that in the eventual dissemination of the results obtained to a wider audience, the use of the same methodology of exposure has undeniable advantages.

8 General Consideration of the Results

The proposed case study aims to assess the social impact of a cultural initiative, and to identify the reasons for considering the capacity, and the expressive adequacy of the model for the evaluation adopted. Specifically, the case study is intended to consider the usefulness of the evaluation results obtained for the determination of further strategic indications for the future, and the verification of the usefulness of the evaluation obtained for the accountability needs.

As highlighted above, the results of the evaluation carried out with the SROI are based on not only hypotheses and variables verified and measured through primary sources, financial and economic data, but also interviews, questionnaires and the analysis of documents.

In particular, the information system of the *Taormina Arte Foundation* from which much data comes is not prepared at the origin for the specific application of this methodology; therefore, consequently the result of applications subsequent to a reorganization of the information system could be even more precise and representative.

The calculation of the SROI was relevant for the understanding of the correctness of the assumptions and strategic assumptions, assumed functional in turn for the analysis of the decisions made which actually had a more significant impact on the total ratio. It is precisely these decisions that can be taken into consideration to identify priority areas for the generation of value.

The SROI calculation model certainly has elements that are the result of discretion and esteem that certainly make it less precise than other impact measurement models; however, from an analytical and "reasonable significance" point of view, the SROI manages to highlight the characters and elements of certain managerial value and of strategic utility. It should be kept in mind that, for example, the balance sheet model derived from the economic and financial accounting (with which the economy from the end of the 1400s to the present day is governed) is structurally plagued by elements of discretion and esteem which, however, in substance do not compromise the scope and the usefulness of the information it generates.

The hypotheses prepared for this type of analysis have shown the same results; that is, the most important figure is made up of economic operators, who, thanks to the creation of the Taormina Film Fest, show a significant increase in the supply of goods and services, consequently increasing their own income. Another element that emerges from the analysis of the outcome data concerns the value of support and collaboration in the realization of the event by Taormina Arte. In fact, thanks to co-planning, it was possible to take advantage of important exchanges of resources between bodies, thus creating added value.

The empirical evidence collected through the SROI assessment process shows how much the intervention of the Taormina Arte Foundation generates a positive and significant social impact for its stakeholders. In fact, the estimates applied showed that the resources invested in the event resulted in returns equal to 2.4 times the capital invested. In particular, the economic operators are the stakeholders that benefit most from a strong impact following the Taormina Film Fest film event. In particular, there is a socio-economic benefit in the sale of goods and services, which, when translated into percentage terms, is equal to 57%, as explained above. To estimate the induced benefits, restaurateurs, hoteliers, transport workers, etc., were interviewed, managing to "profile" the Italian daily users, the users who stay more than 1 day, and the corresponding foreign users with higher propensities to the consumption of food and drinks and hospitality compared to the Italians. The other stakeholders (the organization itself, the city community, national and international cinephile culture, donors/sponsors) also achieved the planned changes overall.

In this context, in an implicit way, the donors act as "business angels" allowing the existence of the event in the initial stages, in exchange for the achievement of objectives over time of general interest related to the development of the territory, and the strengthening of moments of nature, social, and culture, corresponding to its mission. In order to calculate the social return, it was necessary to know the initial investment, i.e., the amount in euros intended for the realization of the film festival. For this reason, all information on the investment components necessary for the assessment was taken from the 2016 and 2017 financial statements.

For the determination of the outcomes, and subsequently of the impact, financial proxies were used for the monetization of non-market goods, for which it is not possible to obtain an exact measurement and, therefore, not always objectively determinable. Of course, every organization involved in the process of creating and distributing the value (therefore benefiting from the social impact generated) has its own information system, which therefore requires careful reunification work. Not so much for accounting matters as for the measurements made through this, it stimulates the use of dialogue with stakeholders, which seems to be the only way (apart from preventive interventions) to arrive at a shared evaluation, reducing the subjective interpretative keys. Then, a dead-weight analysis was performed for each stakeholder, i.e., the calculation of the impact or outcome with the film event and the value of the outcome without the film event.

Similarly, the weight of other factors or actors in determining the outcome (attribution) and the reduction of the impact, or impact on the impact over time (drop-off), was considered, even if, as aforementioned, for the sake of simplification it was decided to consider an evaluation period of only 1 year (12 months), while also foreseeing transformations for longer periods, but which would be difficult to evaluate objectively. The assessment was very prudent, understood from the fact that significant deadweight and attribution have been included in the calculation. Subsequently, the discount and discount rate were assessed at 1.2%, corresponding to the inflation rate for the year 2017, according to the Italian Ministry of Economy and Finance data.

9 Some Conclusions and Implications for Future Development

The attractiveness of the Taormina Arte Foundation for a large and quality public audience, albeit in a very limited period of the year, has strong impacts on the local economy. The intertwining of architectural heritage, landscape and territory value, quality of the artists involved, and local organizational skills constitutes a model of particular significance, capable of enhancing the cross effects of these factors, enhancing economies of scale and economies of purpose. The path taken by Taormina Arte has contributed to highlighting, in an even clearer way, the creation of value in the socio-cultural as well as tourism-economic sphere, and has made it possible to strengthen ties with the other stakeholders who have been involved on this path.

The cinematographic event, therefore, shows the possibility of continuous growth over time in importance and territorial impact.

In summary, in order to strengthen its image, authoritativeness and reliability toward stakeholders, and to facilitate access to public and private financing channels, the Taormina Arte Foundation must always be able not only to adopt strategies of reporting and communication through which they "account" for the activities and results achieved, but also to equip themselves with adequate systems aimed at measuring and evaluating their performance. The results shown also through the calculation summary tables indicate that the SROI seems, despite its methodological limits and a certain laboriousness, to have balanced different needs such as that of sustainability of the calculation, that of simplicity of method, that of considering data coming from by different information systems, and above all, that of acting to identify the value of the strategic choices made and identify new indications of future intervention, together with that of supporting relevant and clear accountability processes.

The impact assessment carried out with a methodology such as SROI therefore has an impact both on the side of calculation sustainability and on the ability to concretely represent a complex reality derived from constantly changing social and economic situations. The trade-off between precision and reasonableness of the measurement must become a constant in the choice of the evaluation methodology. Differently, in accordance with the epistemological premises made above, the risk is to stiffen and inhibit the use of evaluation to actual users. Furthermore impact evaluation of stakeholders involved and the social value of their investments could open a process of new awareness of the impact produced, emphasizing the relevance of the cultural heritage in encouraging community participation and the fruitful intersection of the different spheres of society. In another direction, this research orientation could certainly constitute a precise line of development for research projects and future developments of the study carried out.

References

Arvidson, M., et al. (2010). *The ambitions and challenges of SROI*. Third Sector Research Center, University of Birmingham.

Arvidson, M., et al. (2013). Valuing the social? The nature and controversies of measuring social return on investment (SROI). *Voluntary Sector Review, 4*(1), 3–18. https://doi.org/10.1332/204080513X661554.

Biondi, L., Demartini, P., Marchegiani, L., & Marchioni, M. (2020). Management, participation and entrepreneurship in the cultural and creative sector: An introduction. In M. Piber (Ed.), *Management, participation and entrepreneurship in the cultural and creative sector* (pp. 1–8). Cham: Springer. https://doi.org/10.1007/978-3-030-46796-8_1.

Bourdieu, P. (1991). *Language and symbolic power*. Cambridge, MA: Harvard University Press.

Bourdieu, P., & Passeron, J. C. (1979). *The inheritors: French students and their relations to culture*. Chicago: University of Chicago Press.

Catalfo, P., & Giustra, M. (2020). The value of the relationship between architecture and social engagement: Imre Makovecz's work within the *Faluházak* project. In M. Piber (Ed.), *Management, participation and entrepreneurship in the cultural and creative sector* (pp. 141–159). Cham: Springer. https://doi.org/10.1007/978-3-030-46796-8_8.

Clark, K. (2000). From regulation to participation: Cultural heritage, sustainable development and citizenship. In *Forward planning: The functions of cultural heritage in a changing Europe*. Papers from an expert workshop on cultural heritage in Europe in 2000 to contribute to the 5th European conference of ministers in Slovenia. Council of Europe, pp. 103–113.

Clark, C., Rosenzweig, W., Long, D., & Olsen, S. (2004). Double bottom line project report: Assessing social impact. In *Double bottom line ventures, methods catalog*. University of California. http://repositories.cdlib.org/crb/wps/13

Colbert, F. (1994). *Marketing culture and the arts*. Montréal-Paris-Casablanca: Gaëtan Morin Éditeur ltée.

Cristofoli, D., Macció, L., & Meneguzzo, M. (2014). When civic culture meets strategy: exploring predictors of citizen engagement in participatory strategic plans in Italy. In P. Joyce & A. Drumaux (Eds.), *Management in public organizations: Europeans practices and perspective*. London: Routledge.

Daum, J. H. (2001). *Intangible assets and value creation*. Chichester: Wiley.

Durkeim, E. (1912). *Les formes élémentaires de la vie religieuse*. Paris: Alcan.

Emerson, J., Wachowicz, J., & Chun, S. (2000). Social return on investment: Exploring aspects of value creations in the non-profit sector. In *Social purpose enterprise and venture philanthropy in the new millennium* (Vol. 2). San Francisco: The Roberts Foundation.

Frey, B., & Pommerehne, W. (1989). *Muses and markets: Explorations in the economics of the arts*. Oxford: Basil Blackwell.
Hopwood, A. (1983). On trying to understand accounting in the context in which it operates, accounting. *Organizations and Society, 8*(2–3), 287–305.
Hopwood, A., & Miller, P. (1994). *Accounting as social and institutional practice*. Cambridge: Cambridge University Press.
Husserl, E. (1912). *Les formes élémentaires de la vie religieuse*. Paris: Alcan.
Husserl, E. (1936). *Die Krisis der europäischen Wissenschaften und die transzendentale Phänomenologie: Eine Einleitung in die phänomenologische Philosophie* (1st Edn.).
Longstaff, P. H., Armstrong, N. J., Perrin, K., Parker, W. M., & Hidek, M. A. (2010). Building resilient communities: A preliminary framework for assessment. *Homeland and Security Affairs, VI*(3), 1–23.
Lorusso, S., Cogo, G. M., & Natali, A. (2016). The protection and valorization of cultural and environmental heritage in the development process of territory. *Conservation Science in Cultural Heritage Journal, 16*. https://doi.org/10.6092/issn.1973-9494/7165.
Lorusso, S., Mari Braida, A., & Natali, A. (2018). Interdisciplinary studies in cultural and environmental heritage: History, protection, valorization, management. *Conservation Science in Cultural Heritage, 18*. https://doi.org/10.6092/issn.1973-9494/9234.
Meynhardt, T. (2009). Public value inside: What is public value creation? *International Journal of Public Administration, 32*(3–4), 192–219. https://doi.org/10.1080/01900690902732632.
Myles, J. F. (2004). From Doxa to experience: Issues in Bourdieu's adoption of Husserlian phenomenology. *Theory, Culture and Society, 21*(2), 91–107. https://doi.org/10.1177/0263276404042136.
Nicholls, J., et al. (2009). *A guide to Social Return on Investment*. London: SROI Network.
Peacock, A., & Rizzo, I. (2008). *The heritage game: Economics, policy, and practice*. Oxford: Oxford University Press.
Perrini, F., & Vurro, C. (2013). *La valutazione degli impatti sociali: Approcci e strumenti applicativi*. EGEA spa—CReSV.
Potter, B. (2005). Accounting as a social and institutional practice: Perspectives to enrich our understanding of accounting change. *Abacus, 41*(3), 265–289.
SROI. (2012). *A guide to Social Return on Investment, HF*. www.humanfoundation.com
UN-ECOSOC. (2008). *Achieving sustainable development and promoting development cooperation*. Dialogues at the Economic and Social Council. United Nations Publications. isbn:978-92-1-104587-1.
Watson, K. J., & Whitley, T. (2016). Applying Social Return on Investment (SROI) to the built environment. *Building Research and Information*, 2–5. https://doi.org/10.1080/09613218.2016.1223486
Watson, K. J., et al. (2016). Capturing the social value of buildings: The promise of social return on investment (SROI). *Building and Environment, 103*, 289–301.
Zamagni, S., et al. (2015). Valutare l'impatto sociale. La questione della misurazione nelle imprese sociali. *Rivista impresa sociale, 6*, 77–97.

Pierluigi Catalfo A professor of accounting and management, director of the Research Centre for Territorial Governance (GOT) at the University of Catania. He is a scientific correspondent for the Italian Social Reporting Research Group (GBS), and member and internal auditor of the New Club of Paris (NCP). He was a scientific consultant for the Department of the Italian Presidency of the Council of Ministers and for the Italian Public Sector development Agency (FORMEZPa).

Martina Giustra An architect since 2007, she carries out research on the thought and oeuvre of the Master Imre Makovecz and on Hungarian organic architecture. She is conducting research activity at the Department of Residential Design at the Faculty of Architecture of BME University and at the Balassi Institute, both in Budapest (2012–2014 and 2016–2017). Currently she works as an architect and is a Ph.D. candidate at Marcell Breuer Doctoral School—PTE University of Pécs.

Agata Cardillo She carries out empirical research and consultancy activities and collaborates with the University of Catania for the application of methodology for assessing socio-economic impacts.

A Struggle of Capitals Over the Identity and the Cultural Offering of Festivaletteratura: The Organizational Impact of Audience Development

Luca Pareschi and Noemi Ponzoni

Abstract In this chapter, we explore the case of Festivaletteratura, the most important Italian literary festival, which in the last few years experienced an internal struggle over its cultural offering and cultural identity. Part of the organizational staff, fearing an ageing of the audience, started an audience development project aimed at using the knowledge acquired as a source of cultural capital. Conversely, those organizational bodies not willing to change the cultural offering of the festival relied on symbolic capitals acquired over the years. We make sense of the moves and countermoves of the agents involved in this symbolic struggle through the concept of institutional work, aimed at maintaining or disrupting existing institutions, and show that the most relevant institution in this case is composed by the programme meetings, where the actual cultural offering of each edition of the festival is defined.

Keywords Organizational identity, Cultural offering, Symbolic capital, Cultural capital, Institutional work, Festival

1 Introduction

Festivaletteratura is the most important Italian literary festival, which takes place in Mantova, a UNESCO World Heritage Site, since 1997. Festivaletteratura was a pioneer for the festivalization phenomenon in Italy and kept growing for more than 20 years. The festival was founded by eight citizens of Mantova, the organizing

---

Authors are listed alphabetically and contributed equally to the paper.

L. Pareschi (✉)
Università di Roma 'Tor Vergata', Roma, Italy
e-mail: luca.pareschi@uniroma2.it

N. Ponzoni
Penguin Random House UK, London, UK

P. Demartini et al. (eds.), *Cultural Initiatives for Sustainable Development*,
Contributions to Management Science,
https://doi.org/10.1007/978-3-030-65687-4_12

committee, which are still the heart and brain of the organization. Yet, in the last years, a new generation of managers within Festivaletteratura's organizational bodies attempted to modify the cultural offering of the festival and, consequently, the central and stable characteristics perceived by the audience, which are the festival's identity (Montanari et al. 2013). Indeed, these managers were concerned that the festival's model, although successful in the past, was becoming outdated, as they noticed that Festivaletteratura's audience was ageing. Nonetheless, Festivaletteratura's entrenched power structure was not easy to modify, and a quiet struggle over the identity and cultural offering of the festival began to take place. We present a case study regarding this clash, and we make sense of the moves and countermoves of the agents involved through the theoretical lens of neo-institutionalism, and in particular through the concept of institutional work aimed at maintaining or disrupting existing institutions.

Creative industries are characterized by the need for cultural managers to balance seemingly conflicting logics (Lampel et al. 2000). These logics are socially constructed symbolic systems that order reality and that act at individual, organizational and field levels (Friedland and Alford 1991), as they produce, legitimize and reproduce a system of material practices, assumptions, values and beliefs (Thornton and Ocasio 1999). For creative and cultural industries, a typical example of conflicting logics is the clash between artistic and economic logics (e.g. Eikhof and Haunschild 2006, 2007). Another conflict, which is often overlooked in academic research, is the attitude towards the audience—which is the balance between demand analysis and market construction. Are cultural goods 'an expression of consumers' needs and desires or is 'what consumers want [...] almost entirely shaped by the imagination and creativity of the producers'? (Lampel et al. 2000: 266). This conflict is relevant for our case study, as the fixed cultural offering of the festival, which is a stable feature of the festival, consisting of frontal presentations, started to be questioned by some people working for Festivaletteratura, who proposed new models for the events, hopefully aimed at meeting the preferences of new audiences. But it is not easy to alter the core of the festival, as it is an entrenched power structure.

Institutional entrepreneurs have interest in particular institutional arrangements, and leverage resources to impose specific logics (Hardy and Maguire 2008). Specifically, they can create, maintain, or disrupt institutions through 'institutional work', which is a purposive, 'intelligent, situated institutional action' (Lawrence and Suddaby 2006). Lawrence and Suddaby categorized several forms of institutional work and described how such practices leverage theoretical tools deriving from the sociology of practice (e.g. Bourdieu 1977; Giddens 1984). Following this insight, we will make sense of the institutional work in our case through the concepts of symbolic capital and cultural capital (Bourdieu 1984).

We rely on a range of data sources: one author worked for the editorial staff of Festivaletteratura for 12 years, while the other worked for 6 months in the international authors' hospitality service. During this time, she had the chance to develop an audience-oriented survey, under the guidance of the administrative office of the festival, while both authors realized several interviews to staff members, before and after the survey, and participated in formal and informal meetings. What we argue is that part of the organizational staff of the festival, which we call the 'innovators', aimed at disrupting a specific institution of the festival, which is the 'programme meetings', 'undermining the assumptions and beliefs' (Lawrence and Suddaby 2006: 237) taken for granted within the festival, which are about drafting the programme following the traditional legitimate authority of the organizing committee. Specifically, they aimed at accumulating cultural capital by leveraging an audience development survey, which served as an external form of knowledge (Rindova et al. 2011), used to alter the power structure. Conversely, the organizing committee tried to maintain a tight grip on the programme through the form of institutional work defined 'mythologizing' (Lawrence and Suddaby 2006: 233), which leverages symbolic capital.

The chapter is structured as follows: in the next section, we review the theoretical background which is needed to delve into our case study. Then, we will provide more details regarding the story and the institutions within Festivaletteratura. After introducing our methods and data, we will proceed to findings. In the discussion we will defend our insights. The chapter ends with conclusions.

2 Theoretical Background

Creative industries are generally defined as those 'industries supplying goods and services that we broadly associate with cultural, artistic, or simply entertainment value' (Caves 2000) and for which aesthetic and symbolic meanings are at the core of value creation [DuGay and Prike 2002; Hesmondalgh and Pratt 2005; see Jones and Townley (2016) for a critical analysis of the evolution of the field]. Typically, the definition of 'creative' or 'cultural' industries comprise architecture and design, film, television, video, radio and publishing, fine arts, music and the performing arts, software and computer, gaming, advertising, and crafts (DCMS 2001); but wider definitions may also include theatre and opera, new media, photography, festivals (Power 2002), fashion, tourism, video games, toys, and perfume (for more comprehensive accounts, see Hesmondhalgh 2002; Peltoniemi 2014). Within this context, festivals are often addressed in reference to the festivalisation phenomenon, which points at a soar in the number of cultural festivals in Europe over the last 30 years (Allen and Shaw 2000). Management literature tackled festivals especially in reference to their impact on the territories and their ability to foster urban regeneration (Palmi 2010; Sassatelli 2011; Parente and El Tarabishy 2020), in reference to value creation and marketing (i.e. Chang 2020), in reference to the underlying cultural values (i.e. Knowles 2020), or to the knowledge sharing and diffusion (i.e. Abfalter et al. 2012). Yet, while festivals can provide relevant insights regarding organizing and change (Toraldo and Islam 2019), with few notable exceptions (i.e. Sicca 1998), less attention was devoted to internal dynamics of festivals as peculiar organizations.

Creative industries are characterized by the need for cultural managers to balance seemingly conflicting logics (Lampel et al. 2000), defined as 'the socially constructed, historical patterns of material practices, assumptions, values, beliefs, and rules by which individuals produce and reproduce their material subsistence, organize time and space, and provide meaning to their social reality' (Thornton and Ocasio 1999: 804). In other words, institutional logics are symbolic systems that make human experience meaningful by ordering reality, and they act at individual, organizational, and field levels (Friedland and Alford 1991). For creative and cultural industries, a typical example of conflicting logics is the conflict between artistic and economic logics (e.g. Eikhof and Haunschild 2006, 2007; Pareschi and Lusiani 2020) that was tackled over the years by many authors, which either described a conflict between both logics (e.g. Bourdieu 1984), or a shift of market logic over aesthetic logic (e.g. Thompson 2010), or even a blend of both logics (e.g. Franssen and Kuipers 2013).

While institutional logics shape the structure of a field and of the organizations within it, and they act at the individual level as schemata of interpretation of reality, they are not unalterable or written in stone. On the contrary, DiMaggio (1988) and Oliver (1991, 1992) brought back into the neo-institutional debate discourses about power and agency, which focus on the impact of agents on the institutions characterizing the field in which they operate. Agents in the field can fulfil institutional work, which is a 'purposive action aimed at creating, maintaining and disrupting institutions' (Lawrence and Suddaby 2006) through an 'intelligent, situated institutional action' (Lawrence and Suddaby 2006). Agents who are able to change the institutions in their field are institutional entrepreneurs, which are 'actors who have an interest in particular institutional arrangements and who leverage resources to create new institutions or to transform existing ones' (Maguire et al. 2004: 657). Lawrence and Suddaby (2006) identified different forms of institutional work and grouped them in three sets: nine forms of institutional work aimed at creating institutions, six forms of institutional work aimed at maintaining institutions, and three forms of work aimed at disrupting institutions. We make sense of our case study relying especially on two forms of institutional work: *Undermining assumptions and beliefs* and *Mythologizing*. The former refers to the efforts aimed at disrupting institutions by undermining theirs core practices, beliefs, and rules. This is effective when it can replace existing templates of action, hence lowering the perceived risk of innovation (Lawrence and Suddaby 2006). The latter preserves institutions by rooting their normative underpinnings in their history (Lawrence and Suddaby 2006).

As the analysis of the different forms of institutional work is an analysis of the practice, which describes the intelligent activities of individuals and organizations who are working to achieve a specific outcome (Lawrence and Suddaby 2006), it is inspired by the sociology of practice (e.g. Bourdieu 1977; Giddens 1984). Therefore, we leverage its theoretical tools, and use the concepts of cultural and symbolic capital (e.g. Bourdieu 1984). Bourdieu's framework is useful to tackle the concept of agency within cultural organisations, as it makes power relations explicit (Jones and Townley 2016) and it emphasizes both the material and symbolic aspects of cultural goods (Townley et al. 2009). Central to his theoretical framework is the concept of field, which is each social microcosm defined by a specific stake, own rules, own forms of authority and legitimating process, and a (at least relative) degree of autonomy (Bourdieu 1977). Power within a field depends on the possession of a capital, which is each property that can be used as a resource in the social struggle (Santoro 2009). Among the five kinds of capital, economic, social, cultural, and symbolic (Bourdieu 1984), we focus on cultural and symbolic ones. Cultural capital is linked to the possession of knowledge and information (Bourdieu 1991a; Townley et al. 2009), and symbolic capital is any kind of capital when it is accepted and recognized as a source of prestige and power (Bourdieu 2009): 'The kinds of capital, like the aces in a game of cards, are powers that define the chances of profit in a particular field' (Bourdieu 2005: 30).

Another case of conflicting logic within cultural industries which tends to be overlooked by research is the stance towards the audience: 'There is a longstanding dispute in the cultural industries between those who see cultural goods as an expression of consumers' needs and desires and those who argue that what consumers want is almost entirely shaped by the imagination and creativity of the producers' (Lampel et al. 2000: 266). As reading the market cannot be decoupled from the process of constructing it (Anand and Peterson 2000), several tools aimed at focusing on the audience have been developed, and one of the most successful ones is audience development (AD). From the mid-1990s onwards, this concept became a priority for cultural organizations (Hayes and Slater 2002), due to a change in cultural policies in the United Kingdom that began connecting funding to the engagement of new audiences (Hutton et al. 2004). For policy makers and practitioners, AD has gained increasing popularity, contextualized in a framework that identifies arts and culture as powerful driving forces both from the social and economic points of view (Hayes and Slater 2002). Yet, the concept has been ambiguously defined in the literature as a technical function of marketing management with a wider social remit (Hadley 2017: 275). A clear definition has been (Maitland 2005) and still is a concern in the field: the glossary in the *Study on Audience Development. How to Place Audiences at the Centre of Cultural Organisations*, commissioned by the European Commission in 2015, lists 10 different definitions of AD, with the broader one focusing on the diversification, inclusion, and improvement of the overall experience:

> Audience development is a strategic, dynamic and interactive process of making the arts widely accessible. It [...] can be understood in various ways, depending on its objectives and target groups: i) increasing audiences (attracting audiences with the same socio-demographic profile as the current audience); ii) deepening relationship with the audiences (enhancing the experience [...] thus fostering loyalty and return visits); iii) diversifying audiences (attracting people with a different socio-demographic profile [...]). (European Commission (2015), EAC/08/2015 Tender specifications)

There is an evolution 'from an almost purely marketing-oriented approach to a more holistic and processual vision' of AD (European Commission 2015: 5), which can now be deemed as an 'ideological project' (Hadley 2017: 277) aimed at creating a bond between cultural organizations and their audiences. These relationships should be grounded in the core values of the organization itself (Harlow 2014), which is their identity.

The audience-centric approach has contributed to the development of studies on the topic of cultural audiences per se (i.e. Hayes and Slater 2002; Kawashima 2006; Peterson and Kern 1996), but it is still characterized mainly by a practitioner-oriented set of best practices (i.e. Barlow and Shibli 2007; Maitland 1997; Scollen 2009; Wiggins 2004) or a focus on understanding the reasons for the low levels of cultural attendance in specific segments of the population (i.e. Desai and Thomas 1998).

Indeed, Pegoraro and Zan (2017) highlight that literature falls short in analysing AD's impact on the trade-offs between stakeholders, professionals, and practitioners in cultural fields, all characterized by different capital endowments. Moreover, an audience-oriented strategy 'is likely to impact the organizational structure of cultural institutions' (*Audience development and marketing*, Arts Council England 2018). Nonetheless, research overlooked the organizational impact of the stance towards the audience. We explore this issue through the analysis of the case of Festivaletteratura, and we follow the call of Thompson et al. (2007), which recommend more empirical studies of management practices and work relations within cultural organizations.

3 Festivaletteratura

The first Festivaletteratura took place in the summer of 1997 amid the squares of Mantova, a town located in north-east Italy with a population of approximately 48,000 people. The city's historical centre was declared a UNESCO World Heritage Site in 2007 due to the presence of Renaissance architecture and art. During the last 20 years, Festivaletteratura's events were held in 160 venues around town, realizing the idea that festivals need to be analysed also in terms of spatial engagement, use, and appropriation of key urban spaces (Bollo 2012).

Festivaletteratura came to life thanks to the efforts of eight Mantuan citizens who designed it following the model of Hay Festival, and inspired by a market research project on potential developments in the cultural sector. This was commissioned by *Regione Lombardia* in 1993 to Comedia, a British research and consultancy firm founded in 1978 by Charles Landry, one of the theorists of the now widespread concept of the *creative city*. A focal point of Comedia's final report was the idea of culture as a resource for reviving the local economy, for leveraging the value of cultural assets, and for innovating the town's identity. The report suggested an entrepreneurial attitude towards culture, shifting it from a 'showcasing of heritage' towards the use of existing resources (i.e. publishing) for the creation of something new (e.g. 'a city of books', as Hay-On-Wye had recently become). Born in 1988, the Hay Festival of Literature & Arts is a well-established international literary festival taking place every year in the small town of Hay-On-Wye (Wales, UK) and, now, in several other cities in the world. Festivaletteratura still emphasizes the enduring relationship with the Hay Festival, based on the ideas of attracting the literature audience through a simple participatory formula, a direct and informal relationship with the authors, a wide programme with overlapping events enabling the audience to curate their visit, and the crucial role of volunteers (Polettini 2012). Peter Florence, Hay Festival's director, remains today a regular guest and a consultant of Festivaletteratura.

In the last 20 years, the organizational body of Festivaletteratura has grown considerably, but maintained its original core, the organizing committee: Laura Baccaglioni (cultural manager), Carla Bernini (bookseller), Annarosa Buttarelli (university professor), Francesco Caprini (architect), Marzia Corraini (publisher and gallery owner), Luca Nicolini (president, bookseller), Paolo Polettini (sociologist), and Gianni Tonelli (accountant). The other main organizational body of the festival is the administrative office, whose staff increases from eight employees throughout the year to approximately 40 during the days of the festival. The administrative office has always managed the practical aspects of the festival (i.e. administration, accounting, and fundraising), but over the years it has also taken over some strategic and substantial aspects that were originally managed by the organizing committee. The staff of Festivaletteratura was, and is to this day, mostly composed of relatives and friends, a factor contributing to create the informal and familiar atmosphere that quickly became a defining trait of Festivaletteratura. This aspect is prized by many, authors included: the festival is a case of 'guest loyalty', meaning that many writers, which do not receive any direct compensation as guests of Festivaletteratura, have taken the event to heart and return, becoming involved in a variety of activities and events. Nonetheless, the close and familiar environment is a possible source of controversies when disputes arise.

Festivaletteratura is also composed by Filofestival, the non-profit auxiliary association that supports and promotes the festival. In 2016, Filofestival had 3400 members, which are the only part of the audience that Festivaletteratura has information on. An archive was instituted in 2010, with the aim of structuring and giving value to the unique materials (videos, photos, texts...) gathered since the first edition of the festival. The archive is a precious resource counting up to 140,000 photos, 3400 audio recordings, and about 2300 videos. Currently, three people work on the archives. The press office of the festival is outsourced, but contents are not produced by this agency: during the year contents are produced by the administrative office, while during the festival this massive task passes in the hands of the volunteer editorial staff, which produces video, photos, texts, and a podcast. Volunteers are indeed fundamental for Festivaletteratura, and since 1997 more than 5000 have taken part in the festival (750 just in 2016), taking care of box office, logistics, info points, venue management, and communications. Finally, the festival also relies on a small but loyal network of consultants, with different skills and competences (programme building, relation with publishers, scientific publishers, etc.). While the first Festivaletteratura obtained 62% of its funding from public bodies, in 2016, thanks to a precise strategy implemented over the years, the situation reversed, with 72% of the support that came from more than 150 private sponsors, without any direct contributions from publishers, and only 10% of funding from public bodies.

Table 1 Growth of Festivaletteratura: 1997–2016

	1997	2016
Events	106	404
Guests	140	420
Tickets sold	12,800	70,000
Estimated audience at free events	2000	65,000
Total audience	14,800	135,000
Publishers	14	63
Journalists	130	317
Filofestival members	300	3400
Office (all year long)	3	8
Office (during the festival)	20	40
Press office + editorial staff	2 + 10 volunteers	4 + 120 volunteers
Total volunteers	130	750

Adapted from internal documents of Festivaletteratura

In 1997, the first Festivaletteratura consisted of 106 events with 140 authors and attracted 15,000 attendees, and has quickly grown since then. Festivaletteratura had a pioneering role in the *festivalization* phenomenon in the Italian cultural landscape and is now one of the longest running Italian festivals and was taken as a benchmark by newer events: the most up-to-date study listed 927 festivals in Italy (Guerzoni 2012), signalling how crowded the field has become in the last 20 years. Today, most festivals are geographically situated in the north (79%), in a single town (78%), last between 3 and 5 days (78%), and 32% take place in September, just like Festivaletteratura.

The 20th edition Festivaletteratura took place between the 7th and 11th of September 2016, at a time when Mantova was the Italian Capital of Culture. During these 5 days, 404 events were held (113 of which were free). Table 1 shows how, throughout the last 20 years, the festival has grown to be the largest literary festival in Italy. Nevertheless, data seem to confirm the fears expressed by members of the organization. For example, paid events increased 12% from 2015 to 2016, but the number of tickets sold increased by only 4%. In parallel, free events increased by 19%, but the estimated audience of the free events only grew by 12%.

Festivaletteratura grants access to a large variety of offerings, some of which would be difficult for the audience to find in other circumstances (and some of which possibly would not find an audience elsewhere). The programme of the festival combines 'blockbuster events' in the main square in town, which often have a full house and therefore generate income, with niche events that attract specific clusters of the audience that would otherwise not be interested in attending a literary festival.

The cultural offering of Festivaletteratura is defined each year through the programme meetings, which take place monthly from November to February and then twice a month until May, when the programme for the upcoming edition is almost finalized. In the programme meetings all the organizational bodies are represented, and all the participant can propose new events or new formats which are submitted to a vote. Programme meetings are the forum where Festivaletteratura's cultural offering and identity are discussed, and where the evolution of the festival is defined.

4 Methods and Data

We developed a qualitative case study (Langley 1999; Yin 2014) based on an evolutionary (Abdallah et al. 2019) and inductive approach (Strauss and Corbin 1990). Before describing the data collection process, we must devote some words to the authors' different expertise regarding Festivaletteratura and to their relationship with the festival. One of the authors worked for 12 years as a volunteer in Festivaletteratura's editorial staff, writing articles on the events, conducting interviews, and also working as a photographer. He became acquainted with Festivaletteratura's staff and had several formal and informal discussions on the festival's issues. The other author carried out a 6-month internship within the administrative office in 2016, in the international authors' hospitality service, before and during the 20th edition of the festival. The relationship between the authors and the organization, while providing a valuable source of data and granting an inner knowledge of the festival, deserves some critical reflections.

We think that none of these roles can entitle the authors as 'intern' to the festival, as both positions were carried out for a limited time (few days a year over several years, and few months of one year), and within organizational structures that are peripheral to the core of the organization, which selects the cultural offer. Being not alien to the festival's structure enabled us to interpret our data grounded on our knowledge of the organization. As the focus of the chapter is the organization, we discarded self-reflexive methods, such as auto-ethnography, or self-report, as they focus on, or at least devote reflections to, personal feelings and experiences of the researchers, while we focus on the festival's organizational structure. Also, we wondered if our research can be defined as a sort of action research, but it never falls within the boundary of this practice, as we were never asked by the organization itself to conduct a research aimed at changing Festivaletteratura. One author collected data for her degree's thesis, facilitated by the administrative office, and the research had an impact on the organizational structure, as we detail in the remainder of the chapter.

We moved back and forth from data collection, analysis, theory development, and discussions with Festivaletteratura's staff. For 3 years, we collected data and analysed them following established coding techniques (Miles and Huberman 1984; Saldaña 2016), while at the same time making sense of our findings. A specific complexity of our research, indeed, emerged as analysis and practice became intertwined.

We can divide our data collection and analysis process in three time frames: the first time frame happened in the months before the 20th edition of the festival, when one author did an internship at Festivaletteratura. She had the opportunity to participate in formal and informal meetings, to observe the development phases of the festival and to collect field notes. She also conducted eight semi-structured interviews with three members of the organizing committee, four members of the administrative office (ranging from 6 to 20 years of seniority), and a programme consultant. Interviews were transcribed verbatim and coded (Silverman 1993). Several interviewees revealed a lack of knowledge of the festival's audience, and expressed a concern about a forthcoming generational turnover and a perceived lack of young audiences. Thus, supported by the administrative office, a survey was designed to undertake an audience analysis that would provide the organization with information on demographics, cultural habits, and satisfaction levels of the audience members. In designing the survey, we examined the format of several surveys recently carried out by similar cultural organizations.[1]

The second time frame included the administration of the survey, which took place just before and during the 20th edition of Festivaletteratura: 212 answers were collected online (the link to the survey was added to the on-line ticket booking service), 229 face-to-face, during the festival, to randomly selected people queuing for tickets, and 130 were collected via a link published on Festivaletteratura's Facebook profile. The results of the survey triggered the third phase.

In the final time frame, the authors presented results and analysis in a plenary meeting with Festivaletteratura's staff in January 2018. The results of the survey are not the focus of our research, but they were relevant for the administrative office, which organized the meeting. What was relevant for us was collecting data on the organizational impact of the survey: during the plenary meeting, both authors had the chance to informally discuss with Festivaletteratura's staff the organizational structure of the festival and its relationship with the audience. Finally, in the following months, five more semi-structured interviews were conducted, coded, and analysed to examine the impact of this process on the festival. Specifically, we focused on the strategic moves and countermoves made by different members of the staff, especially leveraging the knowledge acquired. The presentation of the findings mirrors this timeline.

5 Findings

5.1 Internship and First Round of Interviews

In the first phase, interviews focused on Festivaletteratura's identity and cultural offering. A first theme that emerged regards indeed how the success of the festival is grounded in its traditional offering, with authenticity and loyalty to the original

[1]More details on the questionnaire, the items and the procedure followed for the survey are not included in the paper due to space limits, and as they are not the focus of the work. Yet, they are available under request.

formula deemed as central and stable characteristics responsible for success: 'The festival has been growing without ever betraying or losing its true nature, the original spirit has "stuck"' (José, AO[2]). 'Independence and authenticity have always been crucial and have been key in building [the festival's] recognition and appreciation by the audience' (David, AO). However, this faithfulness to the tradition was expressed by members of the administrative office in terms of 'spirit', and 'identity', and not in terms of the actual formula (e.g. type of events). Another trait that was valued was the effort to develop a broad festival programme based on research and quality: '[Over the years] we felt that the festival could increasingly be a festival of discovery, defined by serious research towards less familiar themes, authors belonging to niche literatures, and small and medium-sized publishers' (Jonathan, OC). 'The programme has as its prime objective the search for quality and depth' (Toni, AO).

While pointing at the reasons for success, the interviews realized during the first round also highlighted some concerns about Festivaletteratura's future, which were stronger among younger members of the staff, mostly pertaining to the administrative office: the most mentioned word[3] in the interviews was 'audience', followed by other words pointing at an issue connected to the age of the audience ('youth', 'engagement', 'young', 'fear', and 'aged'.). Other often-mentioned words referred to the actual cultural offering of the festival, as expressed through events and the programme ('programme', 'communication', 'events', 'projects', 'authors', 'rethink', and 'volunteers'). Several members of the administrative office reported, as their main concern, a perceived ageing of the audience, which was only 'perceived' because the festival never attempted at analysing its audience. 'Our audience has been growing with us, ageing with us' (Doris, AO). 'The audience has been getting old. I don't know which paths are best to undertake, but we need to be open-minded about what can attract the attention of a younger audience' (David, AO). 'Compared to other festivals, I perceive an ageing of the audience.' (José, AO). Some of the interviewees indicated, 'People do not choose the festival for its image or brand; they choose it because they are attached and loyal to it. But as we go on, this becomes more and more difficult to sustain' (Dario, OC). Concerns on the audience reverberated also on concerns on the position of Festivaletteratura in the field of festivals: 'The festival has been successful because it was the first of its kind; it managed to build its own position, to intercept a need and satisfy it. Now there are many more similar events' (José, AO). For example, Doris (AO) remarked that even if, on the one hand, the bargaining power of Festivaletteratura has increased, on the other hand, competition has made it more difficult to obtain private funding, as local enterprises split their budgets across a variety of initiatives. When the sector market was getting crowded, 'We had to understand what our unique feature was [...], and even today, we cannot afford to sit back and relax', added Jonathan (OC).

---

[2]Interviews were anonymized, but we indicate the organizational role of the interviewee: AO is administrative office, OC is organizing committee, and C means consultant.

[3]We used Nvivo to perform a wordcount.

While this perceived lack of young audience apparently did not concern the members of the organizing committee, it pushed some of the younger associates to advocate for an evolution of the festival: 'In reference to the cultural systems of production, which evolved in the last 20 years, we also need to update our formula.' (David, AO). After this first round of interviews, an audience-focused survey was developed with the support of (but not asked by) the administrative office, which facilitated the administering of the survey during the 20th edition of Festivaletteratura.

5.2 The Survey

While we focus on the organizational impact of the survey, to describe the processes that it triggered we must first provide some information on the results of the survey, which was the first attempt ever by Festivaletteratura to profile and understand its attendees.

Once the survey was approved by the administrative office, the aim became to collect as many questionnaires as possible. Therefore data was collected (1) online during the purchase of tickets (respondent group A); (2) face-to-face with randomly selected audience queuing for the tickets during the festival (B); and (3) online through Festivaletteratura's Facebook page (C). Demographic data shows that the festival's audience is mostly composed of people belonging to a small and very specific niche: women with a very high level of education aged between 36 and 64. Lower education levels were found in group (C), which was mostly composed of volunteers, some of whom were still students, but the percentage of respondents holding a bachelor's degree or higher was 62.7% for group (A). Overall, almost 4% of the respondents held a Ph.D., and 52% had graduated from university. In Italy, only 15.3% of the men and 19.8% of the women have a university degree (Istat 2016).

In terms of reading habits, while in Italy non-readers are the 59.5% of the population (Istat 2017), at the festival this percentage decreases to 0.5%. Moreover, 43.5% of the festival audience not only reads but also reads more than 12 books a year (only 5.7% of Italians do so). A section of the survey was aimed at understanding the consumption habits of the audience, and their satisfaction. More than 8 out of 10 visitors had attended Festivaletteratura in previous years, and many attendees do not come to the festival for only one event but cherish the experience and are willing to spend more time (and money) at Festivaletteratura. Most of the respondents in category (A) had participated in more than eight events throughout the festival. In general, two prevalent audience groups appeared: one group is composed of people who had attend two to four events and another group of people who had attended more than eight events. The survey also included a series of rating scales regarding atmosphere, programme quality, organization, etc., and two open questions to investigate the festival's strongest assets. All the items directly managed by the festival scored very high in terms of level of satisfaction, with some concerns related to the box office only. The open question inquiring what audiences liked most about the festival confirmed the importance of the 'atmosphere' aspect, which was openly mentioned by almost half of the respondents. Other topics that received a fairly significant number of mentions were the relationship with the city, the quality of the festival offerings (including the discovery of new authors or of niche topics), and the contact with the authors. These are undoubtedly some of the core aspects of Festivaletteratura that are in line with the vision and the values of the festival itself. On the other hand, these results confirmed some of the concerns emerged in the interviews, as the picture of the typical attendee is someone returning to Mantova year after year, perfectly aware of what the festival is and is not, eager to attend as many events as possible, loyal and satisfied. At the same time, the lack of young first-time attendees, maybe just attending one event, is striking.

5.3 *Return of the Results and Second Round of Interviews*

On January 26th 2018, we presented the survey results in the Palazzo Andreani, the historical building that hosts the headquarters of Festivaletteratura. The organizational impact of the survey was discussed in a plenary meeting comprising members of the organizing committee and of the administrative office as well as consultants and archivists. The presentation led to a collective discussion, followed by informal conversations with the staff. In the following months, we conducted five more interviews. In this paragraph, we present the most important organizational issues that emerged in the third step: the difference between innovators and conservatives, the Festivaletteratura's identity, and the different takes regarding the programme meetings. In the next paragraph, we will discuss our interpretation of the institutional work that was performed by different members of Festivaletteratura's staff, leveraging the survey. Also, we will discuss the organizational changes that emerged because of this institutional work.

5.3.1 Innovators Versus Conservatives

Festivaletteratura's staff is split between innovators, who want the festival to evolve and attract new audiences, and a more conservative group that wants to remain loyal to the traditionally successful formula of Festivaletteratura. 'We, and by "we" I mean part of the administrative office, but that's not a completely shared belief, feel the need to question who the Festivaletteratura's audience is. Additionally, we feel the urge to open to new ways of getting in touch with a younger audience. However, this is not a concern for some of the people in this organization' (David, AO). 'The ageing of the audience is more of a concern for the new generation of staff. There are discrepancies in how to address this issue' (Nadine, AO). However, the distinction is not clearcut: innovators do not coincide with the administrative office staff, and conservatives do not correspond with the organizing committee. Additionally, as the staff is composed of relatives, friends and people who have worked together for over 20 years, no one genuinely wants to jeopardize the atmosphere of such a close-knit workplace.

5.3.2 Festivaletteratura's Identity and the Importance of People

The struggles over the audience hide struggles over the identity and the cultural offering of Festivaletteratura: according to someone within the new generation of organizers, 'Maybe it's time for Festivaletteratura to start producing cultural content, instead of just intercepting and re-launching books just because they were published recently. Maybe it's time to change the perspective, so that the festival becomes a place where culture is produced' (Ernest, AO). Indeed, it seems that the festival is just reproducing the same formula of success over and over, as there is no long-term plan for the evolution of the festival: 'We should ask ourselves what the festival will be in five years, but we have never really done that' (David, AO). 'There is no debate and no idea about what the festival will be in ten years.' (Ernest, AO). The lack of such a plan depends on the actual overlap between some members of the organizing committee and the identity of the festival itself, as 'people count more than roles and structures' (Nadine, AO); indeed, 'the identity of the festival cannot be decoupled from the people who created it twenty years ago and since then made it a success' (David, AO). The members of the organizing committee accumulated symbolic capital over the years, and this capital provides a legitimation that makes it difficult to question a structure so tightly linked to the people who developed it. People pertaining to the organizing committee are not formally denying an evolution, but it is very hard for the other members of the staff to lift themselves to a level that allows them to successfully propose new formulas and new formats of cultural offering. This legitimation gap was so high that 'there is nothing, at the moment, that ensures the survival of the festival after the physiological end of the organizing committee' (David, AO). To understand this point, we must directly tackle programme meetings.

5.3.3 The Programme Meetings

The organizing committee's symbolic capital is the dominant capital in programme meetings, where the festival's programme is defined. While these are plenary meetings where supposedly all the participants have the right to speak and propose new events and formats, they are described as an institution which is not really fulfilling its presumed scope: '[The programme meetings] are a festival ritual that must be changed but cannot be changed until [the festival's] structure is changed. Traditionally, some people speak, others do not. There is no real openness to change' (Ernest, AO). The different participants to these meetings have different powers, thanks to their capital endowments, and some participants have a sort of informal veto power (José, AO): 'These meetings are several things at a time: arenas of power, spaces for discussion where ideas can emerge as successful or not. The best ideas are not always approved. These meetings are useful for informing everyone on what's happening, but sometimes, as there is not a clear vote on the ideas, we struggle to understand whether something has been approved or not' (José, AO). Indeed, 'decisions are taken outside, and before' (David, AO). Some innovators even think that these meetings prevent change: even though 'the structure is formally not hierarchical' (José, AO), the organizing committee has informal but strong decision-making power, clearly expressed within a supposedly democratic meeting aimed at defining the programme. This power is based on symbolic capital gathered through the years, leveraging the success of a formula that was proposed by the members of the committee themselves. So, as explicit channels for innovation exist but are not working, the same existence of the programme meeting, as they are structured now, can prevent change: the power structure is almost impossible to challenge or bypass for innovators, unless they are able to change the rules of the game.

6 Discussion

Programme meetings are fundamental practices for the definition of Festivaletteratura's identity and cultural offering. As such, they are a relevant institution for Festivaletteratura, as programme meetings entail 'rules and shared meanings [...] that define social relationships, help define who occupies what position in those relationships, and guide interaction by giving actors cognitive frames or sets of meanings to interpret the behaviour of others', which is the definition of 'institution' provided by Fligstein (2001: 108). Also, program meetings meet the definition of institution provided by those who emphasize the role of agency for conceptualizing institutions: 'humanly devised constraint that structure political, economic, and social interaction. They consist of both informal (sanctions, taboos, customs, traditions and codes of conduct), and formal rules (constitutions, laws, and property rights)' (North 1990: 97). While institutions are by definition associated with mechanisms of self-reproduction, these mechanisms are almost never powerful enough as to avoid the need for maintenance institutional work (Lawrence and Suddaby 2006). Throughout the years, the organizing committee that created the festival accumulated a vast amount of symbolic capital that, before the process described, endowed its members with the legitimation needed to impose their vision of Festivaletteratura as the legitimate one. Thanks to this symbolic capital, they were able to reproduce existing norms and beliefs through an institutional work of 'mythologizing' (Lawrence and Suddaby 2006). The organizing committee used to define how decisions are made, decide which events can be part of the festival or not, and if/how the festival evolves. Thanks to its symbolic capital, the organizing committee had 'a power of constructing reality' (Bourdieu 2009: 166), as 'in the struggle to impose legitimate visions, agents possess power in proportion to their symbolic capital' (Bourdieu 1991b: 106). Indeed, when someone openly challenged this symbolic capital, trying to introduce innovation by questioning the effectiveness of the traditional formula in a programme meeting, the attempt not only failed but risked deteriorating interpersonal relationships.[4] This institutional work, aimed at keeping things as they used to be, roots its legitimation in the past, and in the history of Festivaletteratura, rather than in the present, or in the future: the key way in which the organizing committee works to preserve the normative underpinning of the institution that leads to the development of the programme is by 'mythologizing their history' (Lawrence and Suddaby 2006: 233).

This situation left scarce room for those among the staff who wanted to introduce changes. These members of the staff are the 'actors whose interests are not served by existing institutional arrangements, and who will consequently work when possible to disrupt the existing set of institutions' (Lawrence and Suddaby 2006: 235). Instead of trying a frontal attack to the entrenched power structure, innovators leveraged an opportunity: an audience development process. Yet, the focus on the audience was not only designed to 'develop relationships with new and existing audiences' (*Audience development and marketing* 2018: 3), but was strategically used as a method to alter internal power dynamics. While accumulating knowledge useful for the festival, indeed, these institutional entrepreneurs worked to acquire cultural capital to contrast the existing power structure. In particular, AD was chosen from among the cultural resources available outside of the industry register, as these cultural resources can be used to obtain strategic change and gain a better strategic position (Rindova et al. 2011). As Bourdieu argues, the differential endowments of capitals are the basis for struggles in institutional fields (1993). In the absence of available processes for renewing the festival's formula and stuck by the non by-passable legitimation of the organizing committee, the innovators had to find a creative path for changing the rules of the game and disrupt the programme meeting institution. They did so by 'undermining the core assumption and beliefs' that traditionally lead the decision-making process. Instead of sticking to the taken-for-granted process of tradition as a legitimation, they offered a new way to guide the definition of new events and formats, rooted in the understanding of the audience. They performed thus an institutional work aimed at 'facilitating new ways of acting that replace the existing templates, or decreasing the perceived risk of innovation and differentiation' (Lawrence and Suddaby 2006: 237). By doing so the innovators aimed at changing the identity of Festivaletteratura itself. Being able to manage the audience development process endowed innovators with cultural capital, which they seek to transform into symbolic capital, as the audience development process is recognized as legitimate within Festivaletteratura. Audience development, in this case, is not used as a marketing technique; on the contrary, this supposedly neutral set of techniques is a picklock used to gently change the power structures, avoiding a frontal attack. Indeed interviews conducted in the third phase confirmed that innovators did not really need the survey to learn something new, but to use it as a strategic tool: 'As far as I can remember, several issues that emerged from the survey are in line with evaluations that we qualitatively did' (David, AO). 'There was

[4]Interview with Ernest.

nothing astonishing in the survey: it confirmed our non-systematic and non-quantified intuitions' (José, AO). 'We were not shocked by data, we had confirmations, especially about loyalty and demographics' (Nadine, AO). The process that we described has already produced some results and started disrupting the institution of programme meetings: 'We knew about the audience demographic composition, but it was not perceived as a problem before the discussion [following the survey return] started. And it was really useful to talk about it' (Ernest, AO). '[The] study is recognized as important and is often mentioned in our meetings and informal conversations. It should be repeated more often' (Nadine, AO).

Innovators started changing Festivaletteratura's cultural offering and identity by widening the boundaries of the festival to include spin off projects. Some of these projects take place during Festivaletteratura, while others in the rest of the year, targeting new audiences. Their success can endow innovators with more symbolic capital modifying the power structure within the festival.

An example is the European Project *Read On*, which involves teenagers in co-designing new tools for literature creation and fruition, while *Anthology Lab* invites teenagers to devise their own anthology of novels targeted at their peers. The idea behind targeting teenagers is that 'we should have a long-term project: young people are aware of the festival, but it might be hard [for them] to get close' (Toni, AO). Other new projects targeted at young audiences include *Meglio di un romanzo* and *Prototipi*. The former is an open call for papers for people between 15 and 30 years old to submit a narrative journalism project, while the latter involves young professionals devising projects for new forms of writing and cultural transmission. In this case, a long-term plan related to education and the production of culture seems to be central as well: 'We should believe more in the festival's projects, and in what our core will be in the future. It would be particularly important to develop a vocation towards education, as with [these] projects, Festivaletteratura would attract a younger audience more successfully' (David, AO).

A really relevant new project introduced within Festivaletteratura's programme is *Scienceground*, as it reverses several of the core assumptions of the events traditionally organized within the festival. *Scienceground* is organized by an evolving community composed mostly of university students and researchers, and is aimed at communicating science, and science-related cultural productions. While *Scienceground* organizes events, these are often built with a significant involvement of the audience. Also, the most striking innovation is that *Scienceground* continuously exists for the whole duration of Festivaletteratura: a venue is devoted to the interaction of the audience with the community behind *Scienceground*. This interaction is especially relevant when there are no events, and when the researchers can interact with the audience. As a matter of fact, instead of mono-directional communication, *Scienceground* is a bi-directional participatory initiative, where culture is created together with the audience of the festival. The first edition of *Scienceground* was in 2018. Since then, this part of Festivaletteratura evolved, organizing events in other cities during the rest of the year, and remaining part of Festivaletteratura each September.

7 Conclusion

Festivaletteratura's case shows how a past success formula can be a blessing and a curse. The original format has kept traditional audiences loyal and it has crystallized a power structure that determined Festivaletteratura's success. At the same time, the faithfulness to the original spirit has prevented the festival from attracting younger audiences, thus endangering its survival, and has prevented the natural evolution of the festival. The entrenched power structure reproduces itself, and innovators lack ways to overtly challenge it.

Within Festivaletteratura's staff, no one wants to force fast and abrupt change: staff members know that 'there are different sensitivities among people working for Festivaletteratura [. . .]. Everyone wants the best for the festival, but a clear and shared definition of Festivaletteratura's mission and boundaries is needed' (David, AO). For these reasons, innovators have conducted an institutional work aimed at undermining values and beliefs that underpin the programme meetings. They leveraged audience research as a tool to accumulate cultural and symbolic capital, which has served to gently and smoothly start a process dismantling the existing traditional power structure, thus forcing a gradual change in the festival. This transformation started by widening the festival's boundaries to new events with specific younger target audiences. Hopefully for the innovators, the transformation will continue by integrating these events in the core of the festival. Conversely, the organizing committee worked to maintain the current structure of the institutions by mythologizing the successes of the past. In conclusion, we contribute to the creative industries literature that focuses on the balance between demand analysis and market construction (Lampel et al. 2000) and extend the understanding of institutional work within a cultural organization by analysing the organizational impact of a process of audience development, a concept that was especially successful among practitioners. We described the strategic use of such process, which leveraged audience development as an external cultural resource (Rindova et al. 2011) used to alter internal power dynamics and convert cultural capital into symbolic capital (Bourdieu 1991a, b). Through the use of audience development as a strategic tool to gain knowledge, we connect to the general theme of this book, as innovators within Festivaletteratura had to accumulate cultural capital by relying on a new source of knowledge. The characteristics of this new knowledge enabled them to convert cultural capital into symbolic capital. A consequence of this transformation was the implementation of new initiatives within and beyond the traditional cultural programme of Festivaletteratura, paving the way for an evolution of the festival in terms of innovative models for cultural festivals. Although these recently introduced events are successful, and innovators seem able to change Festivaletteratura's identity and cultural offering, a follow-up of this study in few years will be much needed to understand whether the process that we are describing will keep its promises.

References

Abdallah, C., Lusiani, M., & Langley, A. (2019). Performing process research. In B. Boyd, T. Russell Crook, J. K. Lê, & A. D. Smith (Eds.), *Standing on the shoulders of giants: Traditions and innovations in research methodology* (pp. 91–113). Bingley: Emerald.

Abfalter, D., Stadler, R., & Müller, J. (2012). The organization of knowledge sharing at the Colorado Music Festival. *International Journal of Arts Management, 14*(3), 4–15.

Allen, K., & Shaw, P. (2000). *Festivals mean business. The shape of arts festivals in the UK.* London: British Arts Festivals Association.

Anand, N., & Peterson, R. A. (2000). When market information constitutes fields: Sensemaking of markets in the commercial music industry. *Organisation Science, 11*, 270–284.

Audience development and marketing, Grants for the Arts. (2018). Arts Council England. Retrieved from https://www.artscouncil.org.uk/sites/default/files/download-file/Information_sheets_Audience_development_marketing_Project_grants_170518.pdf on 2018/10/31

Barlow, M., & Shibli, S. (2007). Audience development in the arts: A case study of chamber music. *Managing Leisure, 12*, 102–119.

Bollo, A. (2012). *Il marketing della cultura.* Roma: Carocci.

Bourdieu, P. (1977). *Outline of a theory of practice.* Cambridge: Cambridge University Press.

Bourdieu, P. (1984). *Distinction: A social critique of the judgement of taste.* London: Routledge.

Bourdieu, P. (1991a). Le champ littéraire. *actes de la recherche en sciences sociales, 89*(1), 3–46.

Bourdieu, P. (1991b). *Language and symbolic power.* Cambridge: Harvard University Press.

Bourdieu, P. (1993). *Sociology in question.* London: Sage.

Bourdieu, P. (2005). The political field, the social science field and the journalistic field. In R. Benson & E. Neeveu (Eds.), *Bourdieu and the journalistic field* (pp. 29–47). Cambridge: Polity Press.

Bourdieu, P. (2009). *Ragioni Pratiche.* Bologna: Il Mulino.

Caves, R. E. (2000). *Creative industries: Contracts between art and commerce.* Cambridge: Harvard University Press.

Chang, Y.-C. (2020). Creating value through the performing arts festival: The multi-stakeholder approach. *Journal of Macromarketing, 40*(2), 185–200.

DCMS. (2001). *Creative industries mapping document 2001.* UK Department for Culture, Media and Sport.

Desai, P., & Thomas, A. (1998). *Cultural diversity: Attitudes of ethnic minority populations towards museums and galleries.* BMRB/Museums and Galleries Commission.

DiMaggio, P. J. (1988). Interest and agency in institutional theory. In L. G. Zucker (Ed.), *Institutional patterns and organizations: Culture and environment* (pp. 3–22). Cambridge: Ballinger.

Du Gay, P., & Pryke, M. (2002). *Cultural economy: Cultural analysis and commercial life.* London: Sage.

Eikhof, D. R., & Haunschild, A. (2006). Lifestyle meets market: Bohemian entrepreneurs in creative industries. *Creativity and Innovation Management, 15*, 234–241.

Eikhof, D. R., & Haunschild, A. (2007). For art's sake! Artistic and economic logics in creative production. *Journal of Organisational Behavior, 28*, 523–538.

European Commission. (2015). Glossary to *Study on Audience Development—How to place audiences at the centre of cultural organizations.* Directorate-General for Education and Culture Creative Europe programme. Retrieved October 31, 2018, from https://ec.europa.eu/programmes/creative-europe/sites/creative-europe/files/documents/news/audience-development-study-glossary_en.pdf

Fligstein, N. (2001). Social skill and the theory of fields. *Sociological Theory, 19*, 105–125.

Franssen, T., & Kuipers, G. (2013). Coping with uncertainty, abundance and strife: Decision-making processes of Dutch acquisition editors in the global market for translations. *Poetics, 41*(1), 48–74.

Friedland, R., & Alford, R. R. (1991). Bringing society back in: Symbols, practices, and institutional contradictions. In W. W. Powell & P. J. DiMaggio (Eds.), *The new institutionalism in organizational analysis* (pp. 232–266). Chicago: University of Chicago Press.
Giddens, A. (1984). *The constitution of society: Outline of a theory of structuration*. Cambridge: Polity Press.
Guerzoni, G. (2012). *Effettofestival 2012*. Firenze: Fondazione Florens.
Hadley, S. (2017). European Commission final report: Study on audience development—How to place audiences at the Centre of cultural organisations. *Cultural Trends, 26*(3), 275–278.
Hardy, C., & Maguire, S. (2008). Institutional entrepreneurship. In R. Greenwood, C. Oliver, K. Sahlin, & R. Suddaby (Eds.), *The SAGE handbook of organizational institutionalism* (pp. 198–217). London: Sage.
Harlow, B. (2014). *The road to results. Effective practices for building arts audiences*. New York: Wallace Foundation.
Hayes, D., & Slater, A. (2002). Rethinking the 'missionary position'—The quest for sustainable audience development strategies. *Managing Leisure, 7*(1), 1–17.
Hesmondhalgh, D. (2002). *The cultural industries*. London: Sage.
Hesmondhalgh, D., & Pratt, A. C. (2005). Cultural industries and cultural policy. *International Journal of Cultural Policy, 11*, 1–13.
Hutton, L., Bridgwood, A., & Dust, K. (2004). Research at Arts Council England. *Cultural Trends, 13*(4), 41–64.
Istat. (2016). *Italia in cifre*. Retrieved October 31, 2018, from https://www.istat.it/it/archivio/194899
Istat. (2017). *La produzione e la lettura di libri in Italia*. Retrieved October 31, 2018, from https://www4.istat.it/it/archivio/207939
Jones, C., & Townley, B. (2016). Misfits, mavericks and mainstreams: Drivers of innovation in the creative industries. *Organisation Studies, 37*(6), 751–768.
Kawashima, N. (2006). Audience development and social inclusion in Britain. *International Journal of Cultural Policy, 12*(1), 55–72.
Knowles, R. (2020). *Canada comes to CARIFESTA: The Caribbean International Festival of the Arts, Trinidad and Tobago, 2019 August.*
Lampel, J., Lant, T., & Shamsie, J. (2000). Balancing act: Learning from organizing practices in cultural industries. *Organisation Science, 11*, 263–269.
Langley, A. (1999). Strategies for theorizing from process data. *Academy of Management Review, 24*, 691–710.
Lawrence, T., & Suddaby, R. (2006). Institutions and institutional work. In S. R. Clegg, C. Hardy, T. B. Lawrence, & W. R. Nord (Eds.), *The SAGE handbook of organization studies* (pp. 215–254). London: Sage.
Maguire, S., Hardy, C., & Lawrence, T. B. (2004). Institutional entrepreneurship in emerging fields: HIV/AIDS treatment advocacy in Canada. *Academy of Management Journal, 47*, 657–679.
Maitland, H. (1997). *A guide to audience development*. London: Arts Council of England.
Maitland, H. (2005). Researching audience development: Literature review. *Arts Research Digest, 31*.
Miles, M. B., & Huberman, A. M. (1984). Drawing valid meaning from qualitative data: Towards a shared craft. *Educational Researcher, 13*, 20–30.
Montanari, F., Scapolan, A., & Codeluppi, E. (2013). Identity and social media in an art festival. In *Tourism social media: Transformations in identity, community and culture* (Tourism social science series) (Vol. 18, pp. 207–225). Bingley: Emerald Group.
North, D. C. (1990). *Institutions, institutional change and economic performance*. Cambridge: Cambridge University Press.
Oliver, C. (1991). Strategic responses to institutional processes. *Academy of Management Review, 16*, 145–179.
Oliver, C. (1992). The antecedents of deinstitutionalization. *Organization Studies, 13*, 563–588.

Palmi, P. (2010). L'identificazione del prodotto culturale con il territorio: il festival. In P. Palmi (Ed.), *La Governance dello spettacolo dal vivo. Evoluzione manageriale e competitività: tre sfide organizzative* (pp. 161–201). Bari: Cacucci.

Parente, R., & El Tarabishy, A. (2020). Giffoni film festival: How to morph a rural area in an innovation cluster in the digital arts production. In *SAGE business cases*. London: Sage. https://doi.org/10.4135/9781529722567.

Pareschi, L., & Lusiani, M. (2020). What editors talk about when they talk about editors? A public discourse analysis of market and aesthetic logics. *Poetics, 81*, 101444. https://doi.org/10.1016/j.poetic.2020.101444.

Pegoraro, M., & Zan, L. (2017). Life and death in audience development. The exhibition on Pompeii at the British museum 2013. *Museum Management and Curatorship, 32*(3), 210–231.

Peltoniemi, M. (2014). Cultural industries: Product-market characteristics, management challenges and industry dynamics. *International Journal of Management Reviews, 17*, 41–68.

Peterson, R., & Kern, R. (1996). Changing highbrow taste: From snob to omnivore. *American Sociological Review, 61*(5), 900–907.

Polettini, P. (2012). *Il sistema cultura. In Storia di Mantova. Tra presente e futuro*. Mantova: Tre Lune edizioni.

Power, D. (2002). Cultural industries in Sweden: An assessment of their place in the Swedish economy. *Economic Geography, 78*, 103–127.

Rindova, V., Dalpiaz, E., & Ravasi, D. (2011). A study of organizational use of new cultural resources in strategy formation. *Organization Science, 22*(2), 413–431.

Saldaña, J. (2016). *The coding manual for qualitative researchers*. London: Sage.

Santoro, M. (2009). Introduction. In P. Bourdieu (Ed.), *Ragioni Pratiche* (pp. 7–21). Bologna: Il Mulino.

Sassatelli, M. (2011). Urban festivals and the cultural public sphere. In G. Delanty, L. Giorgi, & M. Sassatelli (Eds.), *Festivals and the cultural public sphere*. London: Routledge.

Scollen, R. (2009). Talking theatre is more than a test drive: Two audience development methodologies under review. *International Journal of Arts Management, 12*(1), 4–13.

Sicca, L. M. (1998). Organizzare i processi nella produzione lirica. Il caso Rossini Opera Festival. *AIDEA, La gestione e la valorizzazione dei beni artistici e culturali nella prospettiva aziendale, atti convegno di Siena*, pp. 30–31.

Silverman, D. (1993). *Interpreting qualitative data*. London: Sage.

Strauss, A., & Corbin, J. (1990). *Basics of qualitative research: Techniques and procedures for developing grounded theory*. London: Sage.

Thompson, J. B. (2010). *Merchants of culture: The publishing business in the twenty-first century*. Cambridge: Polity Press.

Thompson, P., Jones, M., & Warhurst, C. (2007). From conception to consumption: Creativity and the missing managerial link. *Journal of Organizational Behaviour, 28*, 625–640.

Thornton, P. H., & Ocasio, W. (1999). Institutional logics and the historical contingency of power in organizations: Executive succession in the higher education publishing industry 1958–1990. *American Journal of Sociology, 105*(3), 801–843.

Toraldo, M.-L., & Islam, G. (2019). Festival and organization studies. *Organization Studies, 40*(3), 309–322.

Townley, B., Beech, N., & McKinlay, A. (2009). Managing in the creative industries: Managing the motley crew. *Human Relations, 62*, 939–962.

Wiggins, J. (2004). Opportunity to participate: Model of audience development. *International Journal of Arts Management, 7*(1), 22–33.

Yin, R. K. (2014). *Case study research: Design and methods*. London: Sage.

Luca Pareschi (corresponding author) is an Assistant professor in Organization and Human Resource Management at the University of Rome 'Tor Vergata' (Italy). Previously, he worked at University of Venice 'Ca' Foscari', and at the University of Bologna, where he got his PhD in management. His main research interests regard the performative power of words and dictionaries, the management of arts and culture, and the neo-institutional approaches to organizational fields. He conducts research through qualitative and semi-automatic text analysis techniques to extract meanings from texts and reconstruct narratives.

Noemi Ponzoni works in the Audience Insight team at Penguin Random House, UK. She has a 5-year professional experience in audience research in the creative industries. She previously worked as a researcher at research and consultancy charity The Audience Agency (UK), contributing to the evaluation of creative projects and supporting arts organisations in their collection and reporting of audience data to Arts Council England. She holds an MSc in Innovation and Organisation of Culture and the Arts from the University of Bologna.

Participatory Event Platforms in the Urban Context: The Importance of Stakeholders' Meaning of "Participation"

Aura Bertoni, Paola Dubini, and Alberto Monti

Abstract Event organizers tend to seek collaboration with different actors, both as a way to reach legitimization and as a possibility to reach out for resources that would otherwise be unavailable or too costly to achieve. In this chapter, we examine how participation is interpreted and enacted in iterative urban events. We argue that the institutional logics informing stakeholders' views and meaning of *participation* affect the governance of participatory cultural events and the evolution of the activities proposed by event organizers. Based on the longitudinal analysis of an iterative participative cultural initiative (IPCI) taking place in Milano since 2012, our work contributes to the understanding of how IPCIs are structured and evolve. The emphasis on the interplay between different institutional logics associated with the meaning of participation and the governance and activities related to the cultural initiatives sheds light on the specificity and the intrinsic fragility of IPCI where participation becomes overtime a resource—if successfully managed—or liability in itself. At the same time, though, it highlights the incredible potential of participative models in creating distributed value across very different stakeholders.

Keywords Participatory cultural initiatives · Iterative events · Participation · Institutional logics · Milan

1 Introduction

Following the growth and articulation of urban events, literature has become interested in identifying typology of events based on their specificity, their impact, and linking them to urban policies (Richards 2017). In this context, temporary events are viewed as part of the urban cultural landscape; they comprise a variety of initiatives, such as festivals, outdoor performances, programs outside museums (Schuster

A. Bertoni · P. Dubini (✉) · A. Monti
Università Bocconi, ASK research center, Milano, Italy
e-mail: aura.bertoni@unibocconi.it; paola.dubini@unibocconi.it; amonti.man@unibocconi.it

P. Demartini et al. (eds.), *Cultural Initiatives for Sustainable Development*, Contributions to Management Science,
https://doi.org/10.1007/978-3-030-65687-4_13

2001). Some are iterative events (e.g., as festivals); others are pulsar (e.g., Olympics; Richards 2015). In particular, top-down serial events are seen as instances of "eventification" that change the value and the meaning of city spaces (Jakob 2013). Additionally, cities can be classified according to how strategically they use events to support long-term political agendas. In particular, Richards and Colombo (2017, p. 40) consider: "An eventful city purposefully uses a programme of events to strategically and sustainably support long-term policy agendas that enhance the quality of life for all."

However, planned serial events are not directly managed by municipalities, nor are they simply outsourced. Increasingly, cities put in place public-private partnerships as part of a *deétatisation* process (Zan et al. 2007) and as a way to make more efficient use of increasingly scarce public resources. Through partnerships, municipalities seek to orient the trajectory and growth possibilities for this type of initiative (e.g., Jakob 2013; Quinn 2005; Richards 2017) while strengthening the need to manage multiple stakeholders with different needs and logics (e.g., Thornton and Ocasio 1999).

Additionally, research increasingly acknowledges the importance of understanding and managing stakeholders (e.g., Mair and Weber 2019). Event organizers tend to seek collaboration with different actors, both as a way to reach legitimization (e.g., Biondi et al. 2020; Powell and Di Maggio 1991) and as a possibility to reach out for resources that would otherwise be unavailable or too costly to achieve (e.g., Pulh et al. 2008; Richards 2015). Successful multi-stakeholder involvement allows increasing the scope and visibility of the event (e.g., Castells 2009; Piber et al. 2019).

Differences in strategies of stakeholders' management between festivals with different types of ownership and structure are significantly under-researched (e.g., Biondi et al. 2020; Andersson and Getz 2009), as well as the analysis of the empirical findings with sound theoretical lenses (e.g., Mair and Weber 2019). Within this context, participation has gained growing importance as a contemporary issue in cultural management (e.g., UNESCO 1976). At the same time, the notion of participation is multifaceted as it may include different features and meanings, depending on the different paradigms or logic brought in by various stakeholders (Rowe and Frewer 2000; Bonet and Negrier 2018). These elements and the increasing need to incorporate active participation bring in new challenges in the institutionalization process of cultural events (e.g., Piber et al. 2019; Biondi et al. 2020).

In this chapter, we build on and develop previous literature by examining how the way in which different stakeholders interpret *participation* affects the organization of participatory cultural events and the evolution of the activities proposed by event organizers. Following Crowther and Orefice (2014), we look at events as value-creating platforms; this requires looking at how the organization's mission incorporates expectations of a quite diverse set of players and how activities, narratives, and outcomes assessment take into consideration a multifaceted series of stakeholders' goals. This perspective can be complemented by viewing these events as "purposeful spaces," where activities and relationships are developed, producing specific behaviors in a range of stakeholders (Liedtka 2000).

The meaning and the operationalization of participation are crucial in understanding the specificity of the event, its mission, activity, and performance. We claim that the organization of a participatory cultural event reflects and evolves with the different conceptualizations of value (such as economic, social, and cultural) associated with the meaning of participation by the various stakeholders (e.g., Hall and Hall 1990; Bonet and Négrier 2018). We explore our claims through a longitudinal case study on the birth and evolution of a significant participatory cultural initiative in Milano, named BookCity (BCM).

The chapter is organized as follows. First, we introduce the theoretical background with the relevant literature used as our lens to delve into the case. Second, we present the methodology used and the BCM case and its context. We then offer the results and the main insights emerging from the data, as informed by our theoretical lens. We conclude this work with a discussion of the contributions and limitations of the study.

2 Theoretical Background

Temporary events are increasingly part of the urban cultural landscape (Schuster 2001). Some are iterative events, and some are pulsar (Richards 2015): the former brings people together on a cyclical basis to support social ties and generate social capital (e.g., annual festivals), the latter is part of a relatively rare subclass of happenings, marking specific moments in the history of a city (e.g., Capital of Culture). From the city planner point of view, events are not perceived any more as mere tools for tourist promotion or residents get together. Instead, they are increasingly considered relevant cultural tools to mobilize neighborhoods so that different local communities can celebrate their own culture, socialize in new ways, making it possible for passive and active participation to take place (Comunian et al. 2010; Puerari et al. 2017). This new way of seeing is also in line with UNESCO's recommendation to allow concrete opportunities for cultural participation guaranteed for all, to both develop individual personalities and contribute to cultural progress at society level (UNESCO 1976). Finally, this perspective on cultural life entails two dimensions: passive participation, referring to receptive attendance; and active participation, entailing engagement in culture as an activity. This latter facet of culture represents a feature of contemporary culture, promoting individual self-assertion and challenging the top-down cultural approach of established organizations.

In our work, we aim at investigating "what constitutes governance, and how Participatory Cultural Initiatives (PCI) are initiated, governed, and developed over time" (Biondi et al. 2020, p. 2). More specifically, we are interested in exploring processes and mechanisms that, over time, shape the formation and governance of a specific iterative PCI (hereinafter, IPCI), in which very different institutional logics coexist (Battilana and Lee 2014; Wright and Zammuto 2013). We argue that, as all organizations, IPCI thrive to become progressively effective and efficient while pursuing their mission and ensuring necessary resources; as participatory initiatives, they continuously need to mediate among several stakeholders and therefore need to be very adaptive. The tension between increased formalization around a core mission and the adaptation to urban needs (however they are codified and represented) is at the core of their viability.

IPCI address several needs at the city level. They allow diversification and audience development for permanent cultural organizations (Quinn 2005) while allowing smaller independent cultural players to have a say and mingle with more established organizations (Pulh et al. 2008). Moreover, these initiatives may offer the possibility for active participation and event co-creation by different audiences that would typically not get in touch (Crowther and Orefice 2014; Biondi et al. 2020). On the other hand, these events are costly to produce, compete with established cultural organizations for audience attention, urban space, donors' resources, and therefore are intrinsically fragile (Dubini 2016). As it often happens in the cultural fields, their long-term sustainability is partly dependent on the structures and mechanisms put in place at the municipality level.

In this respect, we also assume that policies designed and implemented at the city level can affect the birth and conceptualization of IPCI themselves, and therefore are crucial in determining their trajectory and growth possibilities (e.g., Jakob 2013; Quinn 2005; Richards 2017). In particular, two aspects suggest the municipality as the most critical stakeholder for IPCI organizers to negotiate with. First, given the growing strategic role of IPCI in supporting long-term policy agendas at the city level, the municipality vested interest is that these events will be stable and well-rooted in the city's life. Second, changing political agenda, budget cuts, and emerging social challenges at the city level make IPCI intrinsically fragile and in constant need to legitimize their action (Richards 2015, 2017).

The tension between formalization and adaptability, the need to focus on multiple stakeholders and their objectives suggest looking at in the IPCIs as value-creating platforms (Crowther and Orefice 2014). From this perspective, it is essential to understand what different stakeholders mean for value creation and how their values can be reconciled during the co-creation of an event (Crowther and Orefice 2014). We claim that different views on the meaning of "value creation" are related to the different logics or paradigms from which various stakeholders consider "participation." In particular, we inform our analysis by adapting the framework proposed by Bonet and Négrier (2018, p. 67) that highlights the importance to study in the field how: ". . .the simultaneous coexistence, in one way or another, of all these paradigms results in cultural policies that are complex from an interpretative standpoint. Each paradigm considers citizen participation from a different perspective. At the same time, it is in its practical application that the contradictions (and the convergences among paradigms) shine." The framework identifies four logics: excellence, cultural democratization, cultural democracy, creative economy. Therefore, our work contributes to the understanding of how IPCIs are structured and evolve as the result of the different meanings and values various stakeholders associate with what *participation* means. Empirically, we explore these issues tracking the evolution of a specific IPCI: BookCity Milano.

3 Methods

We carried out a longitudinal exploratory case study research (e.g., Yin 1984) with an abductive approach to analysis (Dubois and Gadde 2002). This choice is coherent with the need to investigate a “contemporary phenomenon within its real-life context, especially when the boundaries between phenomenon and context are not clearly evident” (Yin 1984, p. 23).

Empirical information has been gathered through multiple sources in order to improve the validity and reliability of our findings (Dul and Hak 2007). In particular, interviews as primary data while official documents, social media (e.g., Facebook), websites, and press releases as secondary data. A dataset was created listing locations and events for the eight editions. Considering the categories proposed by Piber et al. (2019), researchers selected 13 people involved at different levels in BCM design and management with a higher incidence of individuals who participated in the organization of all editions. All the people selected accepted to be interviewed for at least 1-h in-depth semi-structured interviews. The interviews were carried out in the first quarter of 2020. We devised a list of scheduled inquiries as an interview guide to ensure a standard approach but, at the same time, allowed the researchers to feel free to pursue other topics that emerged during the interviews (Fontana and Frey 1998). Questions were of an open-ended nature to elicit the most thorough answers from the respondents (Rubin and Rubin 2005). Finally, the interviews were carried out by two of the authors, who also have specific knowledge about BCM, since they are involved with different roles in the organization. In particular, one of the authors had the opportunity to participate in some brainstorming meetings at the genesis of the event, while the other joined the organization in 2018 in the role of a project manager. Having specific knowledge of the organization and its history and being known by the interviewees pave the way to the possibility to increase the quality of the interviews by generating a relaxing climate and also the possibility to better elicit themes from the respondents (Rubin and Rubin 2005). At the same time, the relatively limited participation in the festival design and implementation and the presence of a research protocol reduced the risk of bias by interviewers. The third author acted as an independent judger in triangulating the information and the meaning arisen during the interview with the literature and the secondary data sources. We summarise the information collected in Table 1.

We analyzed the data with a narrative approach to reconstruct the evolution of the festival chronologically. At the same time, as part of the interview process, interviewees were asked to identify the most relevant incidents affecting the progressive structuring of the event and to highlight specific players contributing to the shaping of the event in addition to the formal decision-making process.

In the following analytical section, we have reconstructed the evolution of BCM by putting specific attention on the logics associated with the word “participation” as they emerged in the discourse of the different stakeholders and how they were embedded in the structures and procedures of the organization. We used as a theoretical lens the model proposed by Bonet and Négrier (2018).

Table 1 Source of data analyzed

Source	Volumes
Primary data	
Founders	Three interviews—1 h each
Managers	Four interviews—1.5 h each
Key actors	Six interviews—1 h each
Secondary data	
BookCity press release 2012–2019	857 consulted 75 relevant for the issues analyzed
Facebook official page and website	Available posts

Source: Authors' elaboration on Piber et al. (2019) categories

Table 2 BCM venues by institutional type

N venues/type	2012	2013	2014	2015	2016	2017	2018	2019
No profit	29	45	57	40	63	60	87	88
Private	41	64	65	43	75	96	146	115
Public	55	53	89	86	78	100	114	118
Religious	6	11	17	8	14	12	22	17
Total	131	173	228	177	230	268	369	338

Source: Bocconi ASK centre on BookCity data

4 BookCity Milano: An Event on Reading, an Event for the City

4.1 The Structure of the Event: A Description

BookCity Milano (BCM) is a 4-day "feast" to celebrate reading and books, launched in November 2012. From a governance point of view, BCM is designed and managed by the BookCity Milano Association (from now on, the Association) founded in 2013 by the four leading foundations originating from the four biggest Italian publishing houses—Corriere della Sera Foundation, Giangiacomo Feltrinelli Foundation, Arnoldo and Alberto Mondadori Foundation, Umberto and Elisabetta Mauri School for Booksellers Foundation. The president of each foundation is a member of the BCM board. Every year, one of the four presidents is appointed as the executive president. The management staff includes two event management agencies, with specific skills in the organization and production of cultural events. Since 2013, the municipality signs a 3-year agreement with the Association granting the free use of its civic spaces, some free advertising (e.g., billboards, public transportation), and covering the costs for setting up the temporary bookshop based in the Castello Sforzesco (hereinafter, the Castle) courtyard.

The idea of involving, since the beginning, a wide array of institutions and partners is epitomized in Table 2, which shows the evolution in the number of venues by institutional type involved in BCM over the 8 years.[1]

BCM includes three types of events: self-produced events in which the Association encourages private, public, nonprofit, and religious organizations in the city to design, submit, organize, and deliver presentations, conversations, concerts, and exhibitions around books; co-produced events, where it helps publishers to design events in specific "thematic locations" throughout the city with the support of location providers, including museums, theatres, libraries, and other iconic city places; finally, there are events directly organized and produced by the Association, such as the opening and closing ceremonies and co-designed sponsored special events (produced events).

Among these events, particular attention is devoted to four types of audiences (i.e., schools, universities, community centres, and civically engaged readers) through four dedicated programs (respectively, BookCity for Schools, BookCity University, BookCity for the Communities, and BookCity at Home).

Every year, in early March, a presentation of the incoming BCM edition takes place in the conference room of the city hall. This presentation is the public meeting where the Association delivers available information about the event and communicates deadlines to the different stakeholders involved. Between March and mid-June, the program coordinator collects and matches the co-produced events with public, private, and non-profit organizations that have given their consent to host them. Self-produced initiatives are self-managed. The organizers are invited to provide information about venues, dates and time, guests, books presented or discussed on the BCM website until the end of June. Typically, before the end of July, the overall program is outlined. At the end of the summer, as the management staff starts getting involved with the logistics, the program coordinator clusters groups of events by topic and identifies trending themes. In the meantime, the call for volunteers supporting the management staff has come out, press and communication material is being handled, and special projects have been designed with sponsors.

The city area is divided into 88 neighborhoods called NIL *(Nuclei di Identità Locale)*. Table 3 shows different density indicators as proxies to define the attraction of different neighborhoods for BCM events. Average per NIL shows the average number of venues per NIL; its growth—combined with the growth in the percentage of total NILs covered—indicates the ability of the initiative to spread out throughout the city. Moreover, the growth in the % of NILs with a higher number of venues than the average indicates the ability of the initiative to consolidate its presence in specific NILs. Very recently, BookCity started involving the Metropolitan City of Milano.

[1]The drop in the number of events during 2015 was due to the presence of Expo in Milano and to the occupation of several locations with initiatives associated with it. 2012 opened with 437 events, which rose to 1600 in 2019.

Table 3 BCM geographical diffusion

	2012	2013	2014	2015	2016	2017	2018	2019
N Nil	49	40	48	43	48	60	66	60
% Nil	56%	45%	55%	49%	55%	68%	75%	68%
Average per NIL	2.6	4.2	4.0	3.6	4.4	4.2	5.2	5.3
N NIL > avg	13	8	9	12	10	15	20	14
% NIL > avg	27%	20%	19%	28%	21%	25%	30%	23%
Outside Milano	4	6	34	24	19	17	23	21
In Milano	127	167	194	153	211	251	346	317
Total venues	131	173	228	177	230	268	369	338

Source: Bocconi ASK centre on BookCity data

Table 4 participation over time by type of host

N editions/type	1	2	3	4	5	6	7	8
No profit	112	36	17	10	6	6	5	4
Public private partnership	3	2		4	2		3	1
Private	255	78	28	10	5	9	1	3
Public	351	50	30	16	16	10	17	8
Religious	40	9	3	3	2	3		
Total	761	175	78	43	31	28	26	16

Source: Bocconi ASK centre on BookCity data

Row numbers 7 and 8 in Table 4 show events' relative presence in Milano and the Metropolitan City.

4.2 *The Birth of BCM: Recognizing Opportunity and Assessing (Conflicting) Needs*

The idea for a city event on books originated in 2011 when the publishing industry was looking for an initiative stressing the specificity of Milano as a cluster for book publishing. Besides, the Italian Publishers Association (AIE) and some of the most prominent publishing groups, all located in Milano, were growing dissatisfied with the management of the Turin International Book Fair, the most important book fair in Italy at the time. Lastly, in the aftermath of the 2008 crisis, the book industry was too small and fragmented to be visible as an industrial cluster. As recognized by one of the key actors: "The first goal of BookCity was to bring together the fragmented publishing world in Milan in order to increase its potential" (Key Actor 5). Above all, it was felt that it would have been useful to leverage on an event where publishers could obtain widespread attention on their relevance for the national publishing industry and had the opportunity "to promote reading as a key element of society" (Manager 1). *FuoriSalone*, literally *Outside the Fair* (with reference to the International Furnishing Accessories Exhibition) and now renamed "Milan Design Week," one of the most important international design happenings involving the entire city of Milano in a frenzy of events and installations, was taken as the benchmark for this project.

In 2011, Milano was (and still is) recognized as the city of design and fashion, but not as a publishing one, despite the high number of publishing houses—from books to magazines, newspapers, and journals—bookstores and book clubs. "This was not so much in the agenda of the municipality either" (Founder 1). However, one of the items in the municipality's political agenda was how to contrast the growing differences between the city's central and peripheral areas (Puerari et al. 2017) signaling vested interest in promoting other diffused city events like *FuoriSalone*.

As mentioned by one of our informants: "The newly appointed Mayor and Deputy Major for Culture had built his political vision for the city around the idea of increasing citizen participation to city life, and wanted to develop more inclusive policies involving neighborhoods" (Key Actor 4). This quote shows the recognition by different stakeholders that the periphery was progressively being detached from a highly attractive and heavily gentrified city center (Pulh et al. 2008; Crowther and Orefice 2014). Moreover, the municipality wanted "to give greater visibility to the Castle,[2] and to make Milano attractive from a cultural point of view" (Key Actor 5). At that time, the city was perceived mainly as a business destination and the number of cultural attractors was limited and geographically concentrated. Therefore, it was essential to broaden the number and variety of landmarks to make different parts of the city more visible and attractive.

Nevertheless, early talks lacked momentum: publishing houses had many common interests, which made the idea of BCM attractive, but seemingly competing interests both within the publishers' group (i.e., commercial and visibility issues) and the municipality. Even though both the publishers and the municipality wanted to involve and engage the citizenship, they stood from two different ideas.[3] The publishers wanted to create an event to highlight the economic importance of the book industry for the city and the nation while enlarging the potential audience. The approach proposed was top-down, with the idea of showing the excellence in their portfolio while intending the audience's participation as mainly passive, captive, and potential consumers. Therefore, elements of the creative economy and excellence paradigms were the prevalent views of the publishers (Bonet and Négrier 2018). On the other side, the municipality agenda was in line with the idea that the event should aim at enlarging audiences also in the more peripheral urban areas (i.e., cultural democratization).

---

[2]Castello Sforzesco is a castle built in the fifteenth century by Francesco Sforza, Duke of Milan, on the remnants of a fourteenth-century fortification. Extensively rebuilt by Luca Beltrami in 1891–1905, it now houses several of the city's museums and art collections.

[3]There was a common agreement among interviewees on this narrative (Founder 1 and 2, Key actor 2, 3 and 5).

4.3 *Catalyzing Different Needs: The Mediating Role of the Book Publishing Foundations*

To overcome the roadblocks on the path of BookCity, the four leading foundations originating from the four biggest Italian publishing houses came into play. Whereas the four foundations also had different interests in the project, each of them understood the opportunity to strengthen its autonomy from the mother company and to increase its visibility and mission. Eventually, they did manage to play a valuable role as mediators and proactive coordinators able to synthesize different needs and expectations.[4] The critical element of this achievement "has been their "coopetition": through their cultural arms, the largest publishing groups have found the way to stay together, although they still compete on the commercial dimension" (Manager 1). Coherently, the foundations decided to establish the Association BookCity Milano to create a common platform for all the relevant stakeholders (Crowther and Orefice 2014). The Association became the catalyzer of all discussions on participation, besides having the responsibility to manage BCM.

The final decision to design and organize BookCity Milano took place in early 2012. On the one hand, the association and the municipality became aligned with the idea that BCM is meant to amplify the voice of publishers and not to compete with them. While *what* and *who* are presented is typically up to curators in festivals, BCM should be the place "where publishers could decide which authors or books to bring" (Manager 1). On the other hand, in spring 2012, the municipality welcomed *Piano City*, a participatory format where pianists express their willingness to perform and participants give their consent to host a concert through a dedicated website.[5] Therefore, "Piano City acted as a further driver for the participation component" (Manager 2) in the BCM format. It became a fundamental event that turned the attention of the municipality toward a vision in which participation was characterized by the empowerment of the citizen in the choices of location. Finally, the aim of increasing the diversity of cultural views and institutions involved became relevant and explicit in the political agenda (cultural democracy, Bonet and Négrier 2018).

BookCity Milano owes its name, date, and headquarter at Castello Sforzesco to the then Deputy Mayor for Culture. Since the beginning, the Association endorsed his view (Biondi et al. 2020; Piber et al. 2019), including the proposal of the name "BookCity," and immediately developed a coherent logo. The Association members

---

[4]Here as well interviewees largely agreed (all founders and Key Actors 1 and 5).

[5]Milano has been developing two formats for cultural events: "City" format is generally participatory in nature and involves many non-profit organisations. The benchmark has been *Piano City*, a concept of Berlin-based pianist Andreas Kern brought to Milano in 2012. "Week" format includes "off" initiatives around a relevant fair and the benchmark has been *FuoriSalone*. Born in the 1970s along with a few side initiatives organised during the annual design and furniture fair *(Salone del Mobile)* and held in the showrooms of very few furniture brands, FuoriSalone started growing in the 1980s and had a push in the 1990s.

were skeptical about the other two aspects though: whereas November used to be an "empty" month in the city with no other competing events, it is likely to encounter exposure to the inclemency of the weather. Moreover, the Association did not feel the need for a headquarter since it was meant to be diffused throughout the city. "Admittedly, both ideas turned out to be very wise" (Key Actor 2).

The 13 people we interviewed for reconstructing the story of BookCity Milano represent different interests, roles, and perspectives. Nevertheless, respondents largely agreed on many relevant elements which emerged as the pillars of BCM:

- It has to allow publishers to showcase their authors and books (cultural excellence/creative economy).
- It has to highlight the specificity of Milano as a book cluster (creative economy).
- It has to be diffused throughout the city in order to claim the relevance of reading as an everyday activity to be performed by anyone anywhere (cultural democratization).
- It has to be participatory, meaning that in principle, no genre should be dominant, nor should there be any set "cultural agenda," and anyone should be able to propose an event and any event has to be free of charge. No booking of events is allowed (cultural democracy).

The intervention of the book foundations thus mediated between two different logics of participation, suggesting to leverage on the specificity of Milano as a book cluster and civically engaged city. This left autonomy to publishers to choose titles and authors to present while at the same time encouraged the active participation of different organizations and the valorization of several venues throughout the city.

4.4 Translating Participation in the Event Structure

The Association was able to understand the common ground in the explicit and implicit claims of both publishers and the municipality and to recompose the conflicting ones in an agreed-upon definition of the features of the event. It also had the legitimacy to propose itself as a mediator, given the institutional role and direct personal relationships of the presidents with the founding publishing groups, the publishing industry at large and the city council (e.g., Andersson and Getz 2009; Biondi et al. 2020; Powell and DiMaggio 1991; Suchman 1995). Now the issue was how to translate the corresponding business and culture-oriented instances in the programming and actions of BookCity (e.g., Suchman 1995). To respect the "mandate" of the different stakeholders, i.e., to support the process of turning Milano into a close-knit book cluster and to enable inclusiveness and cultural diversity, "BookCity had to be an open event in an open city" (Manager 2). Under such a vision, BCM should mark the difference from the current book events, and participation had to be central to its conceptualization. The Association defined BCM as a "feast of reading and readers" (Key Actor 2) rather than a literature festival or book fair. In doing so, it set the ground for a bottom-up initiative (cultural democracy) with loose ownership, supporting the entire book value chain of the city (creative economy; e.g., Biondi et al. 2020; Powell and Di Maggio 1991).

Moreover, its contribution to linking culture and welfare, and to adequately develop the less central neighborhoods of the city, was made explicit. Accordingly, "the identity of BCM has become strongly connected with participation" (Key Actor 5). Therefore, BookCity aims at facilitating access to the broadest number of people to discuss books with free and diffused events (cultural democratization). It commits to empower any interested party—regardless it is a company, a parish, a university, etc.—to produce a book initiative during the happening (cultural democracy). As the concept of BCM was taking shape, at the beginning of 2012, the Deputy Mayor for Culture urged the Association to go ahead and organize the first edition in November that year. Thus, in March, the Association gave a presentation in the city hall conference room to "share with the book community what [the Association] had in mind" (Key Actor 6). Plenty of publishing houses, bookstores, libraries, and professionals in the field were invited. The response was positive: "the room was incredibly full" (Key Actor 6) and the Association members had the chance to present the initiative to the city, explain their roles in BCM formation, give dates, and launch the call for event proposals.

Negotiations between the Association, the municipality, and the publishers led to the formalization of the role of BCM, its core activities, and the interaction with other key stakeholders. While BCM would act as a platform to match venues and events, build the calendar of initiatives, and facilitate the relationship between civic and private hosts and publishers, bottom-up participation would be encouraged. Moreover, BCM would have no curatorship, as themes and topics emerge from the collection of bottom-up initiatives submitted by publishers and civic organizations (cultural democracy). Last, promotion of BCM is a shared responsibility: the municipality promotes the initiative through its channels; the Association organizes the overall communication of the initiative via a dedicated press office and digital means; *Corriere della Sera*, the leading city newspaper, prints and distributes the program booklet for the event. Each proponent (either publisher or civic organization) is the owner and promoter of its event.

Nevertheless, as the event grew, other stakeholders (apart from the municipality, the publishing houses, and event organizers) had to be considered. It became apparent that each of them had specific requirements and interpreted participation in a different way.

A clear example of how participation has molded the structure of BCM stays in the evolution of the contribution given by universities and community centers. Whereas universities are part of the BCM line-up since 2012, as the number of the involved entities and organized events grew every year, 2014 saw the creation of a specific section called BookCity University *(BookCity Università)* for events produced by the universities in Milan. Along the same line, readings and other initiatives on books have taken place in prisons and hospitals since the first year. However, only when events systematically addressed to people who cannot attend BCM summed to several initiatives organized by welfare centres (e.g., soup kitchens, shelters for the poor), the Association stepped in. In 2016, a new section called BookCity for the Community (*BookCity per il Sociale*) was created to bring BCM in places that could otherwise not be animated.

4.5 *The Orchestrating Role of the Association*

The bottom-up nature of BCM entails the absence of curatorship; yet, co-production requires "a simple and complicated process at the same time" (Manager 3). First of all, the program coordinator takes care of those events in need of a venue. Each year, the Association receives hundreds of event proposals from publishing houses (cultural excellence). This activity is autonomously managed and the coordinator acts as a platform intermediating between event proponents (mainly publishers) and locations (Crowther and Orefice 2014). This first phase is twofold: the program coordinator tags each event by topic and then evaluates it according to many matching criteria for determining the most appropriate venue.

> Among these criteria—including expected audience, notoriety of guests, size of the location—particular attention has always been given to nature and mission of the location. For instance, events dedicated to history have always been attributed to the Risorgimento Museum. (Manager 3; top-down approach pushing toward cultural excellence)

In a second phase, the program coordinator mediates with event proponents who are put in contact with other relevant stakeholders, including locations, booksellers, and press offices. As everyone has its own logic, objectives, and languages, "other than helping collaborations to speed up, network creation is deliberately facilitated. This is an example of how BCM creates a collaborative culture and a shared sensitivity among players involved within the city, intended as widespread social knowledge and ability to innovate" (Manager 3) (e.g., Mair and Weber 2019).

Self-produced events are self-managed and promoted as a whole by BCM. Once all the events have been inserted in the complete calendar, the coordinator tags them to increase their visibility. Nevertheless, the municipality has been giving to the Association further inputs in order to expand attendance figures (cultural democratization) and to allow diverse social groups to make their cultural interests and practices visible (cultural democracy; e.g., Jakob 2013; Quinn 2005; Richards 2017). Hence, increasingly, the Association "built bridges with some of the most active civic organizations of the city by giving value to their experience and letting them envision and organize their events" (Manager 3).

On the one hand, there is evidence of the goal of mobilizing niches as, "from visual identity of book series to hermeneutic translation and dialect poetry, niches are one of the BCM secrets since the first edition" (Manager 3; cultural excellence). On the other hand, clearly, "the difference between high and low culture appears not relevant for BookCity giving room to citizen initiatives is a substantial and characterizing element for the event" (Manager 3; cultural democracy). In quantitative terms, the extended participation in the project has been visible since the beginning.

Furthermore, "the quality of the proposals was—maybe surprisingly—also high. . . . the combination of these two elements gave a chance for survival and sustainability" (Manager 3; creative economy).

4.6 Making Participation Sustainable

The effort to address several stakeholders needs and to progressively fine-tune processes in dealing with a broader set of audiences and different neighborhoods proved successful in quantitative terms:

> [BookCity] has accomplished the goal it had at its formation, as today it is a highly participated initiative: it offers a broad relational platform to 500 big and small publishers, thousands of professionals working in the field and complementary industries, the Municipality, cultural organisations, readers, and the whole city. (Manager 4)

BCM has "reached and highlighted several places in town, making evident that Milano is indeed a book city" (Key Actor 4). At the time BCM was designed, a few performing arts festivals in Italy changed the traditional relationship with their audiences (Agustì et al. 2018). Conversely, "BCM has been the first one to test participatory programming with books" (Manager 3) and set the ground for further city initiatives in the publishing field. In 2015, Milano launched *Patto per la Lettura*, an alliance among libraries, publishers, bookstores, authors, and committed citizens, aiming to restore value to the act of reading as an essential part of a new idea of citizenship.[6] In 2017, Milano joined the "UNESCO Creative Cities Network for Literature" with the aim to be internationally attractive and locally inclusive at the same time.[7] In this sense, BCM "has become a method for developing participation around books. . .an opportunity to learn mingling and sharing" (Key Actor 2).

However, a point of caution for the Association is to make participation sustainable, in the different dimensions that the mission of inclusiveness entails. First, the logic of cultural democracy cannot but consider 1600 events in 4 days as a huge success. Regardless of their level of formalizations, small groups, associations, and any sort of active collectives can become visible during BookCity. At the same time, this creates on audiences a problem of content overload. As one of our interlocutors recognized:

> BookCity is a mikado. Over the years it has grown incredibly, it has achieved its goals and has managed to create value for Milano and for all the organizations involved, without making anyone rich at the expenses of the others. It is truly noteworthy. Yet, it stands on a very delicate equilibrium: BCM is very fragile as it has grown too much in size and so we all exhausted for higher marginal costs and lower marginal benefits. (Key Actor 3)

[6] *See* https://web.comune.milano.it/wps/portal/ist/st/Patto_Milano/scopri_il_patto (10 June 2020).
[7] *See* https://en.unesco.org/creative-cities/milan (10 June 2020).

The result of years of engaging work on cultural participation is the actual risk of being overwhelmed by an ever-increasing number of submissions for self-produced events. "BCM is a platform without rules. If BCM is endlessly enlarged, inclusiveness becomes exclusivity. In the end, it only works for events that would be successful anyway" (Manager 1). It is difficult for a participatory event like BookCity to discourage participation, "but [the Association] needs to push for more proactivity from local cultural institutions and to be more creative itself" (Founder 1).

All the respondents have raised the issue of over-participation. Under these views, "BCM must find metrics different from quantity" (Key Actor 1). The bottom-up element of BCM provides the opportunity to channel diverse cultures and practices around books and raise awareness toward the least-known ones. Nonetheless, the overload of content can make conversations less meaningful. According to many respondents, "readers are dissatisfied, as it is difficult for them to find what might trigger their interest in such a huge program" (e.g., Key Actor 3).

However, different views have been expressed by the respondents as to the potential solutions for over-participation. Some members of the Association would like BCM to become more curatorial for bounding participation. "If we decide that the Association only admits self-produced event on a given topic, maybe even a broad one, what is worth would not be lost" (Key Actor 4). Concurrently, "self-produced events are vital for BCM. Even those who make noise are important. For this reason, it would be better to reduce events co-produced with publishers and leave the self-produced ones free. This active approach of the Association would help people to significantly engage with the content [but it would also] make BookCity more suited for contributions of sponsors and media partners" (Manager 1).

Secondly, the size reached by the event has stretched the resources available, and not all efforts have been adequately capitalized. Table 4 shows the number of hosts by institutional form and by the number of years participating in BCM. Very few hosts have been mobilized throughout all editions. On the one hand, this is a sign of vitality and it shows how every year new organizations are willing to participate. On the other hand, the high turnover makes it difficult—particularly for smaller organizations—to develop learning mechanisms to become more effective and efficient and to develop a loyal relationship with an audience throughout the years.

Moreover, places may interpret their participation passively. For instance, in principle, all the public institutions involved host events voluntarily. Except for the first year, when the municipality directly selected some locations, public sites receive official communication from the municipality announcing the opportunity to participate. "It is never a problem to find locations. Increasingly, some of them organize initiatives as part of their own activity. I think this is an indication of the fact that BookCity has indeed become part of the city texture" (Manager 2). One crucial aspect of the assessment of these contributions is to take into account the different levels of responsibility and commitment assumed by the hosts. There is the risk that "they perceive themselves only as mere containers; in this sense, the involvement of these institutions is limited to how many days and how many rooms to make available. . .It is not true participation." (Key Actor 5).

Thirdly, the sustainability of BookCity stands on the participation of the private sectors, i.e., publishers and sponsors. The publishing industry likes the format since "its purpose is not to sell books—Milano is already a lively market for books and is luckily full of bookstores—but to promote authors, and to have them part of a conversation" (Key Actor 6; cultural excellence and creative economy). At the same time, publishers tend to prefer central and prestigious locations.

Private financial support is difficult to consolidate. "There are a few sponsors that have generously supported BookCity since the first year and who are still very committed, but the ability to proactively involve sponsors can be improved" (Founder 1). "Maybe this happens because we consider them merely sponsors and not also partners. So, every year we start from zero" (Key Actor 2).

The mediation between the private sector and the municipality is made more difficult as—in the last years—the municipality has stressed the concept of "polycentric city," pushing BCM to develop greater connection and involvement of more peripheral areas. The scattered geography of the events is a vital element of the BCM experience but is interpreted differently by different stakeholders. One of the interviewees stressed the importance of a growing presence throughout the city. She says: "One of the winning elements in the alliance between the Association and the municipality has been the idea of letting culture emerge in every corner of the city and making evident how we all need culture" (Founder 2; cultural democratization). In parallel, a key actor contested the passive involvement of audiences in some of the "top-down events" organized in the suburban areas of the city. She goes: "Genuine engagement requires to understand more in-depth the needs of those who work in the neighborhoods every day. While publishers are not always willing to run the risk of a limited audience because of the not-so-prestigious location, it is also essential to enter in negotiation with the local stakeholders to avoid any form of cultural colonialism" (Manager 3).

Finally, widening and diversifying local audiences is a key factor of the participatory approach taken by BCM. The Association knows remarkably little about its audiences but keeps endorsing the municipality mandate to develop participation among citizens (e.g., Jakob 2013; Quinn 2005; Richards 2017). "Still you know there is a lot of people, you see people chatting while queuing, there are young and old people, families and groups of friends. You see them around, suggesting each other where to go. And on social media too, they are very active. It is fantastic because these people look for book authors, not for rockstars" (Key Actor 6). Despite the high number of initiatives, events succeed in micro-marketing. "Organizers are clearly helped by [the Association], especially through the website, but there is attendance to events because organizers are active subjects and not mere spectators" (Manager 3).

It is hard to say whether BCM has succeeded in making Milano reading more:

> Statistics say that Milano has 3% of Italian population and 15% of the book market, but we cannot claim that this is because of BookCity. Also, we are definitely able to mix people, and trigger curiosity but we did not succeed in persuading all the executives that populate the

> many important companies in town that reading Dante help them making better and wiser decisions. And now we need better and wiser decision-makers. We know and we say that reading a lot helps you in whatever your job is, but we still have a lot to do. (Founder 1)

In the effort of listening to the specific needs of different audiences and of making participation sustainable, the Association has been trying to develop initiatives throughout the year. This approach started as a need to work with schools on an ongoing basis, but it is being also explored as an opportunity to distribute a mass of events that is increasingly more difficult to manage in a 4-day event throughout a longer time frame without limiting the energy of proponents. As far as top-down initiatives are concerned, the possibility to broaden the calendar would help the Association to become a permanent player in the urban cultural scene. The Association "is also becoming the driver of specific initiatives in the peripheral areas of the city that build on its capacity of creating networks and partnerships around books to co-produce social value" (Key Actor 4; e.g., Crowther and Orefice 2014; Richards 2017).

Although this commitment is extensively appreciated within the Association; "committing on literacy for social inclusion requires to build up a dedicated team with specific skills" (Manager 1). More in general, the switch from a temporary to a permanent cultural organization requires a series of organizational changes and new competencies within the Association. Any more prolonged and active participation of the Association—both for permanent activities along the year and during BCM—raises issues on its governance and organization. At the moment, "decisions are based on personal ideas and past choices. This leads all of us to adopt conservative approaches, which makes the innovation process difficult. Moreover, decisions are not contributing to the growth of a collective vision. As a result, choices are truly made at the top or at the operational level" (Manager 1). Consistently, "there are a lot of areas of improvement at governance level. The voice of publishers is not truly heard. While there is a very centralised decision-making process, at the operational level everybody has a say and no one is really owner of her decision" (Key Actor 1).

5 Discussion

In this chapter, we sought to link two streams of literature (one that looks at the interplay of different institutional logics and another that focuses on participation and stakeholder management) around the analysis of IPCI. The main contribution of this paper is to make visible: (1) how different stakeholders can simultaneously hold different perspectives and logics and act accordingly toward the network of players involved in cultural events; and (2) how the organizer can help recompose such plurality of logics (e.g., Glynn 2000; Kerr 1963; Kraatz and Block 2008). This is possible since the Association acts as a platform able to recognize the pressures that each different logic puts on its processes and strategy and to adapt it to the variety of the stakeholders involved. Additionally, the evolution in the meaning attributed to the term "participation" by critical stakeholders is incorporated in the organization throughout the years. This also possible because the Association becomes a space of reflexivity for all the stakeholders involved in the initiative (e.g., Liedtka 2000).

More specifically, the initial vision set forward by the publishing community (a top-down cultural event, striving for excellence with an expected economic outcome for the relevant industry) was not consistent with the political aim of the municipality. The latter was seeking for cultural democratization (audience development) and cultural democracy (audience engagement). The mediating role of the Association has allowed for a reconciliation of these different interests around BCM.

At the same time, a stakeholder can hold different logics simultaneously. For example, the municipality had a directive, top-down approach to selecting relevant institutions to be involved in BookCity as locations. However, it claims for a bottom-up approach to participation, therefore not considering the institutions chosen as part of the relevant audiences to be empowered (see for publishers Pareschi and Lusiani 2020; e.g., Glynn 2000; Lampel et al. 2000).

The organizational solutions put in place over the years have made it possible to simultaneously bring onboard different stakeholders while incorporating their definition of participation and their needs. For example, BookCity for Schools works in the intersection between excellence and cultural democratization and succeeds in growing the captive base of children while making available a broad cultural offering to schools (see Fig. 1 page 68, in Bonet and Négrier 2018). The evolution of more bottom-up experiences, like the self-produced events, represents participation in terms of both cultural democratization and cultural democracy logics. While these are innovative ways to broaden the book audience base, they also encourage existing audiences to assume the responsibility of participatory programming activities. The complexity of this effort has partly to do with the specificity of each stakeholder involved in the participatory process. For example, BookCity University and BookCity for the Community show how specific audiences (respectively, students and people who cannot attend ordinary BCM events) are not necessarily satisfied by the same organizational solutions (i.e., identifying a specific managerial responsibility). In the first case, there is a lack of alignment in the meaning given to participation by universities. In the second one, it stems from the different reasons determining exclusion from cultural participation in diverse settings, including hospitals, prisons, schools, community centers, and shelters for homeless people.

Another relevant element has to do with the evolving meaning attributed to the term "participation" within key stakeholders and the pressure this creates on the organization (e.g., Kraatz and Block 2008). The role of the municipality is an exemplary case. Its original idea was to mark the difference from past policies through citizen empowerment (e.g., Richards 2017). Thus, participation was sought by the Association via geographic diffusion, a variety of proposals, and a lack of curatorial activity. The success of the organizational solutions explains the growth in self-produced events, to the point that cultural democracy is considered by many overwhelming. The next step of participation defined by the municipality as "polycentric city" makes evident the need for BCM to engage with organizations active in the different neighborhoods, in the direction of developing a growing number of critic and engaged communities. Today, "too much" cultural democracy challenges the Association's sustainability and, therefore, its ability to please loyal audiences and, ultimately, the political ambition to create a polycentric city.

Another aspect has to do with considering BCM as a platform. The long-term sustainability of platforms has to do with addressing the needs of different stakeholders simultaneously while creating high switching costs. When this occurs, high network externalities are created within each stakeholder and among various stakeholders. In the case of BCM, this happens only in part. For instance, as to co-produced events, the interplay between BCM, publishers, and venues makes the relationship between BCM and publishers stable and longlasting, while that of BCM and the venues are potentially more problematic. Given BCM visibility and reputation, the relationship between the Association and venues is a win-win for a limited number of organizations. In fact, many perceive themselves as merely "containers of events," despite the dedicated efforts of the program coordinator to match events with the identity of the hosting site. The high volatility of locations suggests to the Association to work more in-depth on the involvement of institutions providing space during BCM. Increasing the loyalty of hosts and engaging them to be proactive would help the Association to benefit as a platform from participation without being overwhelmed, thus transforming participation into a useful resource to increase effectiveness and stability.

While this chapter presents all the limitations typical of case-based analysis, it also opens up relevant streams of further research.

1. Governance. The Association has to be flexible to anticipate and adapt to the multifaceted meanings attributed to the term by the critical stakeholders engaged. Still, strong enough to keep its role as a value-creating platform, bringing different meanings of participation to a synthesis. The Association has successfully developed a method for participation that the city has leveraged in the *Patto per la lettura* and in the UNESCO Creative City application. At the same time, it is a very fragile organization whose value created is mainly immaterial. The relationship between participation and governance is one area worth exploring (Pegoraro and Zan 2017).
2. Ritualization and participation. One of the challenges for cultural events is how to become an integral part of the cultural landscape of the city. Iteration is one key element, but it is not enough. The Association has created and routinized several rituals to become intelligible to its stakeholders and to strengthen its role. Additionally, it started considering "unseasoning" as a way to integrate itself in the city cultural landscape further. The relationship between routinization and ritualization on the one hand and participation on the other is a second area worth exploring (e.g., Kraatz and Block 2008; Lawrence and Suddaby 2006; Oliver 1991)
3. Metrics for performance. One of the characteristics of successful platforms is their ability to grow exponentially. However, the tendency to rely on quantitative measures to assess success might be misleading and counterproductive. BCM

relevance and contribution are overshadowed by the fact that it is depicted as a "big" event (many venues, many events, many attendants). While this is an easily communicated result, it inevitably provides a partial assessment of participation that is not instrumental to its sustainability and stakeholders' involvement. The relationship between performance metrics and managerial practices ensuring participation is a different line of research worth exploring (e.g., Hardy and Maguire 2008; Rindova et al. 2011). Key stakeholders can introduce and use new metrics to strategically orient an institutional change (e.g., Lawrence and Suddaby 2006; Townley et al. 2009).

Acknowledgment The authors acknowledge the support of ASK research center—Università Bocconi.

References

Agustí, L. B., Calvano, G., Carnelli, L., Dupin-Meynard, F., & Négrier, E. (2018). *Be SpectACTive! Challenging participation in performing arts*. Editoria & specttacolo, 425 p., 978-88-32068-01-6. ffhal-02154976f

Andersson, T. D., & Getz, D. (2009). Festival ownership. Differences between public, nonprofit and private festivals in Sweden. *Scandinavian Journal of Hospitality and Tourism, 9*(2–3), 249–265.

Battilana, J., & Lee, M. (2014). Advancing research on hybrid organizing—Insights from the study of social enterprises. *The Academy of Management Annals, 8*(1), 397–441.

Biondi, L., Demartini, P., Marchegiani, L., Marchiori, M., & Piber, M. (2020). Understanding orchestrated participatory cultural initiatives: Mapping the dynamics of governance and participation. *Cities, 96*, 102459.

Bonet, L., & Négrier, E. (2018). The participative turn in cultural policy: Paradigms, models, contexts. *Poetics, 66*, 64–73.

Castells, M. (2009). *Communication power*. Oxford: Oxford University Press.

Comunian, R., Chapain, C., & Clifton, N. (2010). Location, location, location: Exploring the complex relationship between creative industries and place. *Creative Industries Journal, 3*(1), 5–10.

Crowther, P., & Orefice, C. (2014). Co-creative events: Analysis and illustrations. In *Event design* (pp. 138–152). London: Routledge.

Dubini, P. (Ed.). (2016). *Institutionalising fragility. Entrepreneurship in cultural organisations*. Milano: Fondazione Feltrinelli.

Dubois, A., & Gadde, L. E. (2002). Systematic combining: An abductive approach to case research. *Journal of Business Research, 55*(7), 553–560.

Dul, J., & Hak, T. (2007). *Case study methodology in business research*. London: Routledge.

Fontana, A., & Frey, J. H. (1998). Interviewing—The art of science. In N. K. Denzin & Y. S. Lincoln (Eds.), *Collecting and interpreting qualitative materials*. Thousand Oaks, CA: Sage.

Glynn, M. A. (2000). When cymbals become symbols: Conflict over organizational identity within a symphony orchestra. *Organization Science, 11*(3), 285–298.

Hall, E. T., & Hall, M. R. (1990) Understanding cultural differences: Germans, French, and Americans.

Hardy, C., & Maguire, S. (2008). Institutional entrepreneurship. In R. Greenwood, C. Oliver, K. Sahlin, & R. Suddaby (Eds.), *The SAGE handbook of organizational institutionalism* (pp. 198–217). London: Sage.

Jakob, D. (2013). The eventification of place: Urban development and experience consumption in Berlin and New York City. *European Urban and Regional Studies, 20*(4), 447–459.

Kerr, C. (1963). *The uses of the University*. New York: Harper & Row.

Kraatz, M., & Block, E. (2008). Organizational implications of institutional pluralism. In R. Greenwood, C. Oliver, & R. Suddaby (Eds.), *The SAGE handbook of organizational institutionalism* (pp. 243–275). London: Sage. https://doi.org/10.4135/9781849200387.n10.

Lampel, J., Lant, T., & Shamsie, J. (2000). Balancing act: Learning from organizing practices in cultural industries. *Organization Science, 11*(3), 263–269.

Lawrence, T., & Suddaby, R. (2006). Institutions and institutional work. In S. R. Clegg, C. Hardy, T. B. Lawrence, & W. R. Nord (Eds.), *The SAGE handbook of organization studies* (pp. 215–254). London: Sage.

Liedtka, J. (2000). In defense of strategy as design. *California Management Review, 42*(3), 8–30.

Mair, J., & Weber, K. (2019). Event and festival research: A review and research directions. *International Journal of Event and Festival Management, 10*(3), 209–216.

Oliver, C. (1991). Strategic responses to institutional processes. *Academy of Management Review, 16*, 145–179.

Pareschi, L., & Lusiani, M. (2020). What editors talk about when they talk about editors? A public discourse analysis of market and aesthetic logics. *Poetics*. https://doi.org/10.1016/j.poetic.2020.101444.

Pegoraro, M., & Zan, L. (2017). Life and death in audience development. The exhibition on Pompeii at the British museum, 2013. *Museum Management and Curatorship, 32*(3), 210–231.

Piber, M., Demartini, P., & Biondi, L. (2019). The management of participatory cultural initiatives: Learning from the discourse on intellectual capital. *Journal of Management and Governance, 23* (2), 435–458.

Powell, W. W., & DiMaggio, P. J. (1991). *The new institutionalism in organizational analysis*. Chicago: University of Chicago Press.

Puerari, E., de Koning, J. I. J. C., Mulder, I. J., & Loorbach, D. (2017). Shaping spaces of interaction for sustainable transitions. In *30th Annual Aesop Congress 2017*. Universidade de Lisboa.

Pulh, M., Marteaux, S., & Mencarelli, R. (2008). Positioning strategies of cultural institutions: A renewal of the offer in the face of shifting consumer trends. *International Journal of Arts Management, 10*, 4–20.

Quinn, B. (2005). Arts festivals and the city. *Urban Studies, 42*(5–6), 927–943.

Richards, G. (2015). Events in the network society: The role of pulsar and iterative events. *Event Management, 19*(4), 553–566.

Richards, G. (2017). From place branding to placemaking: The role of events. *International Journal of Event and Festival Management, 8*(1), 8–23.

Richards, G. W., & Colombo, A. (2017). Rethinking the eventful city: Introduction. *Event Management, 21*(5), 527–531.

Rindova, V., Dalpiaz, E., & Ravasi, D. (2011). A study of organizational use of new cultural resources in strategy formation. *Organization Science, 22*(2), 413–431.

Rowe, G., & Frewer, L. J. (2000). Public participation methods: A framework for evaluation. *Science, Technology, and Human Values, 25*(1), 3–29.

Rubin, J. R., & Rubin, S. R. (2005). *Qualitative interviewing: The art of interviewing*. London: Sage.

Schuster, J. M. (2001). Ephemera, temporary urbanism and imaging. In L. J. Vale & S. B. Warner (Eds.), *Imaging the City: Continuing Struggles and New Directions* (pp. 361–396). New Brunswick, NJ: CUPR Books.

Suchman, M. (1995). Managing legitimacy: Strategic and institutional approaches. *Academy of Management Review, 20*(3), 571–610.

Thornton, P. H., & Ocasio, W. (1999). Institutional logics and the historical contingency of power in organizations: Executive succession in the higher education publishing industry, 1958–1990. *American Journal of Sociology, 105*(3), 801–843.

Townley, B., Beech, N., & McKinlay, A. (2009). Managing in the creative industries: Managing the motley crew. *Human Relations, 62*, 939–962.
UNESCO. (1976). *Recommendation on participation by the people at large in cultural life and their contribution to it*. https://doi.org/10.1163/ej.9789004164543.1-760.54
Wright, A. L., & Zammuto, R. F. (2013). Creating opportunities for institutional entrepreneurship: The Colonel and the Cup in English County Cricket. *Journal of Business Venturing, 28*(1), 51–68.
Yin, R. (1984). *Case study research*. London: Sage Beverly Hills.
Zan, L., Baraldi, S. B., & Gordon, C. (2007). Cultural heritage between centralisation and decentralisation: Insights from the Italian context. *International Journal of Cultural Policy, 13*(1), 49–70.

Aura Bertoni is an academic fellow at ASK Bocconi centre for research on management and economics of arts and culture institutions. As a teaching fellow, since 2011 she has taught copyright and cultural management in various undergraduate and master programs. She is the author of several peer-reviewed international and national publications. As a researcher, she has participated in many national and international research projects dealing with a wide range of topics including digital technologies for access to knowledge, participatory forms of learning and creation, cooperative models for innovation, culture-led urban regeneration initiatives, and open data for cultural institutions. At the ASK Center she has co-founded and co-directed the mapMI project, which has produced and promoted research about the city of Milan through a digital database of its cultural premises.

Paola Dubini is a management professor at Università Bocconi, researcher at ASK Research Center and visiting professor at IMT Lucca. Her research and professional interests focus on the conditions of the economic viability of cultural organizations, private, public, and no profit, and on culture-led policies according to UN SDGs 2030. She coordinates ASviS—Alleanza per lo Sviluppo Sostenibile—group on culture and sustainable development. She takes part in national and international meetings and conferences on these issues and has published extensively in Italian and English. Her most recent publications include Dubini et al. Institutionalising fragility Fondazione Feltrinelli 2016; Dubini et al. "Management delle organizzazioni culturali" Egea 2017; Dubini Con la cultura non si mangia.falso! Laterza 2018.

Alberto Monti is an adjunct professor and member of the Art, Science and Knowledge (ASK) Center at Bocconi University, Milano (Italy). His research focuses on the relationship between social identity and social networks in organizations and particularly on how organizational identification and informal social relationships affect individuals' attitudes and behaviors such as learning behaviors, knowledge sharing, cooperation, and strategic decision-making. He is also interested in sustainability, innovation, category emergence, and consecration in creative industries. His works have been published in *Research in the Sociology of Organizations, Management Decision, European Management Journal, Computers in Industry, Creativity and Innovation Management, International Journal of Arts Management,* and *the Academy of Management Best Paper Proceedings*.

Cultural Heritage Through the "Youth Eyes": Towards Participatory Governance and Management of UNESCO Sites

Mara Del Baldo and Paola Demartini

Abstract Reinterpreting the Cultural Heritage (CH) in the light of the perceived cultural values and expectations of local communities is a significant challenge for politicians and city managers, particularly in UNESCO sites, due to the universal cultural value that they have embed. From our theoretical perspective, individual cultural values are the basis to build on, to create a shared understanding of CH, which, in turn, drives local communities to safeguard and regenerate their cultural commons.

The Urbino perBene—Urbino forGood project, launched in 2017 by the municipality of Urbino, under the patronage of the Italian Minister for Culture and Tourism and UNESCO, offers a laboratory for the exploration and profound understanding of how a World Heritage Site is experienced and perceived by residents and students and how the public administration has intervened to reconcile multiple and often conflicting interests. Through our research, we have found that without a real understanding of the values and needs of the local communities, in particular the youth, it is difficult to achieve the cultural, economic, social and environmental impacts that institutions expect from investments in CH and that should support the sustainable development and regeneration of the city's fabric.

Keywords Cultural heritage · UNESCO site · Historic urban landscape · Cultural value · Cultural commons · Historic city · Sustainable development

M. Del Baldo
Department of Economics, Society and Politics, Urbino University, Urbino (PU), Italy
e-mail: mara.delbaldo@uniurb.it

P. Demartini (✉)
Department of Business Studies, Roma Tre University, Roma, Italy
e-mail: paola.demartini@uniroma3.it

P. Demartini et al. (eds.), *Cultural Initiatives for Sustainable Development*, Contributions to Management Science,
https://doi.org/10.1007/978-3-030-65687-4_14

1 Introduction

The core themes addressed in this study are the transmission and also the regeneration and enhancement of cultural heritage (CH) through the engagement and participation of younger generations (Kazepov 2005, 2014; Silver et al. 2010; UNESCO 2014; Maggioni 2017; D'Alpaos et al. 2017).

Cultural heritage is regarded by scholars and policy-makers as a driver of development—especially for historic cities—and a key element of urban sustainability (Echter 2015; Nyseth and Sognnæs 2013; Nocca 2017).

The participation of communities and citizens in the management of CH is a theme that has been developed by the EU since the Faro Convention (EC 2005). For this reason, in recent years, there has been a flourishing of studies on the participation of citizens in the co-design and co-management of cultural projects and events (see Demartini et al. 2020). At the same time, to our knowledge, there is a lack of literature in the field of management studies concerning youth engagement in both the safeguarding and regeneration of CH.

The city of Urbino, included in the UNESCO List of World heritage since 1998 for its outstanding universal value (see the site whc.unesco.org), offers an interesting laboratory that allows us to scrutinise and discuss this issue both from a theoretical and practical point of view. Despite being developed based on a process shared at an institutional level, the UNESCO site management plan for the city of Urbino drawn in 2013 did not include the engagement of new generations, and among these, the university students who temporarily live in the historic city centre (Municipality of Urbino 2013a). However, the Urbino perBene project, which was embedded in the UNESCO management plan in 2017, intends to recover youth participation as marked by the UNESCO in its strategy 2018–2021 (UNESCO 2017).

The project's initial motivation was to contrast the phenomenon of young wall writers (for some, understood as a manifestation of the malaise of young people; for others, the expression of creativity) to favour a creative dimension in harmony with the aesthetic canons of the city.

Overall, through this project, the administration aims to understand the value attributed to CH by new generations and their possible contribution to both the preservation and reinterpretation of the site's cultural value which should go beyond the "museum city".

Our paper's contribution is twofold. First, it intends to reflect on the attributes that define the concept of cultural value (Throsby 2003; Klamer 2008) and how it is linked to those of CH (Blake 2000; Rodwell 2003; Vecco 2010) and cultural commons (Rullani 2010; Dameri and Moggi 2019). These reflections will offer a framework for the interpretation of the conflict that has emerged in the historic centre of Urbino between residents and young students who live in this famous Italian campus-city.

Second, our analysis has provided us with food for thought to discuss what the involved municipality and institutions should do to preserve and regenerate the historic city centre.

The paper is organised as follows: Sect. 2 explores the literature on CH and sustainable development; Sect. 3 describes our analytical framework; Sect. 4 presents the research design and methodology followed by the case study's narrative (Sect. 5). Findings and emerging issues are pointed out in Sect. 6. Lastly, Sect. 7 closes the paper with some considerations about limits and a further research agenda.

2 The Debate on Cultural Heritage, Historic Urban Landscape and Sustainable Development

Cultural heritage includes both tangible culture (material aspects of culture, such as buildings, monuments and objects, e.g. books, works of art and artefacts), natural heritage (e.g. landscape) and intangible culture (non-material aspects such as folklore, traditions, language and knowledge), which are embodied in social practices, community life, values, beliefs and expressive forms such as language, arts, handcrafts, music and dance (Throsby 1999; Blake 2000).

Accordingly, several authors have conceptualised and defined CH as the legacy of physical artefacts and intangible attributes tied to a group or society and inherited from past generations, maintained in the present and preserved for the benefit of future generations (Vecco 2010).

Despite its crucial importance, in the past decades, CH has long been absent from the mainstream sustainable development debate (Loulanski 2006; UNESCO 2003, 2013).

More recently, the transformation of CH as a driving force in development has been advised (Rodwell 2008) by leading theorists and practitioners, who have pointed out the benefits of policy interventions for the CH sites' safeguarding and promotion to society (Hosagrahar et al. 2016; Alexandrakis et al. 2019).

Below, the milestones of the institutional debate on the role of CH for sustainable development are briefly outlined.

Since the Paris Declaration "Heritage as a driver of development" (ICOMOS 2011), the key role played by heritage concerning social cohesion, wellbeing, creativity, economic appeal and promoting understanding between communities has emerged (Rypkema and Cheong 2011).

Based on a strong appeal from national and local stakeholders, the 2030 Agenda (UN 2015) has integrated, for the first time, the role of culture, through CH and creativity, across the Sustainable Development Goals. Within the framework of action set by the Agenda, CH development is explicitly mentioned in Goal 11, referring to the need to "make cities and human settlements inclusive, safe, resilient and sustainable through inclusive and sustainable urbanisation, planning and management" (target 11.3) and to "strengthen efforts to protect and safeguard the world's cultural and natural heritage" (target 11.4). This specific target addresses CH protection and safeguarding and does not encompass its valorisation and regeneration. However, as Throsby highlights, the economic, social, cultural and environmental systems are not isolated (Throsby 2003). Rooted in the principle of interconnectedness, CH can be considered as the "glue" for the different dimensions of sustainable development (Srakar and Vecco 2016).

Considering the CH impact of on the economy, society, culture and environment, pointed out by "The Cultural Heritage Counts for Europe" project (CHCfE Consortium 2015), the target mentioned above is a key entry point for integrating CH into sustainable urban development through policies adopted by local institutions. A legal framework is needed for transparent and participatory planning and management, incorporating heritage use and traditional settlement patterns and materials as a core component of liveability and sustainability (ICOMOS 2015).

In this journey, the UN-Habitat New Urban Agenda (NUA) made a step forward in the achievement of many SDGs, highlighting the role of Historic Urban Heritage/ Landscape (United Nations 2016a, b). The NUA recognises both tangible and intangible CH as an essential element, stressing its role in sustainable urban development.

Moreover, the Habitat III Issue Paper on "Urban culture and Heritage" provides a valuable contribution to these questions, considering culture as a key element in the humanisation of cities and human settlements (United Nations 2016a, p. 6), which plays an important role "in rehabilitating and revitalising urban areas, and in strengthening social participation and the exercise of citizenship" (*Ibid.* p. 8). Culture is recognised "as a priority component of urban plans and strategic development policies that safeguard a diverse range of tangible and intangible cultural heritage and landscape" (*Ibid.* p. 17). In a nutshell, culture is considered a lever to increase citizen's quality of life in all its different dimensions.

Accordingly, CH contributes by promoting:

- *Social cohesion and equity*. CH strengthens communities, where citizens associate the historic environment with a shared identity and attachment to the place, including minorities, disadvantaged or socially excluded people.
- *Inclusive economic development*: cultural and creative industries in historic areas have increasingly contributed to revitalising their economies by promoting their intangible heritage.
- *Liveability and sustainability of urban areas*: adaptive reuse of the existing fabric including buildings, historic areas and downtowns can be efficient and ecological.

A separate discussion must be made for historic city centres (UNESCO 2004, 2006, 2008; UNESCO/Un-Habitat 2006; EAHTR 2007; Veirier 2008; Enyedi and Kovacs 2006). In such a context, the Historic Urban Landscape (HUL), defined as "the historic layering of cultural and natural values and attributes", which incorporates the intangible dimension of heritage and the related economic processes (UNESCO 2011, art. 8), is a recent approach to the safeguarding, protection and valorisation of CH that goes beyond the geographical and geomorphological scope, determining the increasing relationships between conservation and development (Greffe 2009). This approach recognises the landscape as a "living heritage"—an "organism" made of complex characters and multidimensional interrelationships

(Veldpaus and Pereira Roders 2014; Fusco Girard 2013; Fusco Girard et al. 2015; Angrisano et al. 2016).

The role of HUL in local development processes has been recognised as "cultural urban heritage/landscape provides quality, sense and meaning to the urbanisation processes, promoting the implementation of "places" as attractive (economic/social/cultural) spaces in the city/metropolitan areas, where many plus values are produced" (UN-Habitat 2014, p. 49).

The enhancement and regeneration of the historic city centre are considered by many to be an opportunity for sustainable economic development based on the spur of culture and creative industries and sustainable tourism (Evans 2002; Stolarick and Florida 2006; Cooke and Lazzeretti 2008; Nakagawa 2010; Sacco et al. 2014). However, there was no lack of criticism because there is also a risk of abuse and the appropriation of common symbolic and material resources (Sacco et al. 2019). Culture-driven gentrification is one example of this type of appropriation (Atkinson and Bridge 2005; Zukin et al. 2015).

Management and interventions in historic environments and CH sites encompass actions to preserve and enhance it, thus preserving and possibly increasing its cultural value. Therefore, new tools and approaches for planning and managing complex and dynamic urban systems are required to improve city governance, fostering creativity, resilience and sustainability, and reduce the conflict between specific interests and general ones. They can be based on pilot experiments, like the one investigated in this paper, whose effectiveness is worth being assessed.

3 Cultural Value, Cultural Heritage and Cultural Commons: Our Analytical Framework

Culture is a substructure of society because it gives meaning to our lives and society. Indeed, through culture, people and groups define themselves, conform to society's shared values and contribute to society. Hence, culture provides people with "meaning" and identity, and this dimension is one of the main reasons to invest in "common culture" (Gielen et al. 2015).

The notion of *cultural value is not straightforward* (Crossik and Kaszynska 2016). It is now part of a common understanding that cultural value expands beyond the limits of inherent value. Indeed, the notion of the cultural value of the historic environment and CH sites is related to a set of characteristics or qualities that society, consisting of various groups with a legitimate interest in the site, attributes to such object or site (De La Torre et al. 2005).

Throsby (2003, pp. 279–280) states that cultural value is "*multi-dimensional, unstable, contested, lacks a common unit of account, and may contain elements that cannot be easily expressed according to any quantitative or qualitative scale*" and is made up of several values:

- the *aesthetic value* refers to "properties of beauty, harmony, form and other aesthetic characteristics" (*ib.*);
- the *spiritual value* is the value given to a cultural artefact or experience either because of its "significance to the members of a religious faith, tribe or other cultural groupings" (*ib.*), or its secular importance;
- the *social value* is the value linked to the sense of connection with others, present and past generations, evoked by artwork or CH;
- the *historic value* relates to the way "it reflects the conditions of life at the time it was created" (*ib.*);
- the *symbolic value* is linked to the capacity of the artwork to convey a particular meaning beyond its material aspects and recall, for example, the sense of belonging, identity and self-awareness;
- the *authenticity value* is the value that comes from the fact that the cultural artefact is original.

Throsby's (2003) decodification of cultural value has been accepted, either implicitly or explicitly, in the cultural economics literature (Snowball 2011; Klamer 2008, 2016), even though some additional components of cultural value have been envisaged by other researchers, such as the experiential value (Hutter and Shusterman 2006, p. 198), that is, the "directly satisfying or pleasurable experience" given by the cultural good/experience and the educational value (Throsby and Zednik 2014, p. 88) that can be identified in terms of the artwork's role in the education of children.

Researchers agree that the attributes of cultural value are not perceived in the same way by each individual; hence, the evaluation of cultural value has a high idiosyncratic component (Hernando and Campo 2017).

All the above dimensions can be applied to cultural goods, services, assets, heritage, and, as far as our case study is concerned, historic towns/cities, in particular. The historic city, indeed, "Is generally built up and closely knit. It is multifunctional, with strong virtues of proximity and accessibility, which limits the need for infrastructure and transport. It is functionally and socially mixed, supporting a wide range of complementary activities. It is human in scale and expresses geo-cultural diversity. This provides a strong sense of tangible and intangible identity" (Rodwell 2008).

The UNESCO clearly stated, "Cultural urban heritage relates the history of the city, its inhabitants, religions and social and cultural transformations. This heritage is deeply anchored in the spatial and economic structure of the cities, their buildings, and monuments. The people living and working in the city identify with it. Today, historic districts are symbols of the city's image; above and beyond their own cultural value they fulfil an important mission in modern urban development: they create the identity and the city's image and are key geographic factors for the local and regional economy" (UNESCO 2008, p. 9).

Furthermore, according to UNESCO, World Heritage properties bear an Outstanding Universal Value (OUV) of cultural and/or natural significance, "Which is so exceptional as to transcend national boundaries and to be of common importance for present and future generations of all humanity. As such, the permanent protection of this heritage is of the highest importance to the international community as a whole" (WH compendium available on UNESCO website).

Therefore, even when considering the list of UNESCO heritage sites, public officers in charge of it should try to balance its safeguard with the needs of the communities that bear legitimate interest in such heritage sites.

Another relevant concept to elucidate is that cultural value is dynamic value. Cultural goods are "discursive constructs," and their values are established through conversation, dialogue and the sharing of meaning (Klamer 2008). The concept of dynamic value can also be considered an explanation of how the value itself is formed through a process of growth/decline which occurs, thanks to the exchange and interaction between the individuals of a certain society along the timeline.

Dynamic value also allows us to introduce the concept very strictly related to cultural commons. Furthermore, it offers a clue about how CH should be managed to preserve it.

Commons are understood as natural and cultural resources accessible to all members of a society inherited from the past. They are organised systems of functions and meanings that are the product of evolutionary learning, realised in the past, sedimented over time and transmitted to us (Rullani 2003).

Commons are linked to particular subjects, that is, interpersonal groups, territorial communities and communities of sense (Ansari et al. 2013). They are rooted in a place because they have a direct relationship with the subjects who share its/their use and have contributed to making its history and meaning.

Core aspects of common goods include the interdependency of individuals in enjoying the good, uniqueness and non-renewability of the resource and almost all the social value of the resources (Arjoon et al. 2018; Euler 2018; Isaac et al. 2010).

Drawing from such a theoretical construct, CH can be conceived as common goods since they belong to a community of people that bear the right to enjoy the cultural content of those assets regardless of their ownership (Dameri and Moggi 2019, p. 1). In other words, cultural assets are commons of a tangible and intangible nature (Rullani 2010) as they are unique, unrepeatable, embody a high social value and enjoyed by a community of individuals (Hess and Ostrom 2007; Riganti and Nijkamp 2016).

CH represents a common because it is (or it should be) available for collective benefit. Cultural assets embody shared content that belongs to a community of people and sometimes to all humanity, for example, UNESCO world heritage properties (Council of Europe 2005; UNESCO 1968). Nevertheless, at the same time, it suffers from vulnerability due to beneficiaries' misuse and/or neglect (phenomenon called the "tragedy of commons" by the ecologist Hardin in 1968). Indeed, vulnerability depends on several factors, ranging from their overexploitation (like in the case of Venice) to underuse/disuse and/or lack of maintenance (Dameri and Moggi 2019).

Specific features of cultural commons are tied to their social and cultural value, rendering them different from traditional commons. Namely, their management requires cooperative and participatory approaches aimed at enhancing their

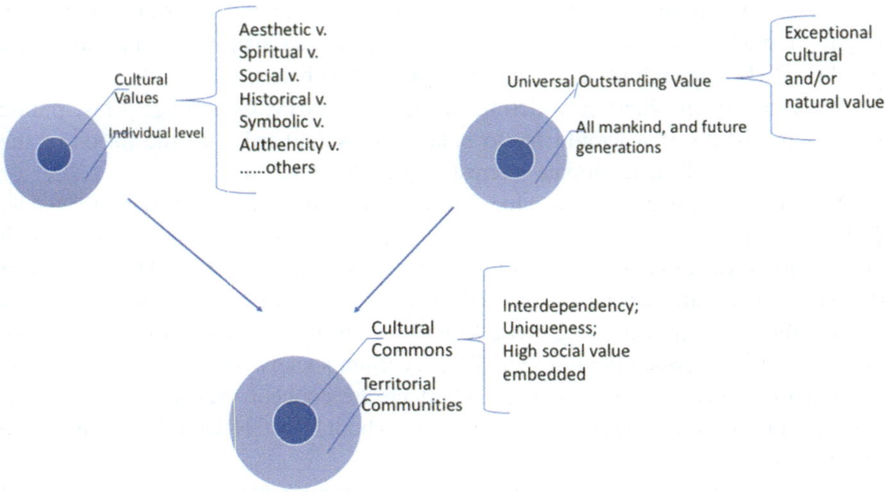

Fig. 1 Attributes of cultural value, universal outstanding value and cultural commons. Source: own elaboration

preservation, economic development and cultural fruition; it is based on the involvement of multiple partners (i.e. public agencies, profit and not-for-profit organisations, local communities, citizens, volunteers) and implies the match between different cultural values and perspectives, including the individual, meso and even universal level.

The vulnerability mentioned above indeed increases in the absence of virtuous governance bodies and collaborative mechanisms aimed at providing long-term sustainable management of the cultural commons (Capello and Perucca 2017; Dubini et al. 2012), especially when they bear a universal outstanding value, as it happens for the World Heritage properties.

Below, Fig. 1 summarises and depicts our framework, linking cultural values attributes to the cultural commons at different levels of analysis.

4 Research Design and Methodology

Drawing from such premises, this work intends to detect and shed light on the perceptions and expectations of the inhabitants and in particular the younger generations regarding the usability and the needs of conservation and enhancement of the cultural capital heritage.

To this end, we have analysed the "Urbino forGood" project, launched by the administration in 2017/2018 (Municipality of Urbino 2013b, 2016).

Urbino is a city of ancient origin that hosts a university campus—whose structures are distributed in prestigious historic buildings in the city centre. A primary component of the community is made up of young people, reaching a population almost equal to that of the residents, so it is vital to find the right balance between these two "souls", affected by forms of malaise and misunderstanding that have arisen.

Young students' experience in the city has been defined as a "temporary journey" of their life (Kazepov 2005, 2014; Silver et al. 2010). Students are considered city users, passing citizens who weave a special relationship with the city, very different from that of the residents (citizens). The coexistence of both old (permanent inhabitants) and young (temporary inhabitant) generates wealth and social capital. At the same time, it causes inconveniences and tensions due to the different ways they understand and experience the city (Ceccarini and Diamanti 2013).

The local administration is interested in understanding their respective expectations and needs and reconciling conflicting interests. Therefore, an exploratory survey was carried out in January 2018 within the project first to understand the expectations and wishes of the local community with a focus on inhabitants of the historic centre and its users, namely the students.

However, this case does not lend itself only to descriptive analysis. Indeed, the second objective is to investigate the value attributed to a world heritage site represented by a historic centre that can be considered a cultural common.

In Urbino, the university plays a leading role and students are a precious resource both economically (because they constitute a vital component for the development of local catering, accommodation activities, libraries, etc.) and socially (for their regenerative power).

Beyond the conflict that emerges between citizens and city users, we will try to understand if the new generations share the universal cultural value attributed to the UNESCO site and if young people are interested in participating in the management of cultural commons.

4.1 *Methods and Data*

The study has been developed following a deductive and inductive approach. The former draws from the literature on commons and CH applied to a historic city. The latter is based on an in-depth analysis of a case study (Yin 2013; Eisenhardt et al. 2016).

The empirical analysis was based on the action research approach since local scholars belonging to the University of Urbino were directly involved in the process of cultural intermediation between the public administration and students. Three researchers from the Department of Economics, Society and Politics took part in the project by collaborating with the municipality and involving young research fellows and a master's degree student in the creation of the questionnaires used for the pilot survey/test. In addition, during the 2019 and 2020 academic years, the researchers provided several seminars and workshops to students and the larger community to sensitise them on the topic.

4.1.1 Data Collection

Data were collected from 2017 to 2019 by one of the authors, who acted as a "participant-observer." Data used for the analysis came from different sources:

1. n. 80 fully completed filled questionnaires addressed to Urbino citizens (n. 30) and students (n. 50).
2. n. 30 semi-structured interviews with university students attending the seminars and workshops organised under the project Urbino PerBene.
3. researchers' notes from n. 10 meetings, seminars and workshops in total.
4. Urbino municipality documents and n. 20 local press articles on the wall writers-graffiti phenomena and the Urbino perBene project.

As far as our research objects are concerned, the most interesting source of information refers to the questionnaires aimed to understand if they perceive the historic heritage of the city as deteriorated and collect their suggestions to increase or generate positive behaviours. The questionnaires were submitted in early 2018 and were organised in four sections, addressing the following aspects:

- Reasons for visiting/using the city centre.
- Problems with enjoying the city centre.
- Suggestions for enhancing how people experience the city centre.
- Perceptions of people's commitment to urban decor.

Although the analysed sample is not statistically representative, the information collected has revealed interesting recurrences for an exploratory investigation.

4.1.2 Data Analysis

As suggested by Miles et al. (2014), data analysis was an ongoing process. Available data were iteratively analysed to allow a progressive elaboration of a general interpretative framework. In the first step, we created a detailed description of the case, putting together interviews and data from all sources. In the second step, we cycled through multiple readings of the data. In line with the purposes of our research, different values and perceptions from Urbino's citizens and city users were identified. With this purpose in mind, each author read the empirical material independently and categorised the stream of words into meaningful categories, via manual open-coding. Subsequently, the results obtained by each author were compared and discussed. In cases of coding disagreements between authors, interviews and other data were jointly reanalysed, and codes were discussed to reach a consensus.

This iterative and interactive analysis process allowed us to describe the conflicts among the two main stakeholders living in the historic centre: the native residents and university students, which can also be considered as a proxy of a conflict among different generations. Furthermore, in our analysis, we found the use of excerpts highly worthwhile as they allow the researchers to focus on the main aspects of the relational capital creation process.

5 The Case Study

5.1 The Setting: Urbino and the UNESCO World List of Heritage Sites

Urbino is a historic city dating back to the fifteenth century located in central Italy (Marche's region). It is a perfect synthesis between landscape and architecture. It represents an urban palimpsest, an example of resilience and transformation, due to the stratification in its historic fabric that allows one to trace different stages of its evolution marked by the supremacy during the Renaissance period. Urbino was the birthplace of the master painter Raphael and attracted some of the most outstanding humanist scholars and artists, eliciting the inclusion of the old city centre into the UNESCO World List of Heritage city in 1999: "Urbino represents a vertex of Renaissance art and architecture, so it has harmoniously adapted to its physical environment and its medieval past, becoming a completely exceptional city" (see Fig. 2).

Fig. 2 The UNESCO Site "Historic Centre of Urbino" coincides with the Area defined by the Renaissance Walls. Source: Siti UNESCO per Bene. World Heritage-The Right Way (www.comune.urbino.pu.it)

However, its tourism vocation, held up in particular by the Duke Palace attraction, does not represent the main economic vocation, since the city's economy mainly relies on the University of Urbino (dating back 1506), located in old buildings inside the walls of the city centre, that had about 13,000 enrolled students (in 2019) and generates jobs and other satellite activities (Maggioni 2017). Currently, the student population is almost equivalent to that of the native inhabitants (14,468). Moreover, while only 1000 of them live in the historic centre, the "campus-city" hosts students from different social and geographical backgrounds who populate the classrooms, libraries and laboratories located in the old city centre. All the architectural heritage is used by them daily, thus risking the safeguarding of the CH, due to the (negative) impact of their presence and conduct throughout the city (D'Alpaos et al. 2017). In this regard, the "studentification" process (Ceccarini and Diamanti 2013; Maggioni 2017) points out how much the university influences the urban life in terms of changes and effects (both positive and negative in nature) related to the settlement and economic, social and cultural dimensions of students in specific contexts.

5.2 The Project Urbino perBene

The "Urbino perBene" project was launched in 2017 by the municipality of Urbino under the patronage of Mibact (The Italian Minister for Culture and Tourism) and UNESCO for the enhancement of youth awareness and urban decorum of the cultural and environmental heritage represented by the historic city centre of Urbino, listed as a UNESCO site.

It was set in motion in 2013, when Urbino and Florence's municipalities approved a memorandum of understanding aimed at tailoring the "Firenze perBene" project to the city of Urbino to push city users to adopt good practices, enhancing their consciousness of being in contact with commons typical of a universal CH, unique and exceptional in value.

In 2017 the "UNESCO forGood" project, concerning Florence, Urbino and Mantua (Florence perBene, Urbino perBene and "Mantova con i tuoi occhi"—Mantua through your eyes) was awarded during the meeting of European World Heritage Associations held in Lübeck (Germany) on the topic of Synergising World Heritage and Sustainable Tourism for Development.

For the implementation of the project, the municipality activated the classification and geo-referencing of the degradation phenomena that have afflicted the city roads and the buildings (i.e. graffiti) through the UNESCO Office—Urban Decoration and Urban Hygiene. A preliminary analysis was carried out through the cataloguing of urban decay phenomena resulting from the soiling caused by wall writers (see Fig. 3).

In parallel, the municipality started the process of involvement and awareness of the various social groups of city users, to set up a process of information, management and maintenance of the site.

Fig. 3 Writings appearing on the walls of the Urbino historic centre of Urbino. Source: Siti UNESCO perBene. World Heritage-The Right Way (www.comune.urbino.pu.it)

A summary of the objectives, steps and outputs of the project is presented in Table 1.

6 Findings and Discussion

The Urbino perBene project is very complex. In this paper, our goal is to dwell on the different perceptions that residents (citizens) and city users (university students) have of the historic city centre to express some considerations in light of our interpretative framework.

6.1 *Conflicts Related to the Coexistence of Citizens and City Users Within the Urbino Historic Centre*

The analysis of conflicts between citizens and city users draws on the questionnaires administered to students and residents.

Our findings confirm a social and spatial separation among students/residents, which significantly affects how the city is experienced and perceived. In other words, there is no real, deep integration between the parties nor an authentic exchange, rather a feeling of mutual tolerance.

Table 1 Urbino perBene—Educare al Bello project (Urbino forGood—cultivating beauty)

Summary of the project	In recent years, vandalism has occurred in the historic centre related to the staining and damage of artistic and architectural works, buildings and urban buildings, which has been accompanied by the abandonment of waste on public land The project supports the importance of culture and education, raising awareness and promoting respect for the environment and the protection of CH to combat degradation behaviour by increasing youth awareness and urban decorum and enhancing the cultural and environmental heritage of the UNESCO historic centre of Urbino site
Aims and objectives	• To enhance the citizens, residents and tourists' awareness and their respectful and attentive attitude towards historic and artistic heritage • To increase awareness of the exceptional beauty of the historic centre, calling for people's involvement in its conservation and enhancement • To promote good practices in favouring the safeguarding of the site for future generations of an invaluable asset
Objectives	• Educate people to understand and respect the CH of Urbino • Educate to have a reciprocal coexistence and good behavioural practices • Raise awareness of the artistic, monumental and landscape beauty of the area • Develop ecological awareness • Extend the knowledge of the CH of the city to all students • Raise awareness of socio-environmental problems to fight the "vandalism" phenomena • Reflect on the condition of the CH so that it can also be transmitted to future generations • To lay the foundations for the direct involvement of young people and students in the implementation of future projects • Promote the participation of young people in the development of fundamental values for the protection of the common good such as respect and civic sense • Encourage students' civic training in the enjoyment and safeguarding of an invaluable asset, heritage of and for all • Make young people protagonists of the process of valorisation and protection of the artistic and environmental heritage
Recipients	• Italian and international students of the University of Urbino Carlo Bo • Middle school students from 11 to 13 years old and high school students from 14 to 18 years • Italian and foreign visitors and tourists • Residents and native inhabitants of the historic centre • Shopkeepers, traders and operators of the city-centre
Methodology	• Collection of information through questionnaires (direct/personal and online submission) • Promotion and support of a set of informative and operative initiatives in collaboration with organisations, institutions and individuals belonging to the local community • Participation in the cultural, social and economic events of the city with the Urbino perBene project
Activities	• Creation of questionnaires to be addressed to the different categories of Urbino's city users • Creation of a decalogue with suggestions (in Italian and English) useful for

(continued)

Table 1 (continued)

	enjoying the city and respecting its places and values • Creation of a map of the centre, indicating its public and tourist services (i.e. drinking fountains, squares and gardens, info points, toilets) • Training courses, laboratories and projects to be promoted in collaboration with students of institutes and high schools of Urbino and the University of Urbino Carlo Bo aimed to sensitise young people to artistic beauty and historical and cultural value • Survey of the quality of the tourist service of the UNESCO site through questionnaires to be submitted to foreign tourists
Tools	• Human resources and materials made available by the schools, universities, volunteers, not-for-profit and civic organisation, local institutions
Collaborations	• Collaborations with ISIA (High School of design) for the realisation of the city map and the decalogue "Urbino perBene" • Collaboration with not-for-profit organisation (i.e. Anteas and Arts Regresso Association) for the organisations of events • Collaboration with the local university department and the local school for the organisation of seminars, conferences, laboratories • Collaboration with volunteers (Volontari per Valbona) and Arts Regresso Association for the removal of graffiti and the restoration of areas and buildings damaged by vandalism
Outputs	• Seminars, congresses, conferences, laboratories and debates in university classrooms and schools • Meetings and social events promoted in favour of the protection of the artistic and architectural heritage, organised in collaboration with local partners (i.e. Anteas) • Elaboration of informative/qualitative questionnaires in Italian and English • City map and decalogue for city users • graffiti cleaning started on the municipal properties defaced and smeared by vandalism; these activities have been accompanied by information and communication campaigns aimed at the various categories of city users

Namely, drawing from the replies to the questionnaire, the following divergent and conflicting perspectives emerged when interviewees were asked about the problems that affect the historic centre of Urbino and how to overcome them.

Students declared that their respect for the city and CH is not waning and that episodes of vandalism are isolated cases. Only 14% of them stated that students care that the local heritage is decreasing. Furthermore, they showed a general tendency to welcome culture-led initiatives. The majority of students believe that the concession of public spaces for cultural, commercial and spontaneous events (64%) and the organisation of "white nights" or museum openings at night (56%) can involve more people to enhance the historic centre.

On the other hand, residents perceived a decrease concerning the urban historic landscape. Their complaints are about defaced buildings (87%), animal excrements (87%) and abandoned waste (80%). They demanded campaigns in favour of the civic sense to preserve the value of the historic centre and the quality of life of residents, which often does not fit in with the expectations of students living on a university campus-city. Sixty percent of citizens reported loud noises, especially during the night, attributed to students partying at pubs. Furthermore, residents believe that greater control regarding the rules, with penalties for offenders (63%), might be more effective to avoid negative consequences concerning poor care of the location. No students indicated rules and sanctions as a viable means for improving the conditions of the city or the quality of the life of citizens.

A series of additional shortcomings emerged, which can be seen as elements of the campus-city's weakness, which tend to produce fractures and affect how the city is actually experienced by students. For instance, the public transport is generally referred to as inadequate. All of the students complain about the lack of night-time transport services, the excessive costs of tickets and the problem of crowded buses.

The relationship between students and the Municipality of Urbino also appears unsatisfactory; students feel economically "useful" to the wealth of the city. However, they believe that the municipal administration is more inclined to respond to voters than to attempt to combine the needs expressed by students and residents. A duality has therefore been generated, igniting potential situations of social conflict.

A distinctive part of the students "feel" the city is transitory and distant, a place where they spend an important period of their life but not a destination where they can settle down in the future. In particular, students asked to use gathering spaces, for cultural events, and greater opportunities for socialisation within the historic centre. Moreover, they expressed a strong desire to participate in the decision/policy-making processes that affect the daily life of the city. Hence, it emerged that implementing and institutionalising bottom-up processes, involving the active engagement of students as primary stakeholders, can represent a first step towards the recomposition of the social and economic fractures of the campus-city.

6.2 Urbino and the Historic Centre, "Through the Eyes of the Youth"

The semi-structured interviews were suitable for collecting personal comments and suggestions useful for the protection and safeguarding of the artistic CH and the improvement of the services located in the centre.

The interviews were also a beneficial source of information that we coded based on our analytical framework to:

(i) detect, through the analysis of personal comments, the cultural value perceived and attributed by individuals to the historic centre of Urbino;
(ii) understand, through the suggestions provided for its management and safeguarding, the value attributed by young people to the historic centre of Urbino—the UNESCO site, as a cultural common.

Table 2 below summarises the most significant excerpts and offers a synopsis to the readers of how we have coded them in light of our interpretative framework.

Table 2 Summary of excerpts coded in the light of our interpretative framework

Name	Excerpts	Throsby's attributes of cultural value (2003)	Excerpts	Cultural Commons' attributes (Rullani 2010; Arjoon et al. 2018; Euler 2018; Dameri and Moggi 2019)
Eleonora	*I imagine Urbino as an elderly lady who can transmit the beauty of her years*	Historical value	*The management of a UNESCO Heritage Site [is]: relationships based on cooperation, trust, reputation and proactivity*	High social value
Serena	*Urbino is a magical city* *Urbino helps you find yourself when looking at the Ducal Palace*	Symbolic, aesthetic, authenticity and social values	*Urbino makes you feel independent but, the same time, never alone* *This City is like a home, and as students, we have the duty to respect and preserve it*	Uniqueness, high social value
Erica	–	–	*The work to raise awareness of the issues of the conservation of cultural heritage and urban decoration is still long, but the involvement of all people is essential, starting with me*	Interdependency, need for proactive involvement of all people
Giuseppe	*[I recognise] the value of the historical, artistic and cultural heritage of the city*	Historical and aesthetic values	*I believe that the commitment of the administration represents the right path to maximise harmony and improve coexistence among the deeply different categories of users of the city. The beating heart of Urbino and the undoubted source of life remain in the students*	Interdependency, focus on young people regeneration energy
Antonio	–	–	*We must sensitise ourselves to respect what is not ours. We, the students, for*	Proactive involvement of the youth as educators

(continued)

Table 2 (continued)

Name	Excerpts	Throsby's attributes of cultural value (2003)	Excerpts	Cultural Commons' attributes (Rullani 2010; Arjoon et al. 2018; Euler 2018; Dameri and Moggi 2019)
			example, could act as educators to help our peers understand the respect for the city	
Federico	*A historic city like Urbino should be subject to particular respect and concern made possible only by a path of awareness, which starts first within the schools*	Historical and symbolic values	*[PA should] not exclusively meet the demands of current residents because it could lead to the progressive and harmful ageing of a city which… has an extreme need for the vitality of young people and their energy*	interdependency, focus on young people regeneration energy
Lorenzo	*Those who gave their contribution by explaining their impressions and their point of view shared with me the love they feel for Urbino*	Spiritual, symbolic and social values	*I understood how important everyone's help and collaboration for the whole city is*	High social value
Martina	*Urbino cannot disregard its history*	Historical value	*Behind graffiti or a mural, we can hypothesise hiding an expressive need and, why not, sometimes, art forms* *Friedrich Nietzsche asserted that it is better to love one's enemies "because they bring out the best" in us*	Interdependency, focus on young people regeneration energy
Vanessa	*We have chosen to study in the city of Urbino, a place rich in history and full of beauty that we breathe every time we walk through the streets*	Historical, aesthetic and symbolic values	*I am very happy with this initiative that … allows us to act as a sort of ambassador for our cities* *We do not have a reserve planet, and there is no "reserve Urbino"; therefore, it is our duty to*	Uniqueness Proactive involvement of the youth as ambassadors of the city

(continued)

Table 2 (continued)

Name	Excerpts	Throsby's attributes of cultural value (2003)	Excerpts	Cultural Commons' attributes (Rullani 2010; Arjoon et al. 2018; Euler 2018; Dameri and Moggi 2019)
			safeguard the beauty that we have received as a gift	
Martina	–	–	*I believe that it is necessary, first of all, to sensitise young people with initiatives that allow them to participate... starting with the management of cultural activities in a "devoted place"*	Interdependency, focus on young people regeneration energy

From the analysis of the interviews, with the most significant excerpts in the table, we have found that youth attribute great cultural value to the heritage of Urbino, collectively recalling all the attributes identified by Trosby (2003), namely: Historical, Aesthetic, Symbolic, Spiritual, Social values.

We deem that this sharing of values, beliefs and feelings is the basis for the development of cultural commons and that the involvement of young people is important to increase their awareness of the outstanding cultural value of the setting in which they live.

The education and awareness process, as the respondents pointed out, takes time and should start right from primary school. Furthermore, to be effective and lasting, the educational content cannot be transmitted only by experts, public officials and/or teachers with a top-down, one-way direction but must be shared, and the young people themselves must communicate this message of care and attention for CH to their peers.

The *awareness-sedimentation-sharing* of individual values is the process that allows for the birth of sedimented collective values that belong to a community of a territory and that we call "commons" (Rullani 2010).

Indeed, in the words of the young people, we found widespread consciousness that CH is conceived as common goods since they belong to a community of people bearing the right to enjoy the cultural content of assets that stem from the past "as a gift". Cultural commons, such as the Urbino historic centre, are unique and non-renewable, and for those reasons, people are disposed to be engaged in safeguarding them. Meanwhile, the interdependency of individuals in enjoying urban CH, and the need to invest in a path of coexistence among the different communities that benefit of them are aspects also emerging from previous research on commons (Arjoon et al. 2018; Euler 2018; Isaac et al. 2010).

What was unexpected is the awareness on the part of many of the interviewees, that youth can bring stimulus and a regenerating power that, if correctly conveyed by the public administration, can help a historic centre like Urbino to avoid remaining crystallised over time as a museum city.

This interesting insight from the interviews was food for thought for some consideration aiming to contribute to the debate on the regeneration and development of historic urban areas (see Sect. 2) in the light of our framework on cultural values and cultural commons.

6.3 Discussion on the Regeneration and Development of Historic Urban Areas and Cultural Commons

By applying the interpretative lens of commons (Rullani 2010), our research highlights that CH is witness to the history and culture of the inhabitants through their practices, activities, memories, desires and identitarian feelings.

In the same vein, its revitalisation means reaching a satisfactory balance between economic development, the needs and the rights of inhabitants and the value enhancement of the city as a public good (Euler 2018). However, in the meantime, as our findings reveal, city users, such as university students in Urbino, live in the same spaces and may have different expectations, roles and cultures. Partnering in managing the revitalisation of historic areas may allow each one to give meaning to the heritage, feeling involved and participating in the projects.

> I believe that collaboration and belonging to a network of partners is a key factor to generate value for all the stakeholders involved. Just as it happens within a company, the same applies to the management of a UNESCO Heritage Site. (Eleonora)

Hence, "Collective cultural heritage management" seems to be the answer to allow all citizens (residents, city users, tourists, migrants) to live and work in the historic urban districts (short- or long-term).

The revival of the feeling of belonging to a history, a culture, a region or a district is symptomatic of the human need to know oneself and for others to recognise one's identity. While cities have become places where cultural expressions are varied and seek to be fully expressed, the role of urban heritage is fundamental in building a common civic sense based on the individual recognition of the cultural values related to CH and the sharing of the concern to safeguard it (UNESCO/Un-Habitat 2006, 2008).

The role of urban policy decision-makers is central in facing this need for the recognition and sharing of values. The value of historic centres as witnesses of the past and laboratories of the present and the future requires approaches and specific regulations designed for cultural commons. In general, sustainable urban development depends on encouraging local democracy, participatory policies (EAHTR

2007) and engagement with the citizens and city users in the regeneration and transformation process (Sacco et al. 2019).

We deem that a good understanding of the attributes of cultural values by stakeholders to these historic areas and social practices in use in these multifunctional, different spaces are essential before embarking on any process of change.

Every historic district, indeed, has its specificity (i.e. the uniqueness of cultural commons), even in the same country, the same region and the same city; thus, no model can claim to replace an analysis of the resources, values and local practices that are subject to constant evolution (Nyseth and Sognnæs 2013).

Concerning our case study, students call for the provision of public spaces, which can be an opportunity to enhance the standing of urban identity through creativity and opening up to cultural diversity.

> One must not to exclusively meet the demands of current residents, because it could lead to the progressive and harmful ageing of a city which, although based on a very important historical legacy, nevertheless, has an extreme need for the vitality of young people and their energy in developing new opportunities and prospects for the future. (Federico)

They call for a (re/new)appropriation that implies a revitalisation process. Such revitalisation relies on governance transformations of established processes based on networks, stakeholders, discourses and practices (Healey 2006) and flexible policies and practices capable of meeting different needs (i.e. regeneration consistent to modern standards of living, youth ideas) and addressing the different forms of challenges that such areas face.

All over the world, many historical areas such as the Museum Quarter in Wien (Biondi et al. 2020), to mention one, have been transformed into cultural, educational, leisure and trade spaces enjoyed by artists, citizens and tourists, often with a reorganisation of the public space.

Art and creativity are an integral part of revitalisation processes in historic districts. Hence, the young people must find a way to express their creative energy in spaces and city events dedicated to them, so that it can be a regenerating force of the historic urban areas.

> I absolutely recommend increasing the promotion of musical events, cultural activities and wider social gatherings, always in a vision of an intimate and transparent collaboration between the students, the university and the municipal administration. (Giuseppe)

The atmosphere, vitality and dynamism of a district are all elements that make people want to live, share and develop an area and, hence, to innovate and create. Thus, creativeness in revitalisation processes often produces projects that involve new methods, new relationships between actors and new ways of working, design, understanding and managing an area.

Concrete results and experiences show that creativity goes hand in hand with innovation and that the enhancement of the attraction of a region and a district can be realised through the improvement of its creative potential, its image and the inhabitants' sense of belonging, and offering broad access to culture for all sections of society.

All this means protecting, developing and giving meaning to tangible heritage by linking traditional and modern knowledge as the young generation calls for.

> Graffiti is an unspeakable action to be fought, but behind graffiti or a mural, we can hypothesise hiding an expressive need and, why not, sometimes, art forms. In my opinion, the most important challenge for Urbino is to reconcile its past with its present and future. (Martina)

Finally, we deem that the Urbino perBene project represents an interesting experimentation, which goes beyond the fight against the phenomenon of wall writers, that is, a first step towards the renewal of the urban context by looking at younger generations.

7 Conclusion

Worldwide examples have illustrated the current trend towards a holistic, interrelated perspective on cultural heritage, which is seen as socially constructed and more dynamic and functional. CH is both an element and a tool for socio-economic development (UN-Habitat 2014; CHCfE Consortium 2015).

Such a holistic approach to heritage and development links sustainability and conservation as two complementary processes that aim to achieve the wise use of resources (Rodwell 2008). It is critical to remember that, "The economic benefits of cultural heritage evaporate if the social, aesthetic, and other values of cultural heritage are not respected and are lost" (Agnew 1999, p. 11).

The attention and debates among theorists and practitioners should be engaged not so much with *if* but rather with *how* CH and development can be integrated successfully and sustainably.

This challenge must integrate cultural and sociopolitical particularities of cities and take into account their history and traditions. To respond to the challenges of long-term revitalisation, all the city actors (i.e. citizens, politicians, public officers, experts, entrepreneurs, researchers, investors, urban planners) are called to promote participatory multidisciplinary sustainable governance (Del Baldo and Demartini 2010).

In this hyper debate, this paper's contribution is twofold. First, the analysis contributes to filling the gap in the management literature, addressing attention to the transmission of the cultural capital heritage among generations and the role of education in "cultivating the beauty", the Urbino perBene project motto, thus providing a basis for further analysis.

Second, the study identifies weaknesses that should be overcome by the public administration to engage the youth in the safeguarding and regeneration of the historic city centre of Urbino and suggest possible directions for successful cultural commons development.

This paper leads the way to future research to study how the public administration of Urbino has worked on urban regeneration projects, both safeguarding the CH in its tangible and intangible dimensions, and, at the same time, making it accessible to the younger generations and empowering them to preserve and enhance the CH they have inherited.

Indeed, the Urbino perBene project marks the importance of the local communities' strong commitment.

In this sense, it also points out the importance of the fruitful contamination of different cultural values belonging to students and inhabitants to nurture shared values, which are the basis for the successful participatory governance of cultural commons.

As a qualitative explorative study, this work is affected by some limitations. Since this study has examined only one case study involving the historic city of Urbino, the analysed process could be impacted by its specific socio-economic conditions, such as being an important campus-city that has a large number of students who temporarily live there. However, we deem that this condition is similar to that of other historic cities in Europe. Therefore, future research could be conducted in other locations to compare our findings. Moreover, the "Urbino perBene" questionnaire was administered to a limited number of residents and students. Hence, our findings do not have statistical significance, though they have provided interesting insights for the politicians, public officers, urban planners and food for thought for researchers.

Acknowledgement While Mara Del Baldo was directly involved in action research, Paola Demartini acted only as a discussant. Authors contributed equally to the drawing of the chapter.

References

Agnew, N. (1999). In R. Mason (Ed.), *Economics and heritage conservation: 11*. Los Angeles: Getty Conservation Institute, J. Paul Getty Trust.

Alexandrakis, G., Manasakis, C., & Kampanis, N.A. (2019). Economic and Societal Impacts on Cultural Heritage Sites, Resulting from Natural Effects and Climate Change. Heritage, 2(1), 279-305. doi:10.3390/heritage2010019.

Angrisano, M., Biancamano, P. F., Bosone, M., Carone, P., Daldanise, G., De Rosa, F., Franciosa, A., Gravagnuolo, A., Iodice, S., Nocca, F., et al. (2016). Towards operationalizing UNESCO recommendations on "historic urban landscape": A position paper. *Aestimum, 2016*, 165–210.

Ansari, S., Wijen, F., & Gray, B. (2013). Constructing a climate change logic: An institutional perspective on the "tragedy of the commons". *Organization Science, 24*(4), 1014–1040.

Arjoon, S., Turriago-Hoyos, A., & Thoene, U. (2018). Virtuousness and the common good as a conceptual framework for harmonizing the goals of the individual, organizations, and the economy. *Journal of Business Ethics, 147*(1), 143–163.

Atkinson, R., & Bridge, G. (Eds.). (2005). *Gentrification in a global context, the new urban colonialism, 2005 housing and society*. London: Routledge.

Biondi, L., Demartini, P., Marchegiani, L., Marchiori, M., & Piber, M. (2020). Understanding orchestrated participatory cultural initiatives: Mapping the dynamics of governance and participation. *Cities, 96*, 102459.

Blake, J. (2000). On defining the cultural heritage. *International and Comparative Law Quarterly, 49*(1), 61–85.

Capello, R., & Perucca, G. (2017). Cultural capital and local development nexus: Does the local environment matter? In H. Shibusawa, K. Sakurai, T. Mizunoya, & S. Uchida (Eds.), *Socio-economic environmental policies and evaluations in regional science* (pp. 103–124). Singapore: Springer.

Ceccarini, L., & Diamanti, I. (2013). Urbino e l'Università: le due città. In Guido Maggioni & Ilvo Diamanti (a cura di), *Studiare a Urbino. Gli studenti, la città, l'Università*. Napoli: Liguori Editore.

CHCfE Consortium. (2015). *Cultural heritage counts for Europe*. Krakow: CHCfE Consortium.

Cooke, P. N., & Lazzeretti, L. (2008). *Creative cities, cultural clusters and local economic development*. Cheltenham: Edward Elgar.

Council of Europe. (2005). *Framework convention on the value of cultural heritage for society*. Strasbourg: Author.

Crossik, G., & Kaszynska, P. (2016). *Understanding the value of arts & culture: The AHRC cultural value project*. Arts and humanities research council.

D'Alpaos, C., Marella, G., Picchiolutto, E., & Stellin, G. (2017). University settlements and host territories: Economic impact evaluation. *Scienze Regionali, 16*(3), 377–400.

Dameri, P. R., & Moggi, S. (2019). Emerging business models for the cultural commons. Empirical evidence from creative cultural firms. *Knowledge Management Research and Practice*, 1–15. https://doi.org/10.1080/14778238.2019.1664945.

De La Torre, M. (2005). In M. MacLean, R. Mason, & D. Myers (Eds.), *Heritage values in site management: Four case studies*. Los Angeles: Getty Conservation Institute.

Del Baldo, M., & Demartini, P. (2010). Values-based public-private networks best practices in Italian local government: The case of "Regione Marche". *Corporate Ownership and Control Journal, 8*(1), 772–784.

Demartini, P., Marchegiani, L., & Marchiori, M. (2020). Culture invites participation. An Inquiry on Matera as European capital of culture 2019. In *Management, participation and entrepreneurship in the cultural and creative sector* (pp. 161–186). Cham: Springer.

Dubini, P., Leone, L., & Forti, L. (2012). Role distribution in public-private partnerships: The case of heritage management in Italy. *International Studies of Management and Organization, 42*(2), 57–75.

EAHTR. (2007). *European Association of Historic Towns and Regions, INHERIT investing in heritage—A guide to successful urban regeneration*. Norwich: European Association of Historic Towns and Regions.

Echter, C. P. (2015). The cultural heritage counts for Europe project and historic cities. *Journal Sustainable Development, Culture, Traditions, 1a*, 49–65.

Eisenhardt, K. M., Graebner, M. E., & Sonenshein, S. (2016). Grand challenges and inductive methods: Rigor without rigor mortis. *Academy of Management Journal, 59*, 1113–1123.

Enyedi, G., & Kovacs, Z. (Eds.). (2006). *Social changes and social sustainability in historic urban centres, the case of Central Europe*. Pécs: Hungarian Academy of Sciences, Centre for Regional Studies.

Euler, J. (2018). Conceptualizing the commons: Moving beyond the goods-based definition by introducing the social practices of commoning as vital determinant. *Ecological Economics, 143*, 10–16.

Evans, G. (2002). Living in a world heritage city: Stakeholders in the dialectic of the universal and particular. *International Journal of Heritage Studies, 8*(2), 117–135.

Fusco Girard, L. (2013). Toward a smart sustainable development of port cities/areas: The Role of the "historic urban landscape" approach. *Sustainability, 5*, 4329–4348.

Fusco Girard, F., Gravagnuolo, A., Nocca, F., Angrisano, M., & Martina Bosone, M. (2015). Towards an economic impact assessment framework for historic urban landscape conservation and regeneration projects. *BDC (Bollettino del Centro Calza Bini, Università degli Studi di Napoli Federico II), 15*(2), 265–293.

Gielen, P., Elkhuizen, S., Van den Hoogen, Q., Lijster, T., & Otte, H. (2015). *Culture: The substructure for a European common*. Rijksuniversitiet Groningen.

Greffe, X. (2009). *Heritage conservation as a driving force for development. Heritage and Beyond* (pp. 101–112). Strasbourg: Council of Europe.

Hardin, G. (1968). The tragedy of the commons. *Science, 162*(3859), 1243–1248.

Healey, P. (2006). Transforming governance: Challenges of institutional adaptation and a new politics of space. *Journal European Planning Studies, 14*(3), 299–320.

Hernando, E., & Campo, S. (2017). An artist's perceived value: Development of a measurement scale. *International Journal of Arts Management, 19*(3), 33–47.

Hess, C., & Ostrom, E. (2007). *Understanding knowledge as a commons: From theory to practice.* Cambridge, MA: MIT Press.

Hosagrahar, J., Soule, J., Fusco Girard, L., & Potts, A. (2016). *Cultural Heritage, the UN Sustainable Development Goals, and the New Urban Agenda.* Retrieved March 15, 2020, from http://www.usicomos.org/wp-content/uploads/2016/05/Final-Concept-Note.pdf

Hutter, M., & Shusterman, R. (2006). Value and the valuation of art in economic and aesthetic theory. In V. A. Ginsburgh & D. Throsby (Eds.), *Handbook of the economics of art and culture* (Vol. 1, pp. 169–208). Amsterdam: Elsevier.

ICOMOS. (2011). *The Valletta principles for the safeguarding and management of historic cities, towns and urban areas.* Adopted by the 17th ICOMOS General Assembly on 28 November 2011 (Online) Valletta Principles GA 2011_CIVVIH_text_EN_FR_final_20120110.pdf.

ICOMOS. (2015, November 20). *Concept note for the United Nations post-2015 Agenda and the Third United Nations Conference on Housing and Sustainable Urban Development (Habitat III).* Cultural Heritage, The UN Sustainable Development Goals, and the New Urban Agenda.

Isaac, J. C., Bermeo, N., Levi, M., Baumgartner, F. R., Keohane, R. O., Axelrod, R., & Mansbridge, J. (2010). Beyond the tragedy of the commons. A discussion of 'Governing the commons: The evolution of institutions for collective action'. *Perspectives on Politics, 8*(2), 569–593.

Kazepov, Y. (2005). Cities of Europe: Changing contexts, local arrangements, and the challenge to social cohesion. In Y. Kazepov (Ed.), *Cities of Europe: Changing contexts, local arrangements and. the challenge to urban cohesion* (pp. 3–43). London: Blackwell.

Kazepov, Y. (2014, February 5). *From citizenship to Cit(y)zenship, cities within multi-level governance arrangements in Europe*, pp. 1–15.

Klamer, A. (2008). The lives of cultural goods. In J. Amariglio, J. W. Childers, & S. E. Cullen-berg (Eds.), *Sublime economy: On the intersection of art and economics* (pp. 250–272). London: Routledge.

Klamer, A. (2016). The value-based approach to cultural economics. *Journal of Cultural Economics, 40*(4), 365–373.

Loulanski, T. (2006). Cultural heritage in socio-economic development: Local and global perspectives. *Environments Journal, 34*(2), 51–69.

Maggioni, G. (Ed.). (2017). *Urbino e le sfide della città-Campus.* Milano: F. Angeli.

Miles, M. B., Huberman, A. M., & Saldaña, J. (2014). *Qualitative data analysis. A methods sourcebook.* Thousand Oaks, CA: Sage.

Municipality of Urbino. (2013a). *Piano di Gestione del Centro Storico di Urbino Sito UNESCO* (UNESCO Heritage site management plan). Urban Planning Office of the Municipality of Urbino minutes and subsequent deliberations. http://www.comune.urbino.pu.it/fileadmin/docs/settori/unesco/ufficio_unesco/attivita-ufficio-unesco-allegati/Piano_di_gestione_del_Sito_Unesco.pdf

Municipality of Urbino. (2013b, November 4). *Memorandum of understanding between the municipality of Florence and the municipality of Urbino.*

Municipality of Urbino. (2016, March 24). *Urbino per Bene (Urbino for Good) project.* Council resolution. UNESCO Office—Urban Decoration and Urban Hygiene.

Nakagawa, S. (2010). Socially inclusive cultural policy and arts-based urban community regeneration. *Cities, 27*, S16–S24.

Nocca, F. (2017). The role of cultural heritage in sustainable development: Multidimensional indicators as decision-making tool. *Sustainability, 9*, 1882, 1–28.

Nyseth, T., & Sognnæs, J. (2013). Preservation of old towns in Norway: Heritage discourses, community processes and the new cultural economy. *Cities, 31*(April), 69–75.

Riganti, P., & Nijkamp, P. (2016). Valuing urban cultural heritage. In L. Fusco Girard & P. Nijkamp (Eds.), *Cultural tourism and sustainable local development* (pp. 75–90). London: Routledge.

Rodwell, D. (2003). Sustainability and the Holistic Approach to the Conservation of Historic Cities. Journal of Architectural Conservation, *1*, 58–73.
Rodwell, D. (2008). *Conservation and sustainability in historic cities*. Hoboken: Wiley.
Rullani, E. (2003). Ritorno al futuro, la ragione riflessiva che intreccia l'economia con la storia. *Annali di Storia d'impresa, 14*, 377–400.
Rullani, E. (2010). *Modernità sostenibile (Sustainable modernity)*. Venezia: Marsilio Ed.
Rypkema, D., & Cheong, C. (2011, November 27–December 2). Measurements and Indicators of Heritage as Development. In *Proceedings of the ICOMOS 17th General Assembly*, Paris, pp. 755–762.
Sacco, P., Ferilli, G., & Blessi, G. T. (2014). Understanding culture-led local development: A critique of alternative theoretical explanations. *Urban Studies, 51*(13), 2806–2821.
Sacco, P., Ghirardi, S., Tartari, M., & Trimarchi, M. (2019). Two versions of heterotopia: The role of art practices in participative urban renewal processes. *Cities, 89*, 199–208.
Silver, H., Scott, A., & Kazepov, Y. (2010). Participation in urban contention and deliberation. *International Journal of Urban and Regional Research, 34*(3), 453–477.
Snowball, J. D. (2011). Cultural value. In R. Towse (Ed.), *A handbook of cultural economics* (pp. 172–176). Cheltenham: Edward Elgar.
Srakar, A., & Vecco, M. (2016). *Culture as a fourth dimension of sustainable development? A statistical analysis*. Retrieved from http://eventos.uva.es/file_manager/get_paper/3421
Stolarick, K., & Florida, R. (2006). Creativity, connections and innovation: A study of linkages in the Montréal region. *Environment and Planning A, 38*(10), 1799–1817.
Throsby, D. (1999). Cultural capital. *Journal of Cultural Economics, 23*(1–2), 3–12.
Throsby, D. (2003). Determining the value of cultural goods: How much (or how little) does contingent valuation tell us? *Journal of Cultural Economics, 27*(3-4), 275–285.
Throsby, D., & Zednik, A. (2014). The economic and cultural value of paintings: Some empirical evidence. In V. A. Ginsburgh & D. Throsby (Eds.), *Handbook of the economics of art and culture* (Vol. 2, pp. 81–99). Amsterdam: Elsevier.
UNESCO. (1968). *Recommendation concerning the preservation of cultural property endangered by public or private works*. In 15th Session of the General Conference, Paris.
UNESCO. (2003). *Text of the convention for the safeguarding of the intangible cultural heritage*. Retrieved from http://www.unesco.org/culture/ich/en/convention
UNESCO. (2004, September 13). *Social sustainability in historic districts, UNESCO round table of experts*.
UNESCO. (2008). *Historic districts for all a social and human approach for sustainable revitalization*. Manual for city professionals, Laure Veirier, UNESCO, Paris. Retrieved May 15, 2020, from https://mirror.unhabitat.org/downloads/docs/10362_1_594123.pdf pp 1-103
UNESCO. (2011). *Recommendation on the historic urban landscape*. Paris: UNESCO World Heritage Centre.
UNESCO. (2013). *Hangzhou declaration. Placing culture at the heart of sustainable development policies*. Doc. CLT-2013/WS/14. Retrieved May 10 2019, from http://www.unesco.org/new/fileadmin/MULTIMEDIA/HQ/CLT/images/FinalHangzhouDeclaration20130517.pdf
UNESCO. (2014). *Operational strategy on youth 2014–2021*. UNESCO. Retrieved July 19, 2020, from https://unesdoc.unesco.org/ark:/48223/pf0000227150
UNESCO. (2017, April 7). *Executive board; 201EX/4.INF.2; implementation of the UNESCO operational strategy on youth (2014–2021): Summary conclusions from the first UNESCO-wide monitoring of the Strategy's implementation, 2014–2016*. Paris. Retrieved July 10, 2020, from https://unesdoc.unesco.org/ark:/48223/pf0000247897
UNESCO/UN-HABITAT. (2006). *Social sustainability in historic districts: Best practices*. Retrieved May 18, 2020, from http://www.tibetheritagefund.org
UN-Habitat. (2014, October 15–18). *Report of the first urban thinkers campus*. Caserta.
United Nations. (2015). Transforming *our world: The 2030 agenda for sustainable development*. New York. Retrieved from https://www.un.org/sustainabledevelopment/32
United Nations. (2016a). *Draft outcome document of the United Nations conference on housing and sustainable urban development (habitat III)*. New York, NY: United Nations.

United Nations (2016b). *Marrakech action proclamation for our climate and sustainable development*. New York, NY: United Nations.
Vecco, M. (2010). A definition of cultural heritage: From the tangible to the intangible. *Journal of Cultural Heritage, 11*(3), 321–324.
Veirier, L. (2008). *Historic districts for all, a social and human approach for sustainable revitalization*. United Nations Human Settlements Programme.
Veldpaus, L., & Pereira Roders, A. R. (2014, July 22–25). The historic urban landscape: Learning from a legacy. In Amoêda, R., Lira, S., & Pinheiro, C. (Eds.), *Proceedings of IV international conference on heritage and sustainable development* (pp. 129–141). Guimaraes.
Yin, R. K. (2013). Validity and generalization in future case study evaluations. *Evaluation, 19*(3), 321–332.
Zukin, S., Kasinitz, P., & Chen, X. (2015). *Global cities, local streets: Everyday diversity from New York to Shanghai*. London: Routledge.

Websites

World Heritage List. Historic Centre of Urbino. Description. Retrieved May 2020, from https://whc.unesco.org/en/list/828/documents/

Mara Del Baldo is an Associate Professor of Business Administration and Economics of Sustainability and Accountability in the Department of Economics, Society and Politics, University of Urbino (Italy),. She was a visiting professor and has given lectures at foreign Universities (the University of Vigo, Spain; the Jurai Dobrila University of Pula, Croatia; the New Bulgarian University of Sofia, Bulgaria; the Corvinus University in Budapest, Hungary; the University of Craiova, Romania). She is currently part of the academic advisory committee in the Global Corporate Governance Institute (USA) and a member of the European Council for Small Business, the Centre for Social and Environmental Accounting Research (CSEAR), the SPES Institute, and the European Business Ethics Network (EBEN) Italia, as well of Italian scientific associations. She serves as a board member and a reviewer of several international scientific journals. Her main research interests include entrepreneurship and small businesses management; corporate social responsibility, sustainability, and entrepreneurial business ethics; sustainable tourism; benefit corporations; SMEs strategies of qualitative development and networking strategies; cultural ecclesiastic heritage management; financial reporting; ethical, social and environmental accounting, and accountability; integrated reporting and accounting; and gender. She published in several Italian and foreign journals (i.e., *Meditary Accountancy Research, Business Strategy and the Environment, Annals of Tourism Research, Journal of Management and Governance*, among others) as well as in national and international conferences proceedings and books.

Paola Demartini is Full Professor of Managerial and Financial Accounting at Roma Tre University, Department of Business Studies. She is the Head of the Corporate Governance Lab, which includes a special section on the Governance of Cultural Heritage and Contemporary Culture. Since 2000, she is member of the supervisory committee of PhD courses on Financial Accounting and Governance. Since 2016, she is involved in the project "Cultural Heritage and Contemporary Culture: Transferring knowledge, Managing Outcomes, and Enabling Sustainability", a collaboration between Corporate Governance Lab (Rome Tre University) and Institut für Organisation und Lernen (Innsbruck University). Currently she is member of the leading research team engaged in a H2020 project titled: SOPHIA—Social Platform For Holistic Heritage Impact Assessment.

Entrepreneurial Cultural Ecosystems in Rural Contexts: Some Insights from Rural Cultural Centers in France

Elena Borin and Cassandre Jolivet

Abstract The debate on cultural entrepreneurship and sustainable development seems to be related mainly to urban spaces, leaving reflections on its role for sustainable development of rural territories rather under-explored. However, there are significant examples of cultural entrepreneurial initiatives in rural areas; the topic seems therefore promising and needs further investigation. The aim of this chapter is to fill this research gap, focusing on a specific type of rural cultural entrepreneurship: the creation of cultural centers in rural territories.

The authors carried out a qualitative empirical investigation based on two significant case studies of rural cultural centers located in the region of Auvergne-Rhône-Alpes in France. The case studies were analyzed according to the case study methodology identified by Yin (2015), based on the triangulation of multiple sources of information.

The results underline that these cultural centers thrived in rural areas, thanks to the creation of site-specific cultural entrepreneurial ecosystems, based on the nurturing of peculiar characteristics related to elements that were both endogenous and exogenous to the territory.

The research highlighted the peculiarities of cultural centers in rural territories and the characteristics of their ecosystems, providing insights into a promising phenomenon and contributing to the debate on the need for a more balanced development between urban and rural regions.

Keywords Business Models · Entrepreneurship in rural areas · Sustainable business models · Sustainable development

---

E. Borin (✉) · C. Jolivet
CEREN, EA 7477 – Burgundy School of Business, Université Bourgogne Franche Comté, Besançon, France
e-mail: elena.borin@bsb-education.com; cassandre.jolivet@bsb-education.com

P. Demartini et al. (eds.), *Cultural Initiatives for Sustainable Development*, Contributions to Management Science,
https://doi.org/10.1007/978-3-030-65687-4_15

1 Introduction

The role of culture for sustainable development has increasingly entered the debate on sustainability, bringing attention to the role of specific cultural and creative industries' subsectors (Duxbury et al. 2017) or as a transversal topic in the Sustainability Development Goals (UN 2015). Within this debate, cultural entrepreneurship has often been studied in its capacity to contribute to regeneration and territorial development (He and Gebhardt 2014), the idea of an entrepreneurial ecosystem has emerged as a key for the enhancement of the territory (Borin 2018; Borin and Donato 2015; Grandadam et al. 2013; Isenberg 2011a, b).

However, these studies have often focused mainly on urban areas, partially disregarding rural cultural initiatives that are nevertheless present and frequently successful (Markusen 2007; Borin and Paunovic 2016). This paper aims at filling this research gap, focusing on a specific category of rural cultural entrepreneurship and its ecosystems: rural cultural centers.

More specifically, the research aims at answering the following main research questions:

RQ1: What are the characteristics of successful cultural centers in rural areas?
RQ2: How are their entrepreneurial ecosystems structured? What are the peculiarities of entrepreneurial ecosystems in the cultural and creative sector in rural territories?

The paper is divided into five sections. After this short introduction, the first section provides an analysis of the scientific literature on the main topics related to this research, dividing the review in thematic subsections. Section 2 presents the research design and methodology, introducing also the selected case studies (*La Gare à Coulisse* and *Les Abattoirs*). Section 3 presents the results of the case study analysis, subsequently discussed in Sect. 4. The last section of the paper summarizes the main points of the research and discusses its contribution to the theoretical and practical debate. The limitations of the study as well as some indications of further avenues of research are also presented.

2 Insights on the Theoretical Debate

2.1 Culture and Sustainable Development

The importance of culture in relation to sustainable development has entered the debate on sustainability quite recently (Soini and Dessein 2016). On one hand, there has been a growing debate on how the different dimensions of culture can interact with the other traditional dimensions of sustainability (Duxbury et al. 2017). On the other hand, culture has been advocated as one of the four pillars of sustainable development equal to social, economic, and environmental dimensions (UCLG

2010; Loach et al. 2017), although the discourse has been sometimes criticized (Isar 2017). Though not explicitly mentioned, the importance of culture as an underlying and transversal topic is evident in the 2030 Sustainable Development Agenda (UN 2015; Yildrim et al. 2019): cultural domains are frequently mentioned or implicitly referred to in several SDGs—Sustainable Development Goals (UN 2015).[1] Recently, the interpretation of culture *in* sustainable development has been criticized as outdated, and the idea to conceive "culture *as* sustainable development" (i.e., constituting the basis for successful reflections on sustainable societies) has emerged (Soini and Dessein 2016). This means that cultural and creative industries are embedded in the whole discourse on sustainable development (Soini and Dessein 2016). Recognizing culture as crucial for development means recognizing the importance of the several fields that belong to the cultural sector: not only cultural heritage (CHCfE 2015) but also cultural and creative industries in general, not just public institutions but also creative entrepreneurial initiatives from the private sector (Aageson 2008; Borin et al. 2018).

With reference to the studies on the contribution of culture and creative subsectors to sustainable development, there has been increasing attention to the need to interpret culture not as a *silo* but as connected to its broader environment, territory, and its stakeholders. This has been linked by the emergence of the "ecology of culture" concept (Holden 2004) and most of all to the "ecosystem approach" (Borin and Donato 2015) that interprets the sector on the basis of its connections with and contribution to the various stakeholders of a territory (Borin 2015).

2.2 Entrepreneurial Ecosystems in Business and Managerial Analysis

This attention given to the ecosystem approach in the cultural and creative sector mirrors the growing consideration of ecosystems by general management and business studies, that draws on the concept of ecosystem in biology (Costanza et al. 1997) and adapts it to the business environment (Iansiti and Levien 2004; Basole 2009; Peltoniemi 2006; Stam 2015). Ecosystems in business are defined as "extended system of mutually supportive organizations; communities of customers, suppliers, lead producers, and other stakeholders, financing, trade associations, standard bodies, labor unions, governmental and quasi-governmental institutions, and other interested parties" (Moore 1996, p. 168). Entrepreneurial ecosystems are dynamic, self-regulating networks of several interdependent actors and factors (Isenberg 2014) that are coordinated in such a way that they enable productive entrepreneurship (Stam, 2015). Literature on entrepreneurial ecosystems has highlighted intentionality, careful programming, and future-making mindsets

[1] See, for example, the Sustainable Development Goals 4 (quality education) and 11 (sustainable cities and communities).

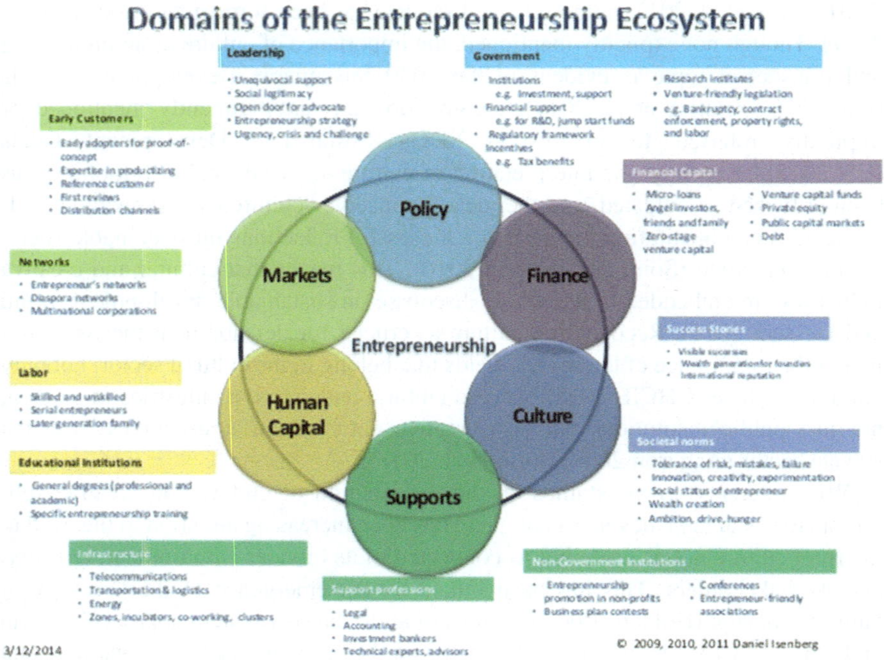

Fig. 1 Domains of the entrepreneurship ecosystem (Source: Isenberg 2010, 2011a, b)

(Peltoniemi 2006; Stam 2015), underlining their location-specific character (Aage and Belussi 2008; Cohendet et al. 2010; Drake 2003) or identifying specific components of successful ecosystem environments (Isenberg 2010, 2011a, b). Among these studies, the work of Isenberg seems particularly significant. Isenberg has classified six specific domains of successful entrepreneurial ecosystems: policy, finance, culture, support, human capital, and markets as highlighted in Fig. 1 (Isenberg 2010, 2011a, b).

The "policy" domain is interpreted as the governmental support given to entrepreneurs (e.g., in the form of regulatory frameworks or tax incentives) as well as the access to basic infrastructure, telecommunication, and transport that might have an impact on businesses' development. The "finance" domain refers to the availability of financial resources (e.g., angel investors, private equity, venture capital, or the access to debt). The "culture" domain includes aspects such as local general tolerance for risk, mistakes, and failures, the abundance of success stories and role models, the positive general opinion on entrepreneurship, or the celebration of innovation. Within the "supports" domain are included the presence of mentors and advisers, as well as professional services like accounting agencies, incubators, or related commercial activities. The "human capital" domain refers to the quality and quantity of the workforce present in the ecosystem. It is related to the availability of factors such as human resources with management and technical talent or outsourcing availability; it is primarily linked to education (e.g., the presence of universities or training centers related to the entrepreneurial domain). Finally, the "markets" domain is related to the presence of customers for the companies' products or services.

2.3 Entrepreneurial Ecosystems in the Cultural and Creative Sector

Isenberg's entrepreneurial ecosystem model is applicable to a large variety of business activities. Entrepreneurial ecosystems in the cultural and creative sector, however, seem to work in a more particular way. A first stream of research noted that entrepreneurs in the cultural and creative sector decide to come together, or cluster, not merely on the basis of the need for physical proximity with customers and competitors, or for policy or financial incentives (Protogerou et al. 2017), but mainly with the purpose to focus on producing collective cultural values (Lash and Urry 1994; Ballico 2017). These studies have therefore addressed the peculiarities and motives of cooperation among the members of cultural ecosystems, reflecting on the need to find appropriate governance systems to ensure fair and balanced collaborations in the long term. Partnerships among public, private, and civic actors are considered as a means of sustainable coordination and exchange between the forces operating in a cultural entrepreneurial ecosystem and as a basis for cultural production (Borin 2017). In these cultural entrepreneurial ecosystems, cultural heritage institutions interact with cultural and creative enterprises, public authorities, and local communities, while at the same time promoting links with other sectors (Borin and Donato 2015). New governance systems and management models should reflect this change and enhance the potential of the cultural sector at large; they are the key to create enabling conditions to promote culture-led sustainable long-term development (Bonet and Donato 2011). Within these frameworks, the need for increasing and structuring the interactions and links between different cultural domains—such as "homemade", "publicly funded," and "commercial culture" (Holden 2015) was also addressed.

Another stream of research focused on the connections between cultural entrepreneurial ecosystems and territories, especially cities. On one hand, they discussed their potential for reviving urban areas (Grandadam et al. 2013; He and Gebhardt 2014). On the other, they claimed the need to create in cities and territories a "fertile soil" in which creativity can flourish on the basis of specific mechanisms, policy tools, and connections between key players that could enhance, nurture, and enact the potential of creative entrepreneurs (Cohendet et al. 2010; Grandadam et al. 2013). In this context, with reference to Isenberg's ecosystem model, the peculiarities of ecosystems in the cultural and creative sector were identified. In particular, specificities are related to the markets domain—traditionally composed by a more active and specialized type of audience, the "cultural prosumers"—and to the culture domain, that is not simply related to the entrepreneurial culture, but also recognizes

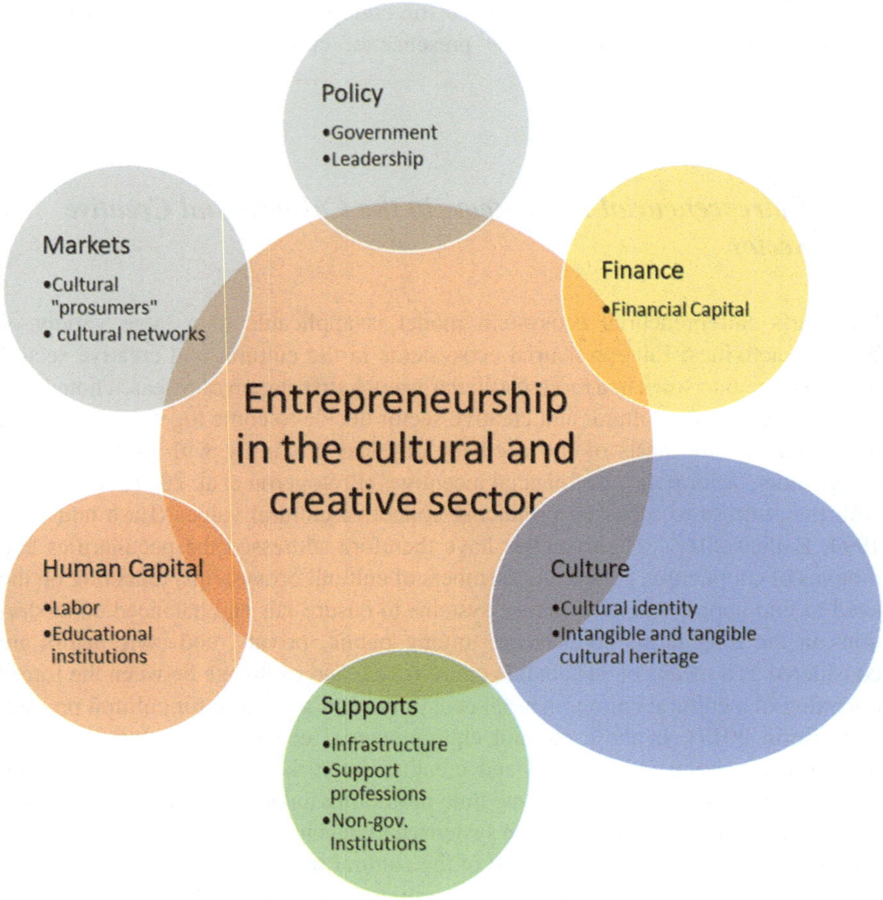

Fig. 2 Domains of the entrepreneurial ecosystems in the cultural and creative sector (Borin 2018)

the importance of local cultural values. Cultural heritage, both intangible and tangible, could influence the development of cultural values and cultural identity of the cities while also being a source of inspiration and development for entrepreneurial cultural initiatives (Borin 2018—see Fig. 2).

The debate on cultural ecosystems and entrepreneurial ecosystems in the cultural and creative sector seems, however, to be related mainly to urban spaces and urban development (Andres and Chapain 2013; Comunian et al. 2010; Cooke and Lazzeretti 2008; Duxbury et al. 2012; Florida 2005, 2017; Hristova et al. 2015; Kagan and Hahn 2011), leaving reflections on the role of cultural ecosystems for sustainable development of rural territories[2] rather under-explored. However, there

[2]Rural areas are defined in opposition to urban areas and are characterized by a low population density and small settlements. According to urban–rural typology statistics (EUROSTAT 2016),

are significant examples of cultural initiatives in rural areas (Markusen 2007; Borin and Paunovic 2016), including the creation of cultural centers.

2.4 Cultural Entrepreneurial Activities in Rural Areas: The Case of Rural Cultural Centers

Among the possible entrepreneurial activities in rural areas, the creation of cultural centers stands out due to its complexity (Delfosse and Georges 2013; Georges 2012; CAE 2019) and could be considered as a relevant example of the potentialities and difficulties of establishing cultural entrepreneurial ecosystems in rural regions. Indeed, in their need to involve stakeholders in the same project and induce them to participate and share, cultural centers have the capacity of creating synergies (especially within the local community, attracting people and services), thus stimulating the creation of cooperation that can be part of broader urban territorial effects (Boix et al. 2015; ENRD 2017). Cultural centers are often created as part of territorial regeneration phenomena in urban areas and can either be the result of top-down policies (Florida 2017; Mulero and Rius-Ulldemolins 2017; Rius Ulldemolins 2014a, b; Vivant 2007) or can be started by local communities and artists. Cultural and local communities are, in both cases, at the core of the process (Florida 2014) and the launching of cultural initiatives should not only retain but also benefit both (Biondi et al. 2018; Comunian et al. 2010; Florida 2017; Sasaki 2010).

The fact that rural cultural centers' projects are less frequent than urban ones (CAE 2019) could be partially due to the fact that the three Ts (Talent, Technology, and Tolerance)—strictly linked to the attractiveness of a place—are more likely to be found in metropolitan areas. Nevertheless, in small rural communities, cultural initiatives could be a way to create local cultural ecosystems, thus stimulating growth and sustainability (Markusen 2007). As a source of local employment and income growth, the development of cultural centers in rural regions could therefore be strategic for policy makers (Hall et al. 2010; Bonerandi and Hochedez 2007; Baudelle and Krauss 2014) and creative hotspots beyond the metropolis that may be the start of regional reinvention (Petrov 2007).

However, in rural areas, there is usually an issue of weak organization by local authorities, which is often an obstacle to local initiatives' development and efficiency (Regourd 2007). With specific reference to cultural entrepreneurship, since rural areas are more extended with fewer inhabitants, artists are often scarcer than in urban areas, causing a sense of isolation, which has a negative impact on the creation of networks of contacts that are frequently needed for artistic creation and dissemination (Delfosse and Georges 2013). This also means that artists frequently struggle to earn a living from working for rural markets and are forced to also work in cities

---

rural areas are "all areas outside urban clusters. "Urban clusters" are clusters of contiguous grid cells of 1 km^2 with a density of at least 300 inhabitants per km^2 and a minimum population of 5000".

(Delfosse and Georges 2013). As a result, their presence in the countryside is usually temporary and they frequently change location. Moreover, there is often a lack of funds, equipment, and training programs for artists in rural areas, which explains why artists tend to prefer cities to rural regions (Delfosse and Georges 2013). Finally, it seems that when artists do manage to make ends meet in rural areas, it is in regions where they are close to a large metropolis or other dynamic regions (Georges 2012).

There are nonetheless several benefits of working in rural territories. For example, rural areas offer a quality of life that could influence an artist's decision to settle on a territory, referring in particular to the fact that peripheral areas are usually less expensive in terms of housing and work spaces (Hall et al. 2010) and offer a natural environment that could be beneficial to the creation process (Delfosse and Georges 2013). Also, in rural areas, artists usually benefit from the possibilities of using unusual spaces for outreach and performances: indeed, such activities frequently happen in open spaces, parks, or outdoor theaters, among others. Moreover, in rural areas, artists usually get more involved in the local social life: they work with schools, local associations, and communities (Delfosse and Georges 2013; Bonerandi and Hochedez 2007) and increasing attention is paid to social engagement (Regourd 2007; Markusen 2007; Bonerandi and Hochedez 2007; Comunian et al. 2010; Delfosse and Georges 2013).

The above-mentioned studies underline some difficulties and opportunities rural areas offer as locations for cultural centers, but there is still a lack of systematization of the success factors of rural cultural centers in order to present a structured model of their ecosystems. There is therefore a need to more effectively investigate the key aspects leading to successful establishment of cultural centers in rural areas as linked to the entrepreneurial ecosystem in which they can thrive. This paper aims at filling this research gap.

3 Research Design and Methodology

As stated in the introductory section, the research is aimed at investigating and identifying the characteristics of successful cultural centers in rural areas and their entrepreneurial ecosystem, thus gathering initial information on the peculiarities of entrepreneurial ecosystems in the cultural and creative sector in rural territories.

Given the complex nature of the topic of investigation and the need to explore it in depth, the authors decided to use a qualitative research approach. Although this research could be regarded as a preliminary investigation of the research questions, this approach was considered as particularly consistent with the objective of shedding light on how the research topics were manifesting (Denzin et al. 2006). The case studies were selected to comply in particular with the principle of representativeness (Patton 2002) of cultural centers located in rural territories.

France was selected as a research area for two main reasons: first of all, the cultural proximity to the researchers, who were either French or working in France, thus avoiding potential cultural biases or misinterpretations; secondly, France seems a particularly significant research scenario since it has over the last three decades implemented a specific policy of "democratization" of culture. Indeed, one of the problems of the country was that most cultural organizations are located in the capital city, Paris, making access to culture problematic in other regions. The central government decided to deal with this issue by implementing policies aiming at making culture also accessible in peripheral areas. These initiatives worked at different levels; for example, allocating incentives in the creation of cultural activities in rural and peripheral regions or promoting the creation of branches of famous Parisian museums in provincial regions and cities—such as the Pompidou in Metz and the Louvre in Lens (Baudelle and Krauss 2014). Rural cultural centers could be considered as a manifestation of this broader policy strategy.

A preliminary mapping of potentially interesting case studies was made in cooperation with three important intermediary organizations (French and European networks of the cultural and creative sector, namely *Art Factories*, *Trans Europe Halles*, and *Le LAB*) that provided important data on the existing cultural centers in cities and rural areas. The authors decided to focus on rural cultural centers that were located in the same geographical region in order to avoid getting results that are biased by different *meso* socioeconomic conditions. It was considered that having an analysis of rural centers in the same territory could lead to better comparison of the results, since they are operating in the same context. The Auvergne-Rhône-Alpes was identified as the most suitable French region for our investigation, since it not only presented a culturally dynamic capital city (Lyon) with a very broad rural territory in which significant cultural centers had been created, but also because, according to the data available from the above-mentioned networks, it was among the most active regions in terms of creation of cultural centers outside Île-de-France (the region in which Paris is located).

After selecting Auvergne-Rhône-Alpes as area of investigation, the authors focused on the section of case studies according to the following criteria: first of all, the cultural centers should have been operating for at least 5 years, in order to avoid too recent or unstable projects; secondly, they should have a multiplicity of activities, not limited to a specific type of artistic production or subsector; thirdly, they should have a certain degree of organization, including a governance structure and a clear operating model. Finally, the centers should have developed activities directed toward the territory, thus allowing the researchers to study the ecosystem in which they were operating. Based on these criteria, two cultural centers in the Auvergne-Rhône-Alpes area were chosen: *La Gare à Coulisses* in Eurre and *Les Abattoirs* in Riom.

These case studies were analyzed according to the case study research protocol identified by Yin (2015), aiming at an explanation-building case description based on the triangulation of multiple sources of information (document analysis, output analysis, and in-depth semi-structured research interviews). The collection of secondary data was based on the analysis of annual reports and press reviews. The collection of primary data was instead carried out through semi-structured research interviews with two categories of interviewees: the directors or managers of the structures and representatives of groups of artists in residence at the cultural centers. The research protocol was designed to understand two main points: first, a preliminary "technical" description of the center, in order to obtain the basic information on the place and an overview of its main features, main projects and activities, and basic vision; secondly, their connection with their territory, their audience, the artists, and artistic communities with whom they were in contact—in other words, their entrepreneurial ecosystem.

The interviews were recorded and transcribed and, after summarizing the general information on the centers, main themes were identified regarding the ecosystem with the aim of shedding light on the main research questions.

The results of the investigation are presented in the following sections. For each of the case studies, the presentation will be divided into two main parts: first, a description of the technical characteristics, vision, main projects, and activities of the cultural center. Then an analysis of the ecosystem will be provided, with reference to the domains of the entrepreneurial ecosystems identified in the literature review [namely, policy, finance, markets, support, human capital, and culture (Borin 2018)]. After an analysis of the case studies, the authors introduced a discussion of the data and a comparative analysis of the two cases, aiming at highlighting the similarities and differences of their ecosystem.

4 Research Results and Discussion

As stated above, the results of the empirical investigation will be presented, focusing first on an analysis of each case study, describing its history, technical characteristics and its values, activities and resources, presenting also a summary of the main features in terms of costs and revenues. Subsequently, the analysis will consider its entrepreneurial ecosystem in the main domains highlighted above. After the case study examination, a comparative analysis will be carried out in order to identify the differences and similarities and answer the research questions.

4.1 La Gare à Coulisses

La Gare à Coulisse's cultural project was initially conceived in a neighboring town, Crest, where the performing arts company *Transe Express* used a pavilion for their rehearsals, signing a partnership agreement with the municipality. This initiative quickly attracted other artists and companies who asked to use the space for rehearsing. After a change in the policy of the town, the project moved to the nearby town of Eurre; the company volunteered for the renovation of a former train warehouse site and the creation of a cultural offer constituted by an "ecosystem of cultural projects." Ultimately, the municipality took charge of the renovation, while the company *Transe Express* was appointed to design the artistic and cultural activities. From 2007 onward, this cultural center used up to three main buildings located on an area of one hectare including spaces for artistic creation and rehearsals, accommodation areas and common spaces including a bar, kitchen, meeting rooms, and offices as well as a pavilion and several caravans.

The mission and values of the company are to establish a "non-conventional relationship with the audiences," creating a less "institutional" place, in which emerging practices could be developed, linked with the local communities. The values of creating a nonconventional relationship with the public and open up the place to the territory are central to the discourses emerging during the interviews. Although the center is "in the middle of nowhere" and does not have the necessary infrastructures to encourage accessibility, the interviewees argued that it is not "conceived as a fortress"; several initiatives are established outside of the center, as public spaces and "third places" intended to sensitize the audience to various artistic practices and stimulate them to visit *La Gare à Coulisses*.

The key activities of the center are related to the main disciplines they are offering (mainly performing arts and circus, dance, and street arts) and consist of residencies for artistic creation and technical works (building up structures for a show for example), amateur workshops, on-site visits for schools, outreach activities with schools, professional training programs for artists, festivals, and cultural and artistic events, which are around 15 shows and four "highlight" activities per year. The "highlights" are particularly significant in their exploration of a different link with the audience. Indeed, they are organized in two phases: in the first, the companies meet and discuss a specific theme; in the second, they exchange with the audience on this same theme.

The interviews with the manager of the cultural center and the artists working there enabled the researchers to have a deeper understanding of the entrepreneurial ecosystem in which they were operating. As explained by one of the interviewees, starting a cultural center in this territory was not easy and it would not have been possible if there had not been the logistic, policy, and financial support of the local authorities. In terms of *policy*, there was a significant help from the local authorities who encouraged *Trans Express* to relocate into the area and were available to offer support for the renovation works for the headquarters of the association and regarding integration in the area. The manager of the association argued that the relationship with the territory was "a trust relationship in the long term."

The fact that *La Gare à Coulisses* was considered a strong asset for territorial development is also demonstrated by the multi-partnership agreement signed with the *DRAC—Direction Régionale des Affaires Culturelles* (Regional branch for cultural affairs), the region, the department, and a group of local municipalities.

In terms of *finance*, the integration with the ecosystem is also strongly visible: Eurre municipality not only provided finance for starting the entrepreneurial initiative, but also continued supporting the operation of *La Gare à Coulisse* together with the other members of the above-mentioned partnership agreement for approximately 50% of the total budget (the remaining 50% comes from self-generated revenues such as workshops, training for professionals, mediation activities, school visits, and from crowdfunding activities). No financial support has come from private sponsors over the 10-year activity period of the cultural center.

However, in terms of *support* (interpreted in cultural entrepreneurial ecosystems mainly as the presence of nongovernmental institutions or local enterprises providing complementary services), the link with supporting companies and associations located in the areas is slowly growing over the past few years. Once more, the municipality has attempted to boost the creation of other commercial activities around the cultural center that could offer complementary services. The overall project includes enterprises, restaurants, the administrative offices of the grouping of municipalities, accommodation for artists with a creative studio, and an organic shop. All those elements have only recently begun to be developed: for the moment, there is no strong link between those elements and the cultural center, but it is trying to promote interactions by keeping the cultural space open and available.

In terms of *markets*, the main traditional audience of the entrepreneurial ecosystems in the cultural and creative sector, i.e., cultural "prosumers" and other stakeholders interested in the cultural field, is described as often lacking. As a result, the cultural center target other diversified segments through shows that are often not performed in their institutional space. Their objective is finding new audiences among local people living in the municipality and department, encouraging them to interact and appreciate disciplines which they would not normally have the opportunity to see. In order to attract these audiences to the performances and cultural offering of the center, *La Gare à Coulisse* attempts to also engage inhabitants of the territory in the daily operations of the center, mainly through volunteering activities.

In terms of *human capital*, the cultural center strongly relies on the work of volunteers, often recruited locally, but most of all upon those artists in residence who could be considered as being its essential resource. The artists involved with *La Gare à Coulisses* are both emerging artists, companies, and students, and a pavilion for the circus school has been installed in the area. As highlighted in the previous sections, in urban spaces, the local cultural and creative milieu, often also rich in arts schools, artists' cooperatives, and collectives or freelance artists, usually provides for a significant number of artists working for or in cooperation with local cultural centers. In the case of *La Gare à Coulisse*, although the managers tried to recruit locally, artists come only partially from the region and are instead more frequently from areas further away or even other countries, bringing with them their equipment and caravans. The process of selection of these artists and companies is rather informal, through word of mouth and especially recruitment during specialized festivals. During an interview, the managers explained that they often work on a system of "cultural barter": they offer to the artists the free use of the spaces and infrastructures to create and rehearse but they ask them to perform for an audience. During an exchange with one of the companies in residence, it transpired that companies particularly like to rehearse in *La Gare à Coulisses* spaces, especially due to the atmosphere of hospitality and support in terms of accommodation, catering, technical assistance, and exchanges with other artists and audience.

In terms of *culture*, the peculiarity of entrepreneurial ecosystems in the cultural and creative sector in urban spaces is usually their strong link with the cultural identity of the territory that is often influenced by its cultural heritage, both tangible and intangible (Borin 2018). Cultural centers often operate in the framework of a specific urban context, and they frequently interact with and are influenced by other cultural and creative activities in the city. As for *La Gare à Coulisse*, the link with the territory comes instead in the form of policy, logistic and financial support, but not in terms of the influence of the cultural and creative milieu or local cultural heritage on the artistic production. As declared by one of the interviewees, the relationship with the territory is not particularly related to its cultural heritage or cultural and creative identity, but it is rather an instrumental one, "a necessary tool, to keep on developing."

4.2 Les Abattoirs

Les Abattoirs is a cultural center located in a former nineteenth-century slaughterhouse that was renovated in 2002 and has been reopened and managed by an association of artists and musicians called *Les Abattoirs*. The association employs one person and around ten collaborators on a temporary basis. The building offers a space of approximately 300 m^2 divided across two floors, including offices, a hall, a bar and a stage, technical rooms, a dressing room, a common room, and a kitchen. Accommodation for artists in residence is organized in external areas. The cultural center is open from September to June.

The mission and values of the cultural center are to encourage a variety of contemporary creation, helping emerging artists and companies to grow. With similarities to *La Gare à Coulisse, Les Abattoirs* also aims also at establishing a diversified and original artistic provision for all types of public reception, stimulating exchanges between artists and audiences. The center does not focus on a specific artistic sector: on the contrary, it encourages the development of all contemporary creative disciplines.

The key activities consist of the organization of outreach activities, shows, and exhibitions, as well as a variety of initiatives aimed at developing the visibility and accessibility of the center. The center also organizes residencies for artists to support them in their artistic creation. The initial idea was to organize residencies and artistic outreach activities mainly involving local artists and companies, but the project grew in a different direction and the center is currently mainly open to companies and artists coming from other regions. Nevertheless, the center attempts to organize most activities both at a local, regional, and national level, for example, a contemporary circus festival.

In terms of the analysis of the entrepreneurial ecosystem, several peculiarities emerged during the interviews with the president of the managing association.

In terms of *policy*, the creation of the center was strongly supported by the municipality of Riom. According to the interviewee, the cultural activities of *Les Abattoirs* are considered “complementary” to the cultural programming of the town, and there are durable links with the municipality, although she argues that there is still insufficient recognition of the center’s contribution to the territory.

In terms of *finance*, *Les Abattoirs* is financially supported by several public authorities: the town of Riom, the Ministry of Culture and Communication, the DRAC Auvergne-Rhône-Alpes, the Regional Council Auvergne-Rhône-Alpes, and the Department Council of Puys-de-Dôme. The public funds allocated to the center cover most of the expenses: sponsorship, ticketing, and the activity of the internal bar provide only limited amounts.

In terms of *support*, as declared by the interviewee, the president of *Les Abattoirs* recognizes the difficulty of having appropriate supporting services in the rural area of Riom although the center was able to promote positive cooperation with other institutional and private, local, and regional partners.

In terms of *markets*, the cultural center attempts to create connections with the local audience that is often not specialized and not familiar with the type of innovative performances and artistic productions that are implemented in the center; indeed, the association wants to offer to both artists and their audience content and activities that are unconventional compared to more institutional spaces. *Les Abattoirs* attempts to connect with them through a multiplicity of activities, such as public rehearsals, outdoor performances, and initiatives targeting young adults, children, and students. Notwithstanding their attention to the territory, the president admits that the center is “still not completely recognized on the local level.” She explains that *Les Abattoirs* has an audience and a recognition on the national level, but that they are sometimes not fully considered locally.

This point also emerges in the analysis of the *human capital* of their ecosystem. As pointed out in the previous case study of *La Gare à Coulisse*, artists and companies in residency are the main human capital asset for the center. Due to the scarcity of local artists and the lack of universities and arts-specialized educational institutions, the center recruits artists and companies mainly during festivals, and more formally by means of an online application process. The artists are offered a space to create and rehearse, as well as technical support and material, free accommodation, meals, and transport. In order to create or reinforce the links with the local community, artists are encouraged to both perform at *Les Abattoirs* and engage with the local audience.

As for the final domain of the cultural ecosystem, that of *culture*, the president does not refer to other cultural and creative activities or local heritage (neither tangible nor intangible) as a source of support or inspiration for artistic production or value creation. Similarly, no reference is made to the cultural identity of the territory.

5 Discussion: The Peculiarities of Entrepreneurial Ecosystems for Cultural Centers in Rural Areas

From a comparative analysis of the two cultural centers, we can interpret their success factors on the basis of the peculiarities of their entrepreneurial ecosystems. As underlined previously, the entrepreneurial ecosystems in the cultural and creative sector mirror traditional components of entrepreneurial ecosystems while maintaining peculiarities related to the strong link with intangible and tangible cultural heritage, other cultural and creative organizations and identity of the territory, often concretizing in the collaboration with other entities of the area such as museums, theaters, performance centers, or other creative companies. Another peculiarity is the need to rely on the presence of cultural networks, "cultural prosumer" audiences, education institutions, universities and research centers, as well as infrastructures. The rural cultural centers analyzed in the previous sections could not rely on some of these ecosystems' characteristics and dealt with these missing links as highlighted in the table below (Table 1).

With reference to *policies*, in both cases the creation of the center was strongly supported by local and regional authorities as part of the enhancement logistics of the territory or within the framework of local cultural policies. Regarding *finance*, since the opportunity to be supported by financial investors or by other resources such as sponsors was barely evident, both centers relied heavily on public funding which, in the case of *Les Abattoirs*, is almost the only source of funds. In terms of *markets*, the traditional audience for cultural events was absent (specialized audience, "cultural prosumers"). This issue was overcome by the creation of a stronger relationship with other stakeholders of the territory, such as schools, local associations, and local communities, that normally constitute just part of the cultural audience; in our cases, these stakeholders were the main recipients of cultural programming, and were involved in the activities of the centers and cooperated with them. This need to reach out to local potential audiences implied adapting artistic programming and performing spaces to attract those audiences. With reference to *human capital*, both cultural centers did not have higher education institutions, universities, or artistic education organizations (e.g., Academies of Fine Arts) in the area, resulting in a scarcity of artists operating in the territory. The lack of local artists was particularly difficult to substitute: in both cases, the centers decided to bring elements that were exogenous from the ecosystems: they recruited artists from festivals and specialized events outside the territory; in the case of *Les Abattoirs*, on a national scale. Hosted artists and companies in residence became temporarily components of their ecosystem. The traditional *culture* domain (cultural identity provided by major cultural heritage assets and cultural values related to the territory) was also lacking, although both Eurre and Riom were characterized by a strong sense of community. Finally, in terms of *support*, both centers were only partially able to find a viable solution to the lack of corollary activities: in the case of *Les Abattoirs*, the center cooperated with local stakeholders and, in the case of *La Gare à Coulisse*, the municipality was

Table 1 Main features of the entrepreneurial ecosystems of La Gare à Coulisse and Les Abattoirs (author's own elaboration)

	La Gare à Coulisse	Les Abattoirs
Policy	Support by local authorities for renovation works; multi-partnership agreement signed with the DRAC, the region, the department, and a group of local municipalities	Strongly supported by the municipality and other local, regional, and national authorities although still some shortcomings in recognizing the contribution of the center to the territory
Finance	Municipality and multi-stakeholder partnership with local, regional, and national public authorities provide around 50% of the total budget. The remaining 50% comes from self-generated revenues; no financial support from private sponsors	Financial support by several public authorities (town of Riom, the Ministry of Culture and Communication, DRAC, Regional Council and the Department Council). Public funds cover most of the expenses. Sponsorship and self-generated revenues contribute marginal funds
Supports	Link with supporting companies (commercial activities, enterprises, restaurants, accommodation for artists, organic shops) and related associations in the areas is not currently present but slowly growing, with the encouragement and support of the municipality. The cooperation is in the development phase	Difficulty of having appropriate complementary services in the rural area, but positive cooperation with various partners that are both institutional and private, local, and regional
Markets	Diversified segments of local inhabitants (e.g., schools), not only the traditional cultural prosumers that are absent in the territory	Attempts to create connections with local (not specialized) audience, by means of "unconventional" activities for the territory. However, the center is "still not completely recognized on the local level"; the place is more recognized at the national level
Human capital	Volunteers, often recruited locally; artists in residency (both emerging artists, students, and companies) who originate only partially from the region and more frequently coming from outside areas or other countries, recruited mainly through WoM and during specialized festivals.	Some inhabitants volunteer for the center. Due to the scarcity of local artists and the lack of universities and arts-specialized educational institutions, the center recruits artists and companies from outside the region, mainly during festivals or through an online application process.
Culture	Inadequate link with the cultural identity of the territory and its cultural heritage. Rather weak influence of the cultural milieu or the local cultural heritage on artistic production	Local cultural heritage and cultural activities are not mentioned (neither as a source of inspiration nor as an essential cultural milieu for the cultural center). No reference is made to the cultural identity of the territory

actively encouraging a broader project for creating infrastructures and connecting other shops and commercial activities in the surrounding area.

Summarizing these results, we noticed that the key to their sustainability was the creation of a more flexible entrepreneurial ecosystem composed of some strong endogenous elements, readapting to the local conditions and based on the insertion of exogenous elements on a temporary basis when the main ecosystem components were absent (e.g., the artists). Despite minor differences, both case studies adopted a similar approach, indicating possibly that cultural entrepreneurial ecosystem models should be more "elastic" in rural communities to compensate for potential drawbacks that operating in rural areas might entail, and at the same time adapt to communities that are slightly diverse from those traditionally related to cultural and creative entrepreneurial activities. Thus, they could not only become more sustainable in the long term, but they could also bring positive results of social cohesion, cultural enhancement, and economic development that could contribute to a more sustainable local development of their territories. Both cultural centers have been operating for several years, testifying that running an entrepreneurial activity in rural areas is possible, despite the difficulties.

6 Concluding Remarks

Research on sustainable development has increasingly underlined the role of culture both as fourth pillar of sustainable development, alongside the economic, social, and environmental one and *as* sustainable development. In this scenario, culture and cultural and creative industries have been related to issues of territorial enhancement, balanced development, and holistic approaches to sustainability. In this framework, growing attention is paid to the context in which cultural and creative industries operate: studies on cultural entrepreneurial ecosystems have emerged as viable interpretations of the link between cultural entrepreneurship and territorial context and as a form of guideline for identifying success factors of entrepreneurial initiatives in relation to a specific area. However, this topic is investigated mainly in urban areas, whereas studies on cultural and creative entrepreneurial ecosystems in rural areas are still scarce. This paper aimed at addressing this topic by focusing on a specific type of entrepreneurship in rural areas: rural cultural centers. Two case studies were selected in the Auvergne-Rhone-Alps region (*Les Abattoirs* and *La Gare à Coulisse*). By means of a qualitative approach, the research focused on their development models and the different components of their cultural ecosystem. The analysis highlighted that both centers encountered difficulties related to the lack of specific ecosystem domains, but they were able to overcome this issue by enhancing some of the strongest local components (such as the support by local authorities and cultivating relationships with local stakeholders such as schools and associations) and bringing in external elements (such as artists and companies) to compensate for the missing ones.

These results indicate that a more flexible ecosystem model for entrepreneurial activities in rural areas is needed, based not just on the presence of territorial assets but also on the idea of opening up the ecosystem to involve important exogenous elements, such as the broader community of artists and cultural companies. Only in this way could entrepreneurial activities in rural areas achieve sustainability in the long term, thus enhancing their potential for increasing the social and economic development of the territory.

These research results could be considered a preliminary investigation of the research questions, due to the restricted geographical scope of the investigation and limited sample of case studies. Further development of the investigation could be the study of other similar centers both in France and other countries, in order to carry out an international comparative analysis.

References

Aage, T., & Belussi, F. (2008). From fashion to design: Creative networks in industrial districts. *Industry and innovation, 15*(5), 475–491.

Aageson, T. H. (2008). Cultural entrepreneurs: Producing cultural value and wealth. In H. K. Anheier & Y. R. Isar (Eds.), *Cultures and globalization: The cultural economy* (pp. 92–107). London: Sage.

Andres, L., & Chapain, C. (2013). The integration of cultural and creative industries into local and regional development strategies in Birmingham and Marseille: Towards an inclusive and, collaborative governance? *Regional Studies, 47*(2), 161–182.

Ballico, C. (2017). Another typical day in this typical town: Place as inspiration for music creation and creative expression. *Australian Geographer, 48*(3), 349–363.

Basole, R. C. (2009). Visualization of interfirm relations in a converging mobile ecosystem. *Journal of information Technology, 24*(2), 144–159.

Baudelle, G., & Krauss, G. (2014). The governance model of two French national museums of fine arts relocated in the province: Centre Pompidou Metz and Louvre-Lens. *Belgeo. Revue belge de géographie*, (1). https://doi.org/10.4000/belgeo.12765.

Biondi, L., Demartini, P., Marchegiani, L., Marchiori, M., & Piber M (2018). The outreach of participatory cultural initiatives: The importance of creating and exchanging knowledge. In *IFKAD Book of Proceedings*.

Boix, R., José Luis Hervás, O., & Blanca De Miguel, O. (2015). Micro-geographies of creative industries clusters in Europe: From hot spots to assemblages. *Papers in Regional Science, 94*(4), 753–773.

Bonerandi, E., & Hochedez, C. (2007). Des machines, des vaches et des hommes. Projets culturels, acteurs et territoires dans un espace rural en crise: l'exemple de la Thiérache. *Norois, 204*, 25–37.

Bonet, L., & Donato, F. (2011). The financial crisis and its impact on the current models of governance and management of the cultural sector in Europe. *ENCATC Journal of Cultural Management and Policy, 1*(1), 4–11.

Borin, E. (2015, October 20). Local participation for the enhancement of cultural heritage: The UNESCO Candidature of the *Climats du Vignoble de Bourgogne*. In *The ecology of culture: Community engagement, co-creation and cross fertilization. Book of Proceedings of the 6th Annual Research Session*. ENCATC Annual Conference, Brussels, pp. 105–115.

Borin, E. (2017). *Public-private partnership in the cultural sector: A comparative analysis of European models*. Bruxelles: PIE Peter Lang.

Borin, E. (2018). Patrimonio culturale e ecosistemi imprenditoriali del settore culturale e creativo. *Urban Design—Paesaggio Urbano, 2*, 102–109.

Borin, E., & Donato, F. (2015). Unlocking the potential of IC in Italian cultural ecosystems. *Journal of Intellectual Capital, 16*(2), 285–304.

Borin, E., & Paunovic, I. (2016). The case of Louvre-Lens: Regional regeneration through cultural innovation. In H. Pechlaner, M. Gon, & M. Valeri (Eds.), *Innovazione, sostenibilità e competitiità. Teoria ed evidenze empiriche* (pp. 67–76). Giappichelli: Giappichelli Editore.

Borin, E., Donato, F., & Sinapi, C. (2018). Financial sustainability of small-and medium-sized enterprises in the cultural and creative sector: The role of funding. In E. Innerhofer, H. Pechlaner, & E. Borin (Eds.), *Entrepreneurship in culture and creative industries. FGF studies in small business and entrepreneurship* (pp. 45–62). Cham: Springer.

CAE—Culture Action Europe. (2019). *Culture crops—Cultural practices in non-urban territories*. Conference Report. Culture Action Europe.

CHCfE Consortium. (2015). *Cultural heritage counts for Europe*. Krakow: CHCfE Consortium.

Cohendet, P., Grandadam, D., & Simon, L. (2010). The anatomy of the creative city. *Industry and Innovation, 17*(1), 91–111.

Comunian, R., Chapain, C., & Clifton, N. (2010). Location, location, location: Exploring the complex relationship between creative industries and place. *Creative Industries Journal, 3*(1), 5–10.

Cooke, P. N., & Lazzeretti, L. (Eds.). (2008). *Creative cities, cultural clusters and local economic development*. Cheltenham: Edward Elgar.

Costanza, R., d'Arge, R., de Groot, R., Farber, S., Grasso, M., Hannon, B., Limburg, K., Naeem, S., O'Neill, R. V., Paruedo, J., Raskin, R. G., Sutton, P., & van den Belt, M. (1997). The value of the world's ecosystem services and natural capital. *Nature, 15*(387), 253–260.

Delfosse, C., & Georges, P. M. (2013). Artistes et espace rural: l'émergence d'une dynamique créative. *Territoire en mouvement Revue de géographie et aménagement, 2*(19/20), 60–76.

Denzin, N. K., Lincoln, Y. S., & Giardina, M. D. (2006). Disciplining qualitative research. *International Journal of Qualitative Studies in Education, 19*(6), 769–782.

Drake, G. (2003). 'This place gives me space': Place and creativity in the creative industries. *Geoforum, 34*(4), 511–524.

Duxbury, N., Cullen, C., & Pascual, J. (2012). Cities, culture and sustainable development. In H. K. Anheier, Y. R. Isar, & M. Hoelscher (Eds.), *Cultural policy and governance in a new metropolitan age* (The cultures and globalization series) (Vol. 5, pp. 73–86). London: Sage.

Duxbury, N., Kangas, A., & De Beukelaer, C. (2017). Cultural policies for sustainable development: Four strategic paths. *International Journal of Cultural Policy, 23*(2), 214–230.

ENRD—European Network for Rural Development. (2017). Re-imagining rural business opportunities. *EU Rural Review, 24*.

EUROSTAT. (2016). *Urban Europe—Statistics on cities, towns and suburbs*. Luxembourg: Publications Office of the European Union. Accessed from https://ec.europa.eu/eurostat/web/products-statistical-books/-/ks-01-16-691

Florida, R. (2005). *Cities and the creative class*. London: Routledge.

Florida, R. (2014). The creative class and economic development. *Economic Development Quarterly, 28*(3), 196–205.

Florida, R. (2017). *The new urban crisis: gentrification, housing bubbles, growing inequality, and what we can do about it*. London: Oneworld.

Georges, P. M. (2012). La friche industrielle en milieu rural, de la marge artistique à l'émergence d'une spatialité créative: le cas de Saint-Julien-Molin-Molette. *Colloque international pluridisciplinaire « De la friche industrielle au lieu culturel »*, pp. 255–262.

Grandadam, D., Cohendet, P., & Simon, L. (2013). Places, spaces and the dynamics of creativity: The video game industry in Montreal. *Regional Studies, 47*(10), 1701–1714.

Hall, J. K., Daneke, G. A., & Lenox, M. J. (2010). Sustainable development and entrepreneurship: Past contributions and future directions. *Journal Business Venturing, 25*(5), 439–448.

He, J. L., & Gebhardt, H. (2014). Space of creative industries: A case study of spatial characteristics of creative clusters in Shanghai. *European Planning Studies, 22*(11), 2351–2368.

Holden, J. (2004). *Creative reading: Young people, reading and public libraries*. London: Demos.

Holden, J. (2015). *The ecology of culture*. Swindon: Arts and Humanities Research Council.

Hristova, S., Šešić, M. D., & Duxbury, N. (Eds.). (2015). *Culture and sustainability in European cities: Imagining Europolis*. London: Routledge.
Iansiti, M., & Levien, R. (2004). *Keystones and dominators: Framing operating and technology strategy in a business ecosystem* (pp. 24–25). Boston: Harvard Business School.
Isar, Y. R. (2017). 'Culture', 'sustainable development' and cultural policy: A contrarian view. *International Journal of Cultural Policy, 23*(2), 148–158.
Isenberg, D. J. (2010). How to start an entrepreneurial revolution. *Harvard Business Review, 88*(6), 40–50.
Isenberg, D. J. (2011a). Introducing the entrepreneurship ecosystem: Four defining characteristics. *Forbes*. http://www.forbes.com/sites/danisenberg/2011/05/25/introducing-the-entrepreneurship-ecosystem-four-defining-characteristics/.
Isenberg, D. J. (2011b). *The entrepreneurship ecosystem strategy as a new paradigm for economy policy: Principles for cultivating entrepreneurship*. Babson Entrepreneurship Ecosystem Project. Babson Park, MA: Babson College.
Isenberg, D. (2014). What an entrepreneurship ecosystem actually is. *Harvard Business Review Blog*. https://hbr.org/2014/05/what-an-entrepreneurial-ecosystem-actually-is
Kagan, S., & Hahn, J. (2011). Creative cities and (un) sustainability: From creative class to sustainable creative cities. *Culture and Local Governance, 3*, 11–27.
Lash, S. U., & Urry, J. (1994). *Economies of signs and space*. London: Sage.
Loach, K., Rowley, J., & Griffiths, J. (2017). Cultural sustainability as a strategy for the survival of museums and libraries. *International Journal of Cultural Policy, 23*(2), 186–198.
Markusen, A. (2007). An arts-based state rural development policy. *Journal of Regional Analysis and Policy, 37*(1), 7–9.
Moore, J. F. (1996). *The death of competition: Leadership & strategy in the age of business ecosystems*. New York: Harper Business.
Mulero, M. P., & Rius-Ulldemolins, J. (2017). From creative city to generative governance of the cultural policy system? The case of Barcelona's candidature as UNESCO City of Literature. *City, Culture and Society, 10*, 1–10.
Patton, M. Q. (2002). *Qualitative evaluation and research methods* (3rd ed.). Thousand Oaks, CA: Sage.
Peltoniemi, M. (2006). Preliminary theoretical framework for the study of business ecosystems. *Emergence: Complexity and Organization (E:CO), 8*(1), 10.
Petrov, A. N. (2007). A look beyond metropolis: Exploring creative class in the Canadian periphery. *Canadian Journal of Regional Science, 30*(3), 451–474.
Protogerou, A., Kontolaimou, A., & Caloghirou, Y. (2017). Innovation in the European creative industries: A firm-level empirical approach. *Industry and Innovation, 24*(6), 587–612.
Regourd, E. (2007). Les associations culturelles, porteuses de projet pour de nouvelles ruralités? *Norois, 204*, 67–78.
Rius Ulldemolins, J. (2014a). Culture and authenticity in urban regeneration processes: Place branding in central Barcelona. *Urban Studies, 51*(14), 3026–3045.
Rius Ulldemolins, J. (2014b). Modelos de política cultural y modelos de equipa-mientos culturales: de los modelos nacionales a los modelos locales. Análisis del caso de Barcelona. *Política y sociedad, 51*(2), 399–422.
Sasaki, M. (2010). Urban regeneration through cultural creativity and social inclusion: Rethinking creative city theory through a Japanese case study. *Cities, 27*, S3–S9.
Soini, K., & Dessein, J. (2016). Culture-sustainability relation: Towards a conceptual framework. *Sustainability, 8*(2), 167.
Stam, E. (2015). Entrepreneurial ecosystems and regional policy: A sympathetic critique. *European Planning Studies, 23*(9), 1759–1769.
UCLG, U. C., & Governments, L. (2010). Culture: Fourth pillar of sustainable development. *UCLG*. http://www.agenda21culture.net/sites/default/files/files/documents/en/zz_culture4pillarsd_eng.pdf (2015-01-01)

United Nations. (2015). *Transforming our world: The 2030 agenda for sustainable development*. New York. https://sustainabledevelopment.un.org/post2015/transformingourworld

Vivant, E. (2007). Les événements off: de la résistance à la mise en scène de la ville créative. *Géocarrefour, 82*(3), 131–140.

Yildrim, E., Baltà Portolés, J., Pascual, J., Perrino, M., Llobet, M., Wyber, S., & Guerra, C. (2019). *Culture in the implementation of the 2030 agenda.* Project report. Paris: Culture2030Goal Campaign. http://openarchive.icomos.org/2167/

Yin, R. K. (2015). *Case study research and applications: Design and methods*. London: Sage.

Elena Borin is Associate Professor, Director of the MSc in Arts and Cultural Management, Coordinator of the Research Axis in Arts and Cultural Management, and Co-coordinator of BSB Centre for Arts and Cultural Management at the Burgundy School of Business (Dijon, France). She holds a PhD in Economics and was awarded the title of *Doctor Europeus* from the University of Ferrara (Ferrara, Italy). Her PhD thesis on "Public-Private partnerships in the cultural sector" (published by PIE Peter Lang) won the *3rd ENCATC Research Award on Cultural Policy and Cultural Management* and the *IUSS Best Thesis award*. She is member of Scientific Committees of several academic programs and also of international committees working on the themes of sustainability and the cultural and creative sector, among which the network "*Routes Towards Sustainability*" and the Scientific Committee of the ENCATC Congress *2019* and *2020*. Among her scientific contributions, her most recent articles address topics related to the themes of partnership, funding, and sustainability have been published by the *Journal of Cultural Management and Policy*, *Journal of Intellectual Capital*, *Urban Design*, and in the *Springer FGF Studies in Small Business and Entrepreneurship*. Her research interests include cultural governance with an international comparative approach, public–private partnerships and multi-stakeholder partnerships, funding and financing for the cultural and creative sector, and the link between cultural and creative industries in the sustainability paradigm.

Cassandre Jolivet, MSc in Arts and Cultural Management, is an independent researcher with research interests in arts, literature, and the socioeconomic trends in the cultural industry. She specialized with a thesis about cultural centers and their interactions with urban and rural territories, focusing on new business practices in the cultural industries (entrepreneurship, third places, models from the SSE...) and accompanying businesses working on environmental and socioeconomic transformations. Her thesis won BSB Best Thesis Award for the academic year 2016/2017 and has been published as a series of articles for the journal *Profession Spectacle*. She has collaborated with the Research team in Arts and Cultural Management of the Burgundy School of Business (France) for an applied research projects on the performing arts sector. Her interest in communication and media has led her to work for the press department of *Atout* France in London and to join the company *Look Sharp* as Press and Public Relation Consultant, where she has been working as a PR consultant since August 2018.

Part III
The Macro Level of Analysis

Culture Indicators for Sustainable Development

Annalisa Cicerchia

Abstract Culture and development are two challenging concepts. They are strongly related, in many and contradictory ways. Despite the fact that a huge literature of well-grounded criticisms of both terms continues to grow from one century at least, the post-modern societies use them and give them important strategic meanings. Culture is recognised as both a driver and an enabler of development. Some circles put forward culture as the ideal fourth pillar of sustainable development. Such recognition calls for appropriate measures and those, in turn, still represent an open issue. In 2019, UNESCO released *The Culture 2030 indicators*, a framework of thematic measures, in four domains (environment and resilience, prosperity and livelihoods, knowledge and skills and inclusion and participation) and 22 indicators. Its purpose is to assess and monitor culture's contribution to the implementation of the Goals and targets of the 2030 Agenda for Sustainable Development. UNESCO has approached such a complex endeavour with an innovative approach. This chapter focuses on both the conceptual and the practical application of the framework to institutions, organisations and enterprises of the cultural sector for planning and evaluation purposes. Its relevance as a compelling stimulus for new targeted statistical data production will also be considered.

Keywords Sustainable Development Goals · Cultural indicators · Cultural statistics · Sustainability of culture

---

A. Cicerchia (✉)
Istituto Nazionale di Statistica (Italian National Statistical Institute), Rome, Italy
e-mail: acicerchia@istat.it

P. Demartini et al. (eds.), *Cultural Initiatives for Sustainable Development*, Contributions to Management Science,
https://doi.org/10.1007/978-3-030-65687-4_16

1 Culture and Sustainable Development: A Challenging Relationship

Culture and development are two complex, inter- or even trans-sectoral, multidimensional and polysemic notions. Their contents change over time and space. Diverse and conflictual views over their definition, place and role coexist at both the theoretical and the practical level at least since the 1950s (Kluchhohn and Kröber 1952; Bauman 1973; Barber and Badre 1998; Cicerchia 2016). However, it is since the 1980s that culture has manifestly acquired a distinctive political dimension; in particular, since then, with the emergence of a new strand of study called culture economics, its impacts on wealth, occupation and business have become apparent.

The relationship between culture and *sustainable development* is of course a more recent topic and indeed less investigated, from a wide range of different approaches. Despite the differences, the majority of these approaches tend to take for granted that any role played by culture in sustainable development is in se positive.

> The exploration of the relationship between culture and sustainable development, including of cultural values and the intrinsic value of culture, and of culture as a potential separate pillar of sustainable development, has been going on for several decades... In other words, culture, by contributing to the intellectual, emotional, moral and spiritual wellbeing of people, and by enabling everyone to exercise their human rights, including their cultural rights, also contributes to sustainable development. (UNESCO 2015)

In recent article, stemming from a COST[1] action targeted at investigating the relationship of culture and sustainability[2] the terms in such relationship are discussed in three different combinations (Soiini and Dessein 2016).

The authors first consider culture as if it played an independent role in sustainability, as a *fourth pillar*—the other three being environment, economy and society. Culture, in this light, is a pillar inasmuch it is a specific *capital,* hence the importance of its conservation, maintenance and preservation in the forms of arts, heritage, knowledge and cultural diversity for the next generations.

The second representation refers to culture having a mediating role in achieving economic, social and environmental sustainability. This representation postulates that material and immaterial culture is an essential resource for local and regional economic development. It also implies that cultural values, *weltanschauung,* models and patterns of behaviour need to be considered when aiming for environmental, economic or social sustainability.

---

[1]The European Cooperation in Science and Technology (COST) is a funding organisation for the creation of research networks, called COST Actions. These networks offer an open space for collaboration among scientists across Europe (and beyond) and thereby give impetus to research advancements and innovation. https://www.cost.eu/

[2]https://www.cost.eu/publications/culture-in-for-and-as-sustainable-development-conclusions-from-the-cost-action-is1007-investigating-cultural-sustainability/

The third representation considers culture as a necessary foundation for meeting the overall aims of sustainability. This representation, labelled *culture as sustainability*, encloses the other pillars of sustainability. The authors conclude that in this third combination, sustainability becomes embedded in culture and "leads to eco-cultural civilisation."

The UNESCO's document and the article by Soiini and Dessein have much in common. In particular, they share two tacit assumptions. The first is that culture is operationally treated as being *one,* only one: a single, compact and comprehensive system of material and immaterial resources that must be channelled towards the achievement of sustainability. The second assumption is that culture is intrinsically *good* and inasmuch a natural ally of sustainability. Both assumptions invite a supplement of reflection.

In the preamble to the Universal Declaration on Cultural Diversity (2001), culture is defined as "…the set of distinctive spiritual, material, intellectual and emotional features of society or a social group, and that it encompasses, in addition to art and literature, lifestyles, ways of living together, value systems, traditions and beliefs." This definition is in line with the conclusions of the World Conference on Cultural Policies (MONDIACULT, Mexico City, 1982), of the World Commission on Culture and Development (Our Creative Diversity, 1995) and the Intergovernmental Conference on Cultural Policies for Development (Stockholm, 1998).[3]

This definition appears consonant with what Zygmunt Bauman (1973) labels *Culture as differential concept:* the term "is employed to account for the apparent differences between communities of people (temporally, ecologically, or socially discriminated). This usage locates the differential concept of culture among numerous 'residue concepts', contrived frequently in the social sciences to explain away the sediment of deviant idiosyncrasies unaccountable for by the otherwise universal and omnipotent regularities (where it shares the ascribed function with ideas, tradition, life experience etc.)." A second premise of the differential concept of culture is "that various, even mutually exclusive, socio-cultural forms may correspond to a single set of non-social (biological, natural-environmental, ecological) conditions". Peter Berger warns us, "one cannot throw a sop to the dragon of relativity and then go about one's intellectual business as usual".[4]

The idea of culture as a single, compact and univocal whole holds only if one speaks, figuratively, at the level of the entire human species and as opposed to *nature: nature vs. culture*. At all the other levels, theory and practice prescribe to acknowledge that culture is intrinsically plural and pluralistic, sometimes even particularistic or partisan. Theory and practice face a multiplicity of cultural and sub-cultural expressions of different human groups, communities, organisations, etc.

[3]http://www.unesco.org/new/en/culture/themes/culture-and-development/the-future-we-want-the-role-of-culture/the-key-ideas/ (Accessed March 11, 2020).

[4]Peter L. Berger 1969. A Rumor of Angels: Modern Society and the Rediscovery of the Supernatural, Garden City, N. Y.: Doubleday and Co.

At core of the idea of cultural diversity lies the notion that human groups express themselves through different cultures, so different that in the most extreme forms of *cultural relativism*, even the boundaries between good and bad seem to blur, as their very definition is rooted in the cultural context. Moral codes, values, rules and the idea of legal vs. illegal are culture-laden, plus they change over time and space and cultural differences mark different historical periods and are geographically distributed. Cultural identities are built on perceived differences (*us* vs. *them*). History is full of examples of conflict, aggression, violence and war originated by cultural identity and differences strongly felt, to the point that one culture is perceived as better than or superior to another. Cultural domination, hegemony and cultural colonisation of the succumbing group and even cultural genocide are monsters generated when we fail to acknowledge and practice the existence of cultural diversity and the right of groups to their own culture. Ethnic group, religion, gender, age, class, place and affiliation are but some of the possible sources of cultural differences.

In short, when we speak of the need of mobilising *culture* for the quest of sustainability, as its fourth pillar, we are evoking a powerful, all compassing, and indispensable dimension. But we should be aware that *out there* we will find a number of different *cultures*. Contents, the rules and the structures of that dimension differ from a human community to another, as their languages differ, and even more. We have to keep in mind that each culture is not a compact block but that it is made of a myriad of components, ordered by cultural models, with varying relevance for fostering or hindering a sustainable development.

My second point is about the postulated natural alliance between sustainability and culture(s). Once we acknowledge that each one of the existing cultures correspond to a specific set of values, norms, rules, attitudes, habits, roles, statuses, symbols, technologies, settlements, etc., we should correspondingly acknowledge that some cultures or some of their contents are more oriented towards sustainability than others. Cultural models—mental representations shared by members of a culture—allow us to conduct our daily business while on "automatic pilot". From marriage to etiquette, from hygiene to health, well-being and comfort and our daily relationship with nature are defined by cultural models. Cultural models orient also the way we employ natural resources, produce goods and services, consume or use them, generate waste and dispose it after use. The needs we feel, the ways we satisfy them and the means we avail of are all culture-laden and culture-specific.

Although with exceptions, the dominant cultural model of well-being and comfort, which is basically a cultural expression of the affluent Western societies of the 1960s, remains built upon possession and conspicuous consumption (Veblen [1899] 1973) of energy and resource-intense, disposable rather than reusable, polluting and waste-generating material objects: from cars to food, from clothes to packaging and—of recent—to an unprecedented amount of digital devices. Indeed, one could hardly define those cultural models as conducive or simply consistent with the idea of sustainable development. The same can be said for those cultural models, still found nowadays in other cultural setting, which deny the rule of law or the idea that all human beings—women, for instance—are made equal and are entitled to the same rights, etc. Truth is that, while indeed we need to summon the strength of key elements within each culture if we want that sustainable ways of life are adopted, take roots and last over time, it is also imperative that other adverse elements are changed, minimised or altogether abandoned. Indeed, we can draw from the frugal traditions of many peasant cultures around the globe an attitude to recycle food, clothes and materials and make them work in new forms in order to reduce waste and energy consumption. On the other hand, we need to get rid of intrinsically unsustainable traditional cultural models, like those, found in many cultures even today, which deprive women from power, freedom of choice or even property rights, or those, even more widespread, which associate high social status with the possession of luxury supercars and other conspicuous goods. Amarthya Sen invited reflection about the "complex epistemic issues involved in identifying the ways in which culture may or may not influence development, and the deeply ethical and political issue of social choice involved in accommodating diverse concerns" (Sen 2000). That complexity manifested itself frequently since the 1960s, when North-South programmes of development aid and cooperation started: "Often development agendas, which are repeatedly built upon the assumption of so-called 'objective' rationality and homogeneity of preferences, do not fully take into account the cultural differences that exist between donors and recipients when designing development strategies" (Minasyan 2014). Many authors in the debate of the 1970s and 1980s contributed to unmasking the ethnocentric character of what—especially in the areas of technologies, energy, production and distribution models—was transferred from the developed to the developing countries and presented as neutrally *modern* and *efficient.* Rostow's Stages of Economic Growth (1960) was the manifesto of a vision of development based on an imitation and reproduction of the economic history of the cultural West. Inglehart and Baker have investigated the controversial relationship of tradition and modernity in development processes "Well into the twentieth century, modernisation was widely viewed as a uniquely Western process that non-Western societies could follow only in so far as they abandoned their traditional cultures and assimilated technologically and morally" superior "Western ways." Indeed, the two authors stress the central claim of modernisation theory is that economic development is linked with coherent changes in culture and social and political life. However, they observe, "cultural change is not linear; with the coming of postindustrial society, it moves in a new direction. Different societies follow different trajectories even when they are subjected to the same forces of economic development, in part because situation-specific factors, such as cultural heritage, also shape how a particular society develops" (Inglehart and Baker 2000). To make a long story short, *tradition*, initially rejected as anti-modern, has come to be recognised as an essential identity component of the different human communities and a key factor in their own individual path to development. The next step is the tacit romantic assumption that tradition, any tradition apparently, is always good, per se and in se.

In the last dozen years or so, a renewed international consensus has been manifested about the need to move from a GDP growth-centred model to a well-being-oriented sustainable model. Indeed, measuring development in terms of GDP can be related to a typically Western cultural pattern, labelled, amongst other things, as *economic reductionism* (apart from the original statement about the conceptual carrying capacity of GDP by one of its very inventors, Simon Kuznet,[5] see also Zamagni 2000; Sen 2009). However, also, the definition of well-being is culture-laden. When, in 2012, the Italian National Scientific Commission on Measures of Wellbeing discussed the direction (positive or negative) to give to indicators concerning social trust and relationships, a discrepancy between a North-European cultural value model favouring reliance on universalistic networks (public services, citizenship rights and so on), on the one hand, and the South-European cultural value favouring reliance on particularistic connections (family and personal friends), on the other, became evident. For those who share the universalistic model, the more the people trust and rely on generalised, non-emotional and non-family-based networks (for instance, public social services addressing various needs), the higher their well-being. Those sharing the particularistic model, conversely, appreciate the contribution to well-being granted by strong family bonds and family-based supporting networks. Development goals in policies and programmes are equally the result of culture-driven values and political choices amongst conflicting alternatives.

So much for the *anthropological* concept of culture: an ocean of meanings. There is however another one, increasingly relevant in the economic and political literature, way less complex in comparison to the first concept, if nonetheless statistically evasive: culture as a—special—sector of economic activity. A concept that entails a myriad of products and services, markets and subsidies, consumers and users, businesses and workers, skills, knowledge, technologies, copyrights, creativity, organisations, processes and infrastructures, resources and investments, monetary and non-monetary value, etc. A considerable amount of studies, starting from the mid-1960s[6] (Baumol and Bowen 1966; Peacock 1969; Robbins 1971), aim at assessing the contribution of that economic sector to development processes. Recently, they have included the notion that the cultural sector is in se a sustainable one, and that it can contribute to a more sustainable economy.

It is against that composite background that we can appreciate how brave the efforts to translate the complex relationship of culture(s) and sustainable development into indicators have been so far. The purpose of the present exercise is: (1) to explore and discuss the operational concepts of culture that have been used for building those indicators; (2) to reflect upon the suggested data sources and methods and their feasibility and (3) to derive from the Culture|2030 indicators ideas for future data collection at a national and local level. The first part of the paper analyses UNESCO's Culture 2030 Indicators and their construction. The second part reflects on the concept of culture, upon which the indicators have been developed. The final

[5]"the welfare of a nation can scarcely be inferred from a measure of national income. If the GDP is up, why is America down? Distinctions must be kept in mind between quantity and quality of growth, between costs and returns, and between the short and long run. Goals for more growth should specify more growth of what and for what". (Kuznets, Report to the US Congress, 1934)

[6]The Journal of Cultural Economics was first published in 1977.

part discussed proposed data sources and their availability and feasibility for orienting future statistical activity in this field.

2 UNESCO's Culture|2030 Indicators

2.1 The Process

"Over the last decade, UNESCO's advocacy for a culture-based approach to development has resulted in several United Nations General Assembly Resolutions that acknowledge the role of culture as an enabler and a driver of sustainable development. This process culminated in the integration of culture in the 2030 Agenda for Sustainable Development, adopted in 2015 by the 193 countries of the UN General Assembly" (UNESCO 2019). The 2030 Agenda is built upon 17 Sustainable Development Goals (SDGs) and 169 related Targets. The 17 SDGs are grouped into "5 Ps" of People, Planet, Prosperity, Peace and Partnerships, reflecting the economic, social and environmental dimensions of sustainability (people, planet and prosperity), as well as its two critical conditions (peace and partnerships).

UNESCO developed the Thematic Indicators for Culture in the 2030 Agenda (Culture|2030 Indicators) as part of their effort "to establish a methodology for demonstrating culture's role and contribution to the implementation of the UN Sustainable Development Goals." This new framework for measuring and collecting data on culture is an instrument aimed at advocating for the role of culture in the SDGs, as well as integrating culture into development plans and policies at the national and local levels. The theoretical and methodological background of Culture| 2030 Indicators includes the Framework for Cultural Statistics (FCS) of UIS, the Culture for Development Indicators Suite (CDIS), the Culture Conventions periodic reporting mechanisms and other monitoring mechanisms and methodologies in the specific context of the 2030 Agenda.

2.2 The Thematic Indicators at a Glance

Basically, Culture|2030 Indicators is a framework made of 22 indicators, grouped into four thematic dimensions, each corresponding to the three pillars of sustainable development: the economic, the social and the environmental. The fourth dimension relates to education, knowledge and skills in cultural fields.

In developing the Thematic indicators, UNESCO starts from the definition of culture adopted in the 2030 Agenda[7]: "a broad view of culture that encompasses the contribution of culture to sustainable development including through cultural

[7]https://sustainabledevelopment.un.org/post2015/transformingourworld.

heritage, the creative industries, local culture and products, creativity and innovation, local communities, local materials, and cultural diversity. At the same time, the experience of development projects and interventions has demonstrated the importance of local knowledge and community participation in order to achieve sustainable development—from health to education". From the start, culture here is a twofold notion: (1) as a sector of activity in itself and (2) as cultural diversity and local culture and products. In terms of the multiple meanings of culture, we have discussed in par.1: the first notion derives from the cultural economics strand and the second reflects the differential quality highlighted by Bauman.

2.3 Purpose and Guiding Principles

The role of culture can be addressed both as a *driver* that contributes directly to bringing about economic and social benefits and also as an *enabler* that contributes to the effectiveness of development interventions. Therefore,

"The UNESCO Thematic Indicators for Culture (Culture|2030 Indicators) is a framework of thematic indicators whose purpose is to measure and monitor the progress of culture's enabling contribution to the national and local implementation of the Goals and Targets of the 2030 Agenda for Sustainable Development. The framework will assess both the role of culture as a sector of activity, as well as the transversal contribution of culture across different SDGs and policy areas. As a set of thematic indicators, it is intended to support and complement the global indicators agreed upon within the 2030 Agenda and foster linkages between different Goals and Targets" (UNESCO 2019, p. 17). Make culture visible, provide a thematic and transversal overview of the role of culture across the SDGs, strengthen advocacy for culture, provide evidence-based results to inform policies and actions, build a knowledge base for action and monitor progress of the contribution of culture to the 2030 Agenda represent the five key aims of the Thematic Indicators.

Their guiding principles include: rely as much as possible on existing data sources, use qualitative and quantitative data to assess the contribution of culture, integrate data from reporting on UNESCO Culture Conventions and programmes, develop instruments to measure culture at both national and urban levels, facilitate cooperation across institutions, propose a framework adaptable to different statistical capacities, provide an aspirational tool for all rather than normative assessment and reflect the Result-Based Management conceptual framework. With reference to the last principle, each thematic dimension combines these different types of indicators to measure inputs (e.g. development of cultural infrastructure), processes (e.g. governance mechanisms) or outputs (e.g. participation in cultural life), with a view to producing an overall understanding of culture's contribution.

2.4 Dimensions and Indicators

The Culture|2030 Indicators are structured into four transversal thematic dimensions: (1) Environment and Resilience, (2) Prosperity and Livelihoods, (3) Knowledge and Skills and (4) Inclusion and Participation. Each dimension combines several Sustainable Development (SD) Goals and Targets.[8] "The framework gives priority to the areas that are relevant to UNESCO's mandate in Culture and where relevant quantitative or qualitative data are already being collected or are likely to be identified" (p. 26). Table 1 summarises the relevance of each dimension to each SDG.

2.4.1 Dimension 1: Environment and Resilience

This dimension provides a framework for assessing role and contribution of culture to sustainable human settlements with a focus on cultural and natural heritage and urban environment (the 'Planet' pillar of the SDGs). It addresses tangible and intangible heritage, and natural heritage, as a lever for sustainable development and as an end in itself. The SDGs related to this dimension are no. 2—*End hunger, achieve food security and improved nutrition and promote sustainable agriculture*, no.6—*Ensure availability and sustainable management of water and sanitation for all*, no.9—*Build resilient infrastructure, promote inclusive and sustainable industrialisation and foster innovation*, no.11—*Make cities and human settlements inclusive, safe, resilient and sustainable*, no.12—*Ensure sustainable consumption and production patterns*, no.13—*Take urgent action to combat climate change and its impacts*, no.14—*Conserve and sustainably use the oceans, seas and marine resources for sustainable development*, no.15—*Protect, restore and promote sustainable use of terrestrial ecosystems, sustainably manage forests, combat desertification, halt and reverse land degradation and halt biodiversity loss* and no.16—*Promote peaceful and inclusive societies for sustainable development and provide access to justice for all and build effective, accountable and inclusive institutions at all levels*.

Safeguarding the world's cultural and natural heritage is a specific target in itself (Target 11.4). Other ways culture contributes to environment and resilience are listed below.

- The integration of intangible cultural heritage and traditional knowledge into policies and strategies encourages sustainable development (sustainable food production, resilient agriculture and conservation of natural resources) (Target 2.4 *Sustainable foodways and agriculture*).

[8]For a complete list of the Sustainable Development Goals and Targets, see: https://sustainabledevelopment.un.org/content/documents/11803Official-List-of-Proposed-SDG-Indicators.pdf. [accessed April 23, 2020].

Table 1 UNESCO's Culture 2030 Dimensions and related SDGs

	Dimensions			
SDGs	1 Environment and resilience	2 Prosperity and livelihood	3 Knowledge and skills	4 Inclusion and participation
1—End poverty in all its forms everywhere				
2—End hunger, achieve food security and improved nutrition and promote sustainable agriculture	X			
3—Ensure healthy lives and promote well-being for all at all ages				
4—Ensure inclusive and equitable quality education and promote lifelong learning opportunities for all			X	
5—Achieve gender equality and empower all women and girls	Transversal			
6—Ensure availability and sustainable management of water and sanitation for all	X			
7—Ensure access to affordable, reliable, sustainable and modern energy for all				
8—Promote sustained, inclusive and sustainable economic growth, full and productive employment and decent work for all		X	X	
9—Build resilient infrastructure, promote inclusive and sustainable industrialisation and foster innovation	X		X	X
10—Reduce inequality within and amongst countries		X		X
11—Make cities and human settlements inclusive, safe, resilient and sustainable	X	X		X
12—Ensure sustainable consumption and production patterns	X		X	
13—Take urgent action to combat climate change and its impacts	X		X	
14—Conserve and sustainably use the oceans, seas and marine resources for sustainable development	X			
15—Protect, restore and promote sustainable use of terrestrial ecosystems, sustainably manage forests, combat desertification, halt and reverse land degradation and halt biodiversity loss	X			

(continued)

Table 1 (continued)

SDGs	Dimensions 1 Environment and resilience	2 Prosperity and livelihood	3 Knowledge and skills	4 Inclusion and participation
16—Promote peaceful and inclusive societies for sustainable development, provide access to justice for all and build effective, accountable and inclusive institutions at all levels	X			X
17—Strengthen the means of implementation and revitalise the global partnership for sustainable development				

- The conservation of natural heritage directly contributes to environmental sustainability. Intangible Cultural Heritage and traditional knowledge are also significant components of ecosystem management by local communities and in safeguarding plans and mechanisms for natural heritage conservation (Target 6.6 *Water-related ecosystems*, Target 14.5 *Marine area conservation*, Target 15.1 *Sustainable terrestrial ecosystems* and Target 13.1 *Climate and disaster resilience*).
- Natural, historically derived and local building practices, and intangible cultural heritage can help mitigating the risks of climate-related disaster, support resilience and enhance the adaptation capacities of communities (Target 13.1 *Climate and disaster resilience*).
- Cultural tourism and eco-tourism are central to sustainable tourism and play a primary role in protecting the environment. Policies and measures on sustainable tourism can be integrated into national, subnational and local development plans, mechanisms and strategies (Target 12.b *Sustainable tourism management*).
- In order to improve sustainable management of heritage, cultural policies and strategies must reduce illicit trafficking and encourage the recovery of stolen assets (Target 16.4 *Recovery of stolen assets*).
- Cultural facilities form part of quality, reliable, sustainable and resilient infrastructure for cities. Historic buildings, spaces and urban areas as well as careful and compatible new designs rooted in local materials and contexts enhance the urban space and reinforce cultural identity. Cultural facilities integrated into territorial planning enhance the diversity of public space and citizen well-being (Goal 11, several Targets). Similarly, public green spaces available to cultural activities stimulate social cohesion and function as meeting points, thus contributing to a quality environment (Target 11.7 *Inclusive public spaces*).

2.4.1.1 The Indicators of Dimension 1

Indicators of Dimension 1 assess the dimension of spending on heritage protection and conservation, the distribution of cultural indoor and outdoor facilities and the environmental quality of the management of heritage. While theoretically appropriate for representing heritage "as an end in itself", the three indicators on spending and endowment are in practice scarcely applicable to intangible heritage and quite difficult to be covered by good quality comparable data (cultural expenditure data are often aggregated with other budget items). The measurement of how heritage is a lever for sustainable development appears weaker, as it relies only on two indicators about the environmentally correct management of heritage sites, where the *development* component seems scarcely covered.

Indicator 1: Expenditure on Heritage

> Total expenditure (public and private) per capita spent on the preservation, protection and conservation of all cultural and natural heritage, by type of heritage (cultural, natural, mixed, World Heritage Centre designation), level of government (national, regional, and local/municipal), type of expenditure (operating expenditure/investment) and type of private funding (donations in kind, private non-profit sector, sponsorship).

This indicator illustrates how financial action by public authorities, at the local, national and international levels, alone or in partnership with civil society organisations (CSO) and the private sector, to protect and safeguard cultural and natural heritage has a direct impact on safeguarding heritage and in making cities and human settlements more sustainable.

Sources

- UNESCO data: UIS
- National and local sources: National Statistical Institutes, Administrative data, Specific national surveys and Information systems for culture when available.

Method

Disaggregation:

- By type of heritage: cultural, natural and mixed World Heritage properties
- Public expenditure by level of government (national, regional and local/municipal)
- Type of public expenditure (capital expenditure and operating expenditure)
- Private funding: donations in kind, private non-profit sector and sponsorship.

The Authors stress that it might be difficult to calculate, amongst other things, because countries' national accounting frameworks may not clearly separate cultural, natural and other activities; financial transactions may be channelled away for different uses and may be double counted at different levels of public administration. Also, the indicator does not measure non-monetary factors such as national regulations or national/local policies (e.g. fiscal incentives, tax benefits for donations or sponsorships).

Indicator 2: Sustainable Management of Heritage

Checklist[9] for the sustainable management framework to safeguard and manage cultural and natural heritage, practices, knowledge, and movable historical artefacts.

This indicator offers a general picture of the strengths and shortcomings of public action to protect and promote heritage sustainability through the analysis of three components:

- national and international registers and inventories;
- action to protect, safeguard and manage heritage involving all stakeholders and fostering sustainability;
- the level of support mobilised to safeguard and revitalise heritage.

Data Sources

- UNESCO data: periodic reports of the 1972, 1970 and 2003 Conventions as well as the survey data from the 2011 and 2015 Recommendations
- National and local sources: administrative data, specific national surveys and information systems for culture when available

Method
Checklist (numerical and Yes/No).

Indicator 3: Climate Adaptation and Resilience

Checklist for the climate adaptation framework, particularly including traditional practices for resilience.

This indicator aims to assess measures to foster climate change mitigation and adaptation and enhance resilience through sustainable safeguarding and management of tangible and intangible cultural heritage as well as natural heritage.
Data Sources

- UNESCO data: periodic reports of the 1972 and 2003 Conventions.
- National and local sources: administrative data, specific national surveys and Information systems for culture when available.

Method
Checklist (numerical and Yes/No).

The indicator includes "Evidence of policies/measures to take into account traditional and local community knowledge in assessing the possible impact of climate adaptation on heritage elements and practices"; "Evidence of integrating cultural factors, including knowledge, traditions and practices of all people and communities, into local strategies on environmental sustainability"; "Evidence of integrating cultural factors, including knowledge, traditions and practices into agricultural strategies" amongst the items in the list.

[9]The complete checklist is in UNESCO (2019, pp. 38–46).

Indicator 4: Cultural Facilities

The distribution of cultural facilities through spatial mapping.

The indicator aims to assess the diversity of cultural facilities (libraries, museums, galleries, performance venues, cinemas, *traditional cultural spaces*,[10] creative hubs, educational institutions and cultural internet sites) and their territorial distribution. It enables the identification of cultural areas in relation to the population, transport, administration and economic centres. This indicator aims to assess how cultural facilities are integrated in the urban landscape and provide an enabling environment for communities to attend cultural events, practice and participate and for culture professionals and businesses to thrive. It also enables the spatial assessment of areas better served or in greater need of cultural facilities.

Data Sources

- UNESCO data: UIS
- National and local sources: administrative data

Method

Spatial Analysis focuses on diversity, capacity, physical area and network analysis. Indicators also include measures of the number of facilities, their economic output, the public finance if applicable and the number of visitors or users.

Analysis of the gender dimension is also recommended.

Indicator 5: Open Space for Culture

Number and size of open spaces used for cultural purposes by type of use.

The indicator aims to assess the extent of public open spaces, the nature of the spaces and the degree of public use (including traditional markets[11]).

Data Sources

- UNESCO data: UN-Habitat—Public open space strategies SDG 11.7.1
- National/Local sources: administrative data

Method

Amongst the potential metrics that can be used here, the Thematic Indicators list:

- The area (m^2, ha) of ‘cultural open space’ as a percentage of all public open space
- Number of ‘cultural open spaces’ as a percentage of all open spaces

[10]Traditional cultural spaces are defined areas used as traditional village meeting/socialising spaces, e.g. ‘kiva’—a semi-underground walled area with built hearths and furnishings (Hopi), ‘maneva’ roofed open-sided platform used for socialising sleeping eating and village meetings (Kiribati), ‘marae’ open-enclosed area associated with ancestors used for meetings prayers, etc. (New Zealand).

[11]The definition of Traditional market adopted in the UNESCO text is ‘a market which is built and managed by government, private, cooperative or local people’s self-support with business places such as stores, kiosks, stalls and tents, or some other similar names, which is owned/ managed by small-medium traders, with small scale business and small capital, of which its buying and selling process is done through bargaining’ (UN Habitat).

- Percentage distribution of cultural events in open space by domain as defined by the Framework of culture statistics.[12]

The gender dimension of this indicator should also be taken into account.

2.4.2 Dimension 2: Prosperity and Livelihood

This dimension proposes an assessment framework for the contribution of culture in driving and enabling more inclusive and sustainable economies, in line with the 'Prosperity' pillar of the SDGs, by generating income, employment and revenue, through cultural goods, services and enterprises. The SDGs addressed here are no.8—*Promote sustained, inclusive and sustainable economic growth, full and productive employment and decent work for all*, no. 10—*Reduce inequality within and amongst countries* and no.11—*Make cities and human settlements inclusive, safe, resilient and sustainable.*

The contribution of Culture to prosperity and livelihoods is declined in terms of direct impact of the sector on GDP and job and business creation (e.g. heritage conservation, heritage tourism and in the creative sector—Target 8.3 *Jobs, entrepreneurship and innovation*). Heritage tourism can support job creation and promote local culture and products, contributing to sustainable development (Target 8.9 *Policies for sustainable tourism*). Public policies can encourage economic activities and employment opportunities through increased investment in cultural and natural heritage and infrastructure (Target 11.4 *Cultural and natural heritage*). The governance of culture creates the enabling conditions that allow cultural activities and forms to thrive. Policies and regulations may also favour more equitable international trade (Target 10.a *Differential treatment on trade* and Target 8.a *Increase Aid for Trade*).

2.4.2.1 The Indicators of Dimension 2

The indicators of Dimension 2 are drawn from the classical culture economics. They are theoretically irreproachable, but they nonetheless have two important limits. One is that four of them are based on monetary transactions, which, especially in the digital era, are increasingly less adequate to grasp in full the economic dynamic of the sector. The other is that the statistical information on the field described by Dimension 2, even in those countries with advanced statistical systems, still resents from uncertain definitions (e.g. what is a cultural enterprise or a cultural occupation)

[12]Central cultural domains are defined as common sets of culturally productive industries, activities and practices directly associated with the creation, production, distribution and enjoyment of central cultural content. They are Cultural and Natural Heritage, Performance and Celebration, Visual Arts and Crafts, Books and Press, Audio-visual and Interactive Media, Design and Creative Services and Intangible Cultural Heritage (transversal domain).

and insufficient granularity and detail to describe, for instance, households' cultural expenditure as separated from other leisure time expenditure.

Indicator 6: Culture in GDP

Percentage of Gross Domestic Product attributable to private and formal cultural production.

The indicator aims to assess the overall contribution of the culture sector to the economy in a given territory. One limitation of this indicator is that it is not able to take into account all cultural activities including those that are informal and unpaid. It aligns with the international classification of the Framework for Cultural Statistics.

Data sources

- National and local sources: National Accounts, Business surveys and censuses, Service and commercial surveys, Government records, Cultural special surveys, artist registers, etc., Private sector sources (e.g. special surveys done by trade unions or chambers of commerce).

Method

To obtain the ratio of cultural GDP, add the values obtained using the ISIC statistic codes included in the UIS Framework for Cultural Statistics (UNESCO-UIS 2009, pp. 52–64) and then compare this sum with the gross domestic product (GDP) of the local economy.

Indicator 7: Cultural Employment

Number of people employed in the cultural and creative sectors and cultural occupations as a percentage of overall employment for the latest year.

The indicator aims to assess the role of culture as an employer at the national and local level.

Data sources

- UNESCO data: UIS
- National and local sources: National Accounts, Population Census, Labour Force surveys (LFS), Administrative records (e.g. social security registers) and Professional associations.

Method

Cultural employment includes: A. People with a cultural occupation and who work in cultural businesses (e.g. an actor in a theatre); B. People with a cultural occupation but who work in a business not engaged in cultural activity (e.g. a designer in the motor industry) and C. People who work in cultural businesses but who do not have a cultural occupation (e.g. an accountant working in a theatre). The indicator is calculated as the sum of all these three groups as a percentage of all employed persons. Disaggregation by gender or by ethnic group is suggested.

Indicator 8: Cultural Businesses

Trends in cultural businesses as a percentage of all businesses.

The indicator aims to assess are the enabling conditions offered to cultural businesses (especially small and micro-enterprises).

Data sources

- National and local sources: Business surveys and Registers of businesses (such as the Chamber of Commerce and sectoral bodies).

Method

Level of change as measured by annual percentage points difference in cultural businesses as a percentage of all businesses for the given area. Gender differences in the ownership of businesses are also suggested for investigation.

Indicator 9: Household Expenditure

Percentage of total household expenditure devoted to cultural activities, goods and services.

This indicator aims to assess how households of a given territory value cultural goods and services through market transactions and to obtain insight into the size and the potential of the local market for cultural activities, goods and services.

Data sources

- National and local sources: Industry surveys and censuses, Service surveys, Small establishment surveys and household expenditure surveys.

Method

CHFC is the household final consumption expenditure on cultural activities, goods and services set against total household final consumption expenditure; HCSCOICOP codes is the total amount of household expenditure in the selected COICOP codes; HFC is the total household final consumption expenditure. The authors underline that many transactions involving cultural goods and services are non-market and therefore escape this proposed measure.

Indicator 10: Trade in Cultural Goods and Services

Exports of cultural goods and services as a percentage of all exports.

This indicator aims to assess the degree to which products as expressions of culture are exported reflecting the economic demand, the international profile of the country/city's cultural products and services and the regulatory environment to enable this.

Data sources

- UNESCO data: UIS
- National and local sources: National Customs and Revenue reports preferred, otherwise the international COMTRADE database (comtrade.un.org).

Method

(a) The value of physical cultural exports using the codes (Harmonised System where possible) as listed in FCS as a percentage of the value of all exports of physical goods.

(b) The value of exports of cultural services using the codes as listed in FCS as a percentage of the value of all service exports.

Indicator 11: Public Finance for Culture

Proportion of public expenditure devoted to cultural and creative activities and the annual public budget and expenditure for the cultural and creative sectors.

This indicator aims to monitor the amount of actual public spending on cultural and creative activities.

Data sources

- National and local sources: Administrative data, Specific national surveys and Information systems for culture when available.

Method

For public expenditure for culture, the following disaggregation would be required:

- Public expenditure by administrative unit
- Type of public expenditure (capital investment and operating expenditure)
- Public expenditure per inhabitant
- By sector of intervention
- Public expenditure by source.

Indicator 12: Governance of Culture

Checklist of the governance framework to support culture and creativity.

This checklist aims at:

- representing an overall picture of the government policies and regulatory frameworks in place to support activities in the culture sector, as well as the decision-making processes in cultural domains;
- assessing the regulation of the Culture sector and promoting better working and trade conditions for better livelihoods;
- assessing the degree of development of the governance framework at national/local level for culture in general and by cultural domains specifically.

Data sources

- UNESCO data: Periodic reports of the 1954, 1970, 1972, 2003 and 2005 Conventions.
- National and local sources: Administrative data, Specific national surveys and Information systems for governance culture when available.

Method

Checklist (numerical and Yes/No).

2.4.3 Dimension 3: Knowledge and Skills

This dimension "highlights the contribution of culture in building knowledge and skills and focuses on the contribution of culture to the transmission of local cultural values, knowledge and skills and fostering empowerment through education and training processes and policies. It emphasises the role of cultural diversity in education and vocational training". The Dimension refers to the SDGs no. 4—*Ensure inclusive and equitable quality education and promote lifelong learning opportunities for all*, 8—*Promote sustained, inclusive and sustainable economic growth, full and productive employment and decent work for all*, 9—*Build resilient infrastructure, promote inclusive and sustainable industrialisation and foster innovation*, 12—*Ensure sustainable consumption and production patterns* and no.13—*Take urgent action to combat climate change and its impacts.*

2.4.3.1 The Indicators of Dimension 3

The four indicators of this dimension confirm the unresolved definition of culture at the root of the—however commendable—exercise, as they swing from cultural diversity to the arts and back to the intangible heritage and back again to tangible heritage. This is probably the dimension where the effort to grasp and quantify the multidimensional content of the notion of culture shows its most significant limits and tries overcoming them by adding layer-to-layer.

Indicator 13: Education for Sustainable Development
The indicator is still being developed. Once defined, it aims to assess the extent to which global citizenship education and education for sustainable development, with a particular emphasis on cultural diversity, are mainstreamed at all levels.

Data sources

- UNESCO data: UIS
- National and local sources: Administrative data, Specific national surveys and Information systems for culture when available.

Indicator 14: Cultural Knowledge

Checklist on cultural education and capacity building

This indicator aims to assess the contribution of cultural knowledge to sustainable development practices.

Data sources

- UNESCO data: Periodic reports of the 1972, 2003 and 2005 Conventions, International Bureau of Education.

Method
Checklist (numerical and Yes/No).

Indicator 15: Multilingual Education

Percentage of instructional hours dedicated to multilingualism in relation to the total number of instructional hours dedicated to languages in: (a) primary (ISCED 1) (b) lower secondary school (ISCED 2)

The indicator aims to assess the extent to which multilingualism is promoted in primary and secondary education as an approximation of the levels of promotion of intercultural dialogue, safeguarding and understanding of cultural diversity within the education system.

Data sources

- UNESCO data: Education Sector of UNESCO, IBE
- National and local sources: Official school curriculum obtained from the Ministry of Education

Method

$$\text{Indicator} = \text{ILR} + \text{II} + (1 - 1/\text{B}) \times \text{ION}$$

ION is the annual percentage of instructional hours dedicated to official or national languages during a particular level of schooling (ISCED 1 or ISCED 2), in relation to the total number of hours dedicated to teaching languages; ILR is the annual percentage of instructional hours dedicated to local or regional languages during a particular level of schooling, in relation to the total number of hours dedicated to teaching languages; II is the annual percentage of instructional hours dedicated to international languages during the same level of schooling, in relation to the total number of hours dedicated to teaching languages; B is the number of official or national languages taught.

Indicator 16: Cultural and Artistic Education

Percentage of instructional hours dedicated to cultural education in the first two years of secondary school (ISCED 2), in relation to the total number of instructional hours.

The indicator aims to assess the degree to which cultural studies are included in the secondary school curriculum, as an approximation of the levels of encouragement in the education system of creativity, the promotion of the appreciation of cultural expressions and "the desire for culture" amongst young people.

Data sources

- UNESCO data: Education Sector of UNESCO, IBE.
- National and local sources: National Official school curriculum obtained from the Ministry of Education.

Method

Obtain the number of hours by subject from curriculum-based sources and calculate the percentage of hours devoted to culture.

Indicator 17: Cultural Training

> "Number of students enrolled in post-secondary and tertiary education in the field of culture during the reference year as a percentage of all students enrolled in these levels of education" and "Number of students graduating from post-secondary and tertiary education in the field of culture during the reference year as a percentage of all students graduating in these levels of education".

This indicator aims to assess the extent of participation in culture and creative studies at the post-secondary level.

Data sources

- UNESCO data: Education Sector of UNESCO, IBE.
- National and local sources: Ministries of Technical and Higher Education, Ministry of Culture.

Method

- % of all students in post-secondary education who are enrolled in programmes with culture and creative industry Fields of Study
- % of all students enrolled in tertiary education who are enrolled in programmes with culture and creative industry Fields of Study
- % of all students graduating in post-secondary education programmes with culture and creative industry Fields of Study
- % of all students graduating in tertiary education programmes with culture and creative industry Fields of Study.

2.4.4 Dimension 4: Inclusion and Participation

The last dimension aims at assessing the contribution of culture in building social cohesion and fostering inclusion and participation. It focuses on access to culture, the right of all people to participate in cultural life and their freedom in cultural expression, including artistic and creative freedom. The SDGs involved are no. 9—*Build resilient infrastructure, promote inclusive and sustainable industrialisation and foster innovation*, 10—*Reduce inequality within and amongst countries*, no.11—*Make cities and human settlements inclusive, safe, resilient and sustainable* and 16—*Promote peaceful and inclusive societies for sustainable development, provide access to justice for all and build effective, accountable and inclusive institutions at all levels.*

2.4.4.1 The Indicators of Dimension 4

The five indicators address relevant aspects of inclusion and participation: intercultural tolerance, artistic freedom, physical access to cultural facilities, cultural participation and participatory processes. The Achilles' heel of this set is represented by both the chosen variables of such abstract and complex phenomena and their proposed sources. In particular, the mere availability of infrastructures per capita does not represent access, as barriers may be architectural, economic and cognitive. The indicator about participatory processes is hardly an indicator in the proper sense. About sources, national surveys on specific phenomena, often mentioned for this dimension, do not exist in many countries.

Indicator 18: Culture for Social Cohesion

This indicator is an aggregate of three indicators:

- Intercultural tolerance: Percentage of people who do not object to having a neighbour from another culture.
- Interpersonal trust: Percentage of people reporting that other people can be trusted.
- Perception of gender equality: Degree of positive assessment of gender equality (subjective output).

This composed indicator aims to assess the degree of inter-cultural understanding, to measure the degree of personal acceptance of people from other cultures and to measure the gaps between women and men with respect to their opportunities and rights to take part in the cultural, social, economic and political life of their country.

Data sources

- National and local sources: Administrative data, Specific national surveys (including the Rosenberg question) and Information systems for culture when available.
- World Values Survey (WVS); Latino Barometer: Interpersonal Trust (A60112); Asian Barometer: Most people can be trusted (Q024); Afro Barometer: Most people can be trusted or trust others.

Method

The calculation varies according to the different data sources.

Indicator 19: Artistic Freedom

Checklist on the level of support for artistic freedom and to identify the status of the artist.

This indicator aims to assess the level of a sustainable environment for artists and creators.

Data sources

- UNESCO data: 2005 Convention periodic reports.
- National and local sources: Ministry of Culture, Administrative data, Specific national surveys and Information systems for culture when available.

Method

Checklist (numerical and Yes/No).

Indicator 20: Access to Culture

Availability of cultural infrastructure in relation to the distribution of the population.

This indicator aims to assess the degree to which different people have access to cultural facilities. This measure complements indicator 4 'Cultural facilities' as it aims to assess the number of cultural facilities in a city or country in relation to the size of the population. Where data are available, it may be disaggregated by types of cultural infrastructure.

Data sources

- National and local contributions: Administrative data and Information systems for culture when available.

Method

This indicator aims to assess the availability of cultural facilities in relation to the population that might be expected to use them. For each administrative district or province, take the number of each type of facility (e.g. museum) and the total resident population. Calculate the standard deviation of each facility across the country/city as an indicator as to the extent to which each facility is evenly distributed across the territory.

Indicator 21: Cultural Participation

The composite measure combines three indicators:

1. Cultural site visits: Trends in the number of visits to selected cultural sites or performances.
2. Cultural attendance: Percentage of the population who have participated at least once in a going-out cultural activity in the last 12 months.
3. Individual cultural activities: Percentage of households reporting practicing cultural activities at home in the last 12 months (including: Using the internet for cultural purposes).

The main purposes of the measure are:

- To assess the overall number of visits to cultural sites or facilities and, through time series, whether interest to particular types of facility is increasing or declining.
- To assess the proportion of the population who attend a cultural event or facility and, through time series, whether the proportion of the population attending cultural events outside the home is increasing or decreasing.
- To assess the extent to which people engage in cultural activities or skills at home and to monitor the role of cultural activities online.

Data sources

- National and local sources: Administrative data, specific national surveys and Information systems for culture when available. Data from Internet service providers.
- Regional surveys such as Eurobarometer and Latinobarometer.

Indicator 22: Participatory Processes

The extent of participation of all stakeholders including local communities in the processes for developing and implementing cultural policies, programmes, and initiatives that concern them.

The indicator aims to assess the opportunities open to civil society and to cultural sector professionals and minorities to participate in the formulation and implementation of cultural activities, policies, measures and programmes that concern them.

Data sources

- Periodic reports of the 1972, 2003 and 2005 Conventions.
- National and local sources: Administrative data, Specific national surveys and Information systems for culture when available.

Method

Checklist (numerical and Yes/No).

3 Discussion

3.1 The Concept of Culture

Definitions and purposes of the 22 indicators show that the concepts of culture that they convey are different and vary from a Dimension to another. Basically, there are three concepts involved: culture as a sector of economic activity (in 18 of them), culture in the anthropological sense, as tradition (in three of them, combined with other concepts) and culture in the anthropological sense, as cultural diversity (in four indicators, plus one, combined with other concepts). Table 2 shows how the different concepts are distributed in the four dimensions and their indicators.

3.1.1 Environment and Resilience

All the five indicators in this Dimension are built on the concept of culture as a sector of economic activity, addressing Cultural heritage and cultural venues. Three of them also include references to *tradition,* as a source of practices for sustainable management and climate adaptation for (tangible) heritage. But how is *tradition* operationally defined? How old a practice must be, for being listed as traditional? Is being traditional automatically equal to be sustainable? Evidence shows that this is not so. Examples are the variety of disruptive, albeit traditional, fishing techniques, including trawling, by-catch and even explosives or the disposal of solid waste into large open-air dumps, traditional mineral and stone extraction techniques, like open-air quarries, etc.

3.1.2 Prosperity and Livelihood

The totality of the Indicators in this Dimension refer exclusively to culture as a sector of economic activity, seen as source of occupation and income and therefore worth receiving adequate public support and investment and governance. The

Table 2 Concepts of culture adopted in the dimensions and indicators

Dimensions and indicators	Concepts of culture		
	Sector of economic activity	Anthropological sense, tradition	Cultural diversity
Environment and resilience			
1. Expenditure on heritage	X		
2. Sustainable management of heritage	X	X	
3. Climate adaptation and resilience		X	
4. Cultural facilities	X		
5. Open space for culture	X	X	X
Prosperity and livelihood			
6. Culture in GDP	X		
7. Cultural employment	X		
8. Cultural businesses	X		
9. Household expenditure	X		
10. Trade in cultural goods and services	X		
11. Public finance for culture	X		
12. Governance of culture	X		
Knowledge and skills			
13. Education for sustainable development			X
14. Cultural knowledge			X
15. Multilingual education			X
16. Cultural and artistic education	X		
17. Cultural training	X		
Inclusion and participation			
18. Culture for social cohesion			X
19. Artistic freedom	X		
20. Access to culture	X		
21. Cultural participation	X		
22. Participatory process	X		

anthropological concept and the notion of cultural diversity are not included in the proposed measurements.

3.1.3 Knowledge and Skills

The majority of indicators in the third Dimension, including one still in process of being finalised, convey the value of *cultural diversity*, which the educational system is called upon to protect and nurture from a generation to another. The other two indicators address artistic and creative education, a key function in the cultural sector framework.

3.1.4 Inclusion and Participation

In this Dimension, the proportion of indicators addressing the consumption and practice of cultural goods and services prevails over those designed for measuring cultural diversity. Cultural diversity is mostly measured as inclusion in decision-making processes of a variety of stakeholders, including cultural minorities, while participation is identified with cultural consumption and practice of the conventional kind (visiting museums, attending cultural events, etc.).

3.2 Data and Sources Criticalities

The Thematic Indicators are conceived as a toolkit for global application, to a scale ranging from the city level to the national level. They therefore should be viable for countries with different statistical capacity. At a first glance, the framework appears to be data driven. That is, indicators are preferably built on existing datasets. However, very few countries, in the so-called developed world, are able, today, to feed with data all the proposed indicators. In this sense, Culture|2030 Indicators represent a target for the development of an integrated system of culture-for-sustainable-development statistics. To reach that target requires huge investments. Moreover, it requires progressive and repeated refinement of the definitions, because the UNESCO indicators, while an extraordinary step forward, still contain blurred areas and vagueness. In some respects, like, for instance, the measures of cultural participation, the semantic load of the indicators is too big, and the proposed proxies are still not fully convincing.

3.3 Conclusions

In this exercise, I have tried to explore and discuss the operational concepts of culture that have been used for building the UNESCO Culture 2030 indicators, to reflect upon the suggested data sources and methods and their feasibility and to derive from them ideas for future data collection at a national and local level.

For over 40 years, UNESCO has guaranteed a tremendous effort for the production of a solid system of culture statistics. Culture is a tricky subject, and its very definition can be hardly pinned down to a single concept. From the first, innovative Framework (1986), to the sophisticated Culture for Development Indicator Suite in the 2010s,[13] the path followed is clear and consistent and aims to take into statistical account not only the economic dimension of culture but also cultural human rights and the depths of the anthropological meanings of culture. If in theory the UNESCO's set of indicators should reflect this complexity, in practice, at least so far, the majority of the proposed measures concern the cultural sector of economic activity. Culture in the anthropological sense inspires just three indicators, two of which express a rather rosy vision of tradition, a bit too simplistically identified with "all things good".

Apart from their inevitable limits, largely due to data constraints, the Thematic Indicators discussed here add something new and important to the cultural studies: they are inspired by and tailored to the 17 Goals of the 2030 Agenda for Sustainable Development. This is a game-changer as compared to the past.

Of course, there are open issues, and the future culture statistical activity will be called to deal with them. Data sources are an issue. Standards and norms are also an issue, since the Framework does not propose optimal values or benchmarks. Mere incremental or decremental trends are not enough, if indeed this toolkit must monitor the progressive contribution of culture to development. But, at a different level, together with a general improvement of the statistics, the project will succeed only if some political sore spots will be politically tackled and solved. Culture is never mentioned openly in the 2030 Agenda. The UN hesitates to rule about which is the proper course of cultural policies to support sustainable development. Hence, establishing targets and reference values is probably still too delicate and embarrassing a task to be completed at this stage.

A final remark about the use of *contribution* instead of *impact* to describe the role of culture in the attainment of the SDGs is in order. It is a practical decision, one that shifts the focus from the universe of meaning of *impact*, which can be direct or indirect, but also planned and unexpected as well, to the universe of meaning of a deliberate, planned and carefully constructed *contribution,* which deserves appreciation.

References

Barber, W., & Badre, A. (1998). *Culturability: The merging of culture and usability*. Presented at the Conference on Human Factors and the Web. Basking Ridge, NJ: ATandT Labs. Retrieved from http://zing.ncsl.nist.gov/hfweb/att4/proceedings/barber/

Bauman, Z. (1973). *Culture as Praxis*. London: Routledge and Kegan Paul.

Baumol, W., & Bowen, W. (1966). *Performing arts: The economic dilemma*. New York: Twentieth Century Fund.

Cicerchia, A. (2016). Culture and development. The global quest for indicators. *Economia della Cultura, 2/2016*, 249–260. https://doi.org/10.1446/84503.

Inglehart, R., & Baker, W. (2000). Modernisation, cultural change, and the persistence of traditional values. *American Sociological Review, 65*, 19–51. https://doi.org/10.2307/2657288.

[13]https://en.unesco.org/creativity/activities/cdis. Accessed 10 August 2020.

Kroeber, A. L., & Kluckhohn, C. (1952). *Culture: A critical review of concepts and definitions*. Harvard University Peabody Museum of American Archeology and Ethnology Papers 47.

Minasyan, A. (2014). *Your development or mine? Effects of donor-recipient cultural differences on the aid-growth nexus*. University of Goettingen and University of Heidelberg: http://ices.gmu.edu/wp-content/uploads/2014/10/Your-Development-or-Mine-Effects-of-Donor-Recipient-Cultural-Differences-on-the-Aid-Growth-Nexus-by-Minasyan.pdf

Peacock, A. (1969). Welfare economics and public subsidies to the arts. *Manchester School of Economic and Social Studies, 4*(December), 323–335.

Robbins, L. (1971). Unsettled questions in the political economy of the arts. *Three Banks Review, 91*(September), 3–19.

Sen, A. (2000, December 13). *Culture and development*. World Bank Tokyo Meeting. Retrieved September 10, 2019, from http://info.worldbank.org/etools/docs/voddocs/354/688/sen_tokyo.pdf

Sen, A. (2009). *The idea of justice*. Cambridge: Harvard University Press.

Soiini, K., & Dessein, J. (2016). Culture-sustainability relation: Towards a conceptual framework. *Sustainability, 8*, 167. https://doi.org/10.3390/su8020167. www.mdpi.com/journal/sustainability

Veblen, T. (1899). *The theory of the leisure class: An economic study in the evolution of institutions*. New York: Macmillan.

UNESCO. (1986). *The UNESCO Framework for cultural statistics (mimeo)*. Retrieved August 18, 2020, from http://uis.unesco.org/sites/default/files/documents/unesco-framework-for-cultural-statistics-1986-en_0.pdf

UNESCO. (2015). *UNESCO's work on culture and sustainable development: Evaluation of a policy theme*. IOS/EVS/PI/145 REV.5. Paris: UNESCO.

UNESCO. (2019). *Culture 2030 indicators*. Paris: UNESCO.

UNESCO-UIS. (2009). *The 2009 UNESCO framework for cultural statistics (FCS)*. Montreal, Quebec: The Unesco Institute of Statistics.

Zamagni, S. (2000). Economic reductionism as a hindrance to the analysis of structural change: Scattered notes. *Structural Change and Economic Dynamics, 11*(1–2), 197–208.

Annalisa Cicerchia is a senior culture economist at the Italian National Statistical Institute since 2011. She served as a researcher at the National Institute for Economic Planning—ISPE (1995–1998), then as a senior researcher at the National Institute for Economic Studies and Analysis—ISAE (1999–2010).

She is an expert in cultural policy strategic planning and evaluation. Her research areas include culture economics, culture statistics, cultural indicators, the social impact of culture, culture and well-being, cultural participation, cultural and creative economy and enterprises, cultural sustainability.

Member of Editorial Board of the journal Economia della cultura, of ICOM-Italia, she takes part in the Eurostat Working Group on culture statistics. Professor of Management of the creative enterprises and vice director of the Master on Economics and Management of Tourism and Cultural Activities at the Faculty of Economics of the Roma University Tor Vergata. Consultant for planning and evaluation of culture-based international development cooperation projects for the European Commission and the Italian Ministry of Foreign Affairs (Egypt, Lebanon, Syria, Kosovo, Cape Verde).

Member of the Evaluation Unit of the Italian Ministry of Cultural Heritage (2004–2008). Member of the Managing Committee of the COST ACTION IS1007-Investigating Cultural Sustainability (2013–2015). She is a founding member of the Cultural Welfare Centre.

Currently she is a member of the leading research team of the Horizon 2020 project SoPHIA-Social Platform for Holistic Heritage Impact Assessment.

The Digitalisation of Cultural Heritage for Sustainable Development: The Impact of Europeana

Emanuela Macrì and Concetta Lucia Cristofaro

Abstract The chapter analyses the digitalisation process as a driver of sustainable development in the cultural field, with particular attention to cultural heritage. Cultural heritage is among the most evident cases of relevance to sustainability; in fact, it risks a gradual extinction, at least in the units whose fragility may be irreversible (Macrì and Trimarchi 2015).

The exploratory research is divided into two phases: the first, on the desk, explores the theme of the digitalisation of cultural heritage in national and international literature as a tool to guide sustainable development. The second phase, on the job, analyses the management and growth of cultural heritage through the Europeana portal, as well as the contribution of a short interview with privileged observers from the sector.

This chapter helps improve the understanding and evaluation of digitalisation practices in the cultural heritage sector in Europe. We believe that our research may highlight some interesting elements that can contribute to the efficient use of technology in the cultural heritage sector.

This work is characterised by the offer of innovative profiles because the enhancement of cultural heritage through digitalisation represents an area of activity that, at present, does not find substantial systematisation in the Italian academic debate.

Keywords Digitalisation · Cultural heritage · Sustainable development · Europeana

---

E. Macrì (✉) · C. L. Cristofaro
Department of Law, Economics and Sociology, University of Magna Graecia, Catanzaro, Italy
e-mail: emanuela.macri@unicz.it; concetta.cristofaro@unicz.it

P. Demartini et al. (eds.), *Cultural Initiatives for Sustainable Development*,
Contributions to Management Science,
https://doi.org/10.1007/978-3-030-65687-4_17

1 Introduction

In the literature, many scholars agree that digital technology plays an important role in the innovation of the cultural sector and the valorisation of cultural heritage. The digitalisation and publication of collections online can potentially allow access to content across the globe, and as such, liberate this huge knowledge (Borowiecki and Navarrete 2017). In this sense, digitalisation is a means for safeguarding cultural heritage for future generations and providing access to cultural values (Marcum 2007). Digitalisation applied to cultural heritage transposes the essence of cultural objects presented in their physical form into the digital world, making this essence available to visitors from anywhere and at any time effectively and efficiently.

Europe needs a new action plan to make the best use of information and communication technologies (ICT); this plan could contribute to accelerating economic recovery and laying the foundations for a sustainable digital future.

The Agenda 2030 for sustainable development mentions cultural heritage as one of its indicators because culture has been recognised as a vector of economic and social development. UNESCO has also stated that culture should be at the heart of sustainable development policies, which is why it has launched the Culture Program for Sustainable Urban Development. This program stresses the role of culture and highlights the potential, positive contributions that the proper management of cultural heritage can make to sustainable development.

The theories that considered heritage and its conservation as opposed to economic development have been exhausted because they are currently considered effective partners in development (Loulanski 2006). For example, recent technological innovations and applications have had an impact on traditional sectors, such as the cultural heritage sector, creating new opportunities for the valorisation of cultural heritage and above all arousing particular interest because they are considered capable of attracting a wider public (Marchegiani and Rossi 2016).

This chapter analyses the characteristics of digitalisation as a tool for sustainable development in the cultural sphere with particular attention to cultural heritage. Europeana supports the analysis of the digitalisation process of cultural heritage as an example of a digital platform for the management of cultural heritage. Therefore, this study is based on explorative qualitative research that has been divided into two main phases: the first aims to explore the theme of digitalisation as a tool for the sustainable development of cultural heritage in the national and international literature to provide a theoretical framework. The second step aims to analyse a series of information concerning the management and enhancement of cultural heritage through the Europeana portal. Furthermore, to provide added value to our research, expert interviews were proposed to understand the importance of digitalisation in the cultural sector and Europeana's contribution to the sustainable development of cultural heritage.

The results of the present study provide further implications for future research on the role and impact of innovation in the cultural sector, which could be used in a refined framework with high practical relevance.

The originality of this chapter is represented by the provision of innovative profiles because the enhancement of cultural heritage through digitalisation is an area that, at its current stage, has not found substantial systematisations in the Italian academic debate. Finally, our work represents one of the first attempts to analyse the digitalisation of cultural heritage through planning processes that involve several considerations in the economic, social and knowledge management field.

This chapter is structured as follows: Sect. 1 is the introduction, followed by the glossary in Sect. 2 and the research methodology in Sect. 3. Section 4 presents the theoretical framework, and Sect. 5 analyses the case study: the experience of Europeana. Section 6 discusses the topic Europeana against social media, Sect. 7 describes the impact assessment and Sect. 8 confirms the interview results. Finally, Sect. 9 presents the conclusions.

2 Glossary

In this chapter, many keywords are used to describe the paper's content, as well as its subject matter. Some of these terms are often abused and turned into labels, due to custom of simplifying complex concepts. Although the literature has recently begun to deepen the meaning of concepts such as cultural heritage, digitalisation and sustainable development, their meanings tend to be quite diverse. This happens because the topics taken into consideration can be interpreted from different points of view, and socio-economic and legal aspects contribute to reconstructing the mosaic of interpretations that then create cross-fertilisation. A preliminary glossary could facilitate the interpretation of our research also in light of the growing attention to shared metrics whose need for universal precision ends up draining the multiple meanings of some crucial concepts.

2.1 Towards a Definition of Cultural Heritage

The term cultural heritage has been and is often misused, producing incorrect semantic expansions and transfers. These misuse stems from the generalised use of the term, often used to describe other contexts, such as monuments or cultural heritage. However, all these objects cannot include the full meaning of the term "cultural heritage", which could take on specific meanings depending on the context in which it is used.

The General Conference of the United Nations Educational, Scientific and Cultural Organisation (UNESCO), at its Paris Convention in 1972, considered cultural heritage only as monuments and groups of buildings and sites, thus considering only material goods. In the following years (2003), attention was also given to the immaterial aspects of culture, to promote the richness of cultural diversity, in any form and expression. As a result, we now have its current definition, which includes several main categories: tangible cultural heritage, movable, immovable and underwater. The International Council on Monuments and Sites (ICOMS) also provided a similar definition in 2002: Cultural heritage is the expression of ways of life developed by a community and passed down from generation to generation, including customs, practices, places, objects, artistic expressions and values. Cultural Heritage is often expressed as Intangible or Tangible Cultural Heritage.

Vecco (2010) stresses that the concept of cultural heritage has undergone a continuous evolution in history, as well as that of culture. Finally, Daniel et al. (2012) have considered heritage as the legacy of physical artefacts and intangible attributes of society that are inherited from past generations, maintained in the present and donated for the benefit of future generations, thus using a more pragmatic vision.

2.2 *Sustainable Development*

In 1987 the Brundtland Report[1] entitled "Our Common Future" was published. It defines sustainable development as "development that meets the needs of the present without compromising the ability of future generations to meet their own needs" (WCED 1987).

Sustainable development is not only determined by politics and public order; the whole society plays an important role, as concepts have to be adopted both for day-to-day planning and political and economic decisions, requiring profound changes in economic and social structures and consumption and production patterns.

As part of a new roadmap for sustainable development, the United Nations have endorsed Agenda 2030, which contains the Sustainable Development Goals, a call for action to protect the planet and ensure the global well-being of all people. These common goals require the active involvement of individuals, businesses, administrations and countries around the world. The Sustainable Development Agenda 2030 contains 17 Sustainable Development Goals (SDGs),[2] which have given new impetus to global efforts to achieve sustainable development. The EU is committed to play an active role in maximising progress towards the sustainable development goal (European Union 2019), also through a new action plan to accelerate economic recovery and lay the foundations for a sustainable future.

---

[1]The report that follows the work of the 1983 World Commission on Environment and Development (WCED) chaired by Gro Harlem Brundtland.

[2]17 Goals: 1. No poverty; 2. Zero Hunger; 3. Good health and Well-Being; 4. Quality Education; 5. Gender Equality; 6. Clean Water and Sanitation; 7. Affordable and Clean Energy; 8. Decent work and Economic Growth; 9. Industry, Innovation and Infrastructure; 10. Reduced Inequality; 11. Sustainable Cities and Communities; 12. Responsible Consumption and Production; 13. Climate Action; 14. Life below Water; 15. Life on Land; 16. Pace and Justice Strong Institutions; 17. Partnerships to achieve the Goal.

2.3 *Digitalisation*

While some authors use the terms digitisation or digitalisation interchangeably, others offer distinct but different definitions for the two terms, such as Brennen and Kreiss (2016), who build their distinction on a broad and interdisciplinary literature review. For them, digitisation is "the material process of converting analogue information flows into digital bits" and then action. At the same time, digitalisation is "the way in which many areas of social life are restructured around digital media and communication infrastructures", thus a broader meaning not relegated only to the technological aspect.

Borissova (2018) also refers to digitisation as the process by which, through scanning or some other method of reproduction, an electronic copy of a specific analogue information medium is created. Alternatively, Preuss (2016) argues that there are specific requirements that a good digitalisation process must have and that it should ideally be followed by all institutions working in the field. There are six steps,[3] but the most important is the first one concerning the specification of content, which includes: identifying the purpose and object of digitalisation; anticipating the type of digital use; finding ways to digitise content without destroying or damaging it; analysing the legal framework for digitalisation, publication and use; assessing the resources needed to complete the tasks involved.

Today, however, digitalisation and digital transformation are described as a "global megatrend that is radically changing existing value chains across industries and public sectors" (Collin et al. 2015) or as "the catalyst, enabler and driver of society's development over previous decades" (Bengtsson et al. 2014).

Finally, Lévy (2002) highlights digital networks by specifying that information is physically present somewhere, for specific support, but is also virtually present in every point of the network where requested.

The importance of the issue was also underlined by the Digital Agenda presented by the European Commission for the Europe 2020 strategy, which set the objectives for the EU's growth by 2020, in which it was proposed to exploit the potential of information and communication technologies better to promote innovation, economic growth and progress.

3 Research Methodology

The research presents the results of an exploratory and descriptive study that aims to understand and reconstruct the general dynamics of the phenomenon (Eisenhardt 1989; Yin 1994; Sinkovics et al. 2016). This study is based on a research method

[3]In addition to the first there are: digital cataloguing, digitalisation, digital presentation, digital backup, digital preservation.

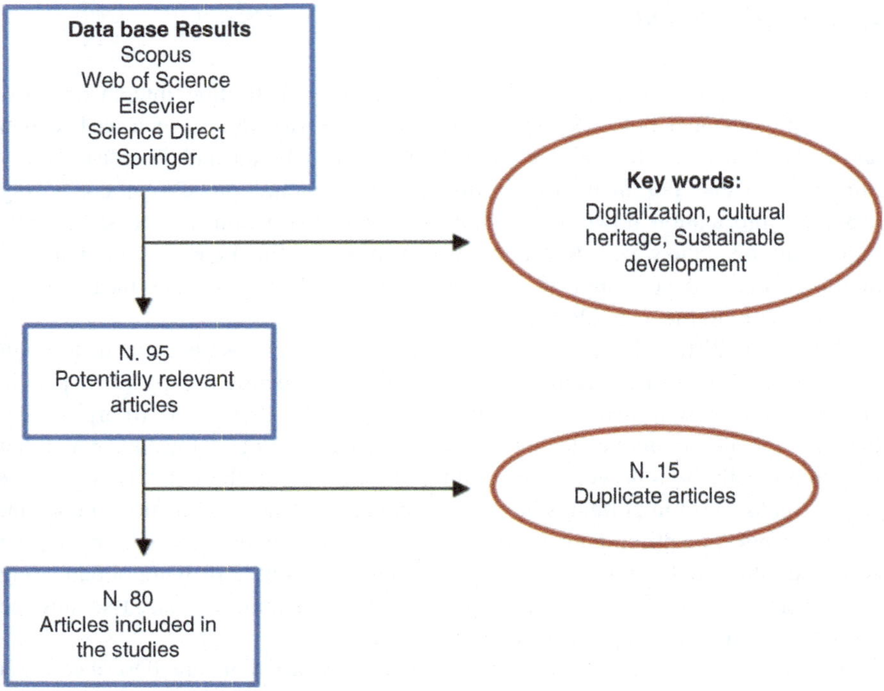

Fig. 1 Literature review. Source: Our elaboration

developed over two phases. In the first phase, the literature is explored, and the characteristics of the digitalisation process of cultural heritage are highlighted.

The second step, on the job, aims to analyse a series of information relating to the management and enhancement of cultural heritage, through the Europeana portal. In addition, to provide added value to the research, expert interviews were proposed to understand the importance of digitalisation in the cultural sector and the contribution of Europeana to the sustainable development of cultural heritage.

In this phase, we used a systematic review (PRISMA) proposed by Moher et al. (2009) (see Fig. 1), to improve the quality of systematic reviews and provide substantial transparency in the selection process. The analysis process developed was as follows:

- Four keywords have been identified: “digitalisation”, “cultural heritage”, “sustainable development”, “Europeana”;
- Five relevant databases were used to identify the studies, such as Scopus, Elsevier, Science Direct, Springer and ISI Web of Science;
- Only articles from economics, management and social science journals published in English were considered from the identified databases (e.g. books, book chapters/reviews, working papers).

At this point in our literature review process, 156 results were obtained.

From the reading of the abstracts, 95 articles were considered because they were deemed useful for our research. Eventually, they were reduced to 80 after deleting the duplicates and reading the rest completely.

The second phase of the research, on the job, took place between October 2018 and January 2020. This phase was carried out with the use of a case study method (Yin 2002). In this research, the case study is represented by the Europeana platform. The analysis process of this second phase was to examine the following aspects of Europeana: ability to innovate cultural heritage in a digital environment; basic platform services for data partners, end-users and re-users; data and functionalities offered; communication with social media, platforms and sustainability; impact assessment. The analysis was supported by the study of information, institutional documents and publications available on the web. In this phase, we used an interview as a research tool to gather the necessary information on the research topic. According to Gaskell (2000), the use of the interview in qualitative research is important because it allows us to obtain a detailed description of a phenomenon. In particular, we used a semi-structured interview to propose a common frame of reference for all interviewees. The outline of the interview constitutes the perimeter of the contents to be treated (Corbetta 1999). In fact, our interview was composed of several questions and was divided into two parts. Within the first part, there were questions on the impact of digitalisation in the cultural context. In the second part, there were questions about Europeana as a driver for sustainable development of cultural heritage. Our interviews were initially directed at members of the Management Board of Europeana Network Association (ENA),[4] and also other personalities involved in various ways in the platform. Following practical impediments, we decided to interview subjects who, despite not being part of the studied phenomenon, are experts and connoisseurs of the phenomenon, i.e. the "privileged observers". They are experts of the topics covered in the present chapter, thanks to their role within particular organisations and institutions. The general study approach used in the chapter is summarised in the following Fig. 2.

4 Theoretical Framework

4.1 The Current State of Digitalisation in Europa

Assessing the position of European countries in the digitalisation arena is an essential and crucial issue in the context of cultural digitalisation. Here, strategic and political plans must be based on the existence of solid and adequate infrastructures, especially when European cultural resources can be transformed into an important pillar of the digital and real economy. To emphasise these aspects, we have looked to the Digital Economy and Society Index (DESI), an indicator that

[4]Group of 2500 cultural heritage professionals.

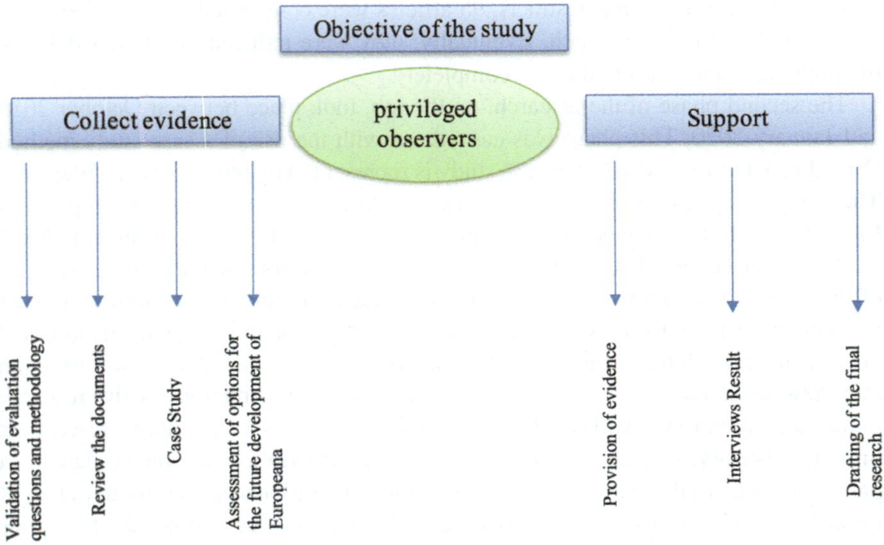

Fig. 2 General study approach. Source: Our adoption CARSA 2018

Table 1 Five components of DESI

Connectivity: deployment of broadband infrastructure and its quality
Human capital: skills needed to take advantage of the possibilities offered by digitalisation
Use of the Internet: consumption of online content video calls as well as online shopping and banking
Integration of Digital Technology: digitalisation of businesses and e-commerce
Digital Public Services: digitalisation of public services, focusing on e-Government and e-Health

Source: Our processing

measures the performance of European countries regarding digital performance. The exploration and analysis of the components of DESI are very important because it allows us to understand if Europe has a solid basis to undergo a radical digital change. There are five components, which are shown in Table 1.

Compared to the results of DESI 2018, there is a change from the first position. In fact, Finland takes first place and moves Denmark to fourth place. Sweden is in second place, while Holland climbs one position to third place. The last four positions are occupied respectively by Poland, Greece, Romania and Bulgaria. Over the past year, all EU countries have improved their digital performance, but they still have a long way to go, and the EU as a whole needs to improve in order to compete on the global stage. The results of DESI 2019 are shown in Fig. 3.

Although some countries are still lagging behind on the components of digitalisation in recent years, great efforts have been made throughout Europe by public institutions and private entities to convert "physical cultural heritage" into "digital". The European Commission's Directorate-General for Communication Networks, Content and Technologies has conducted a wide-ranging action to

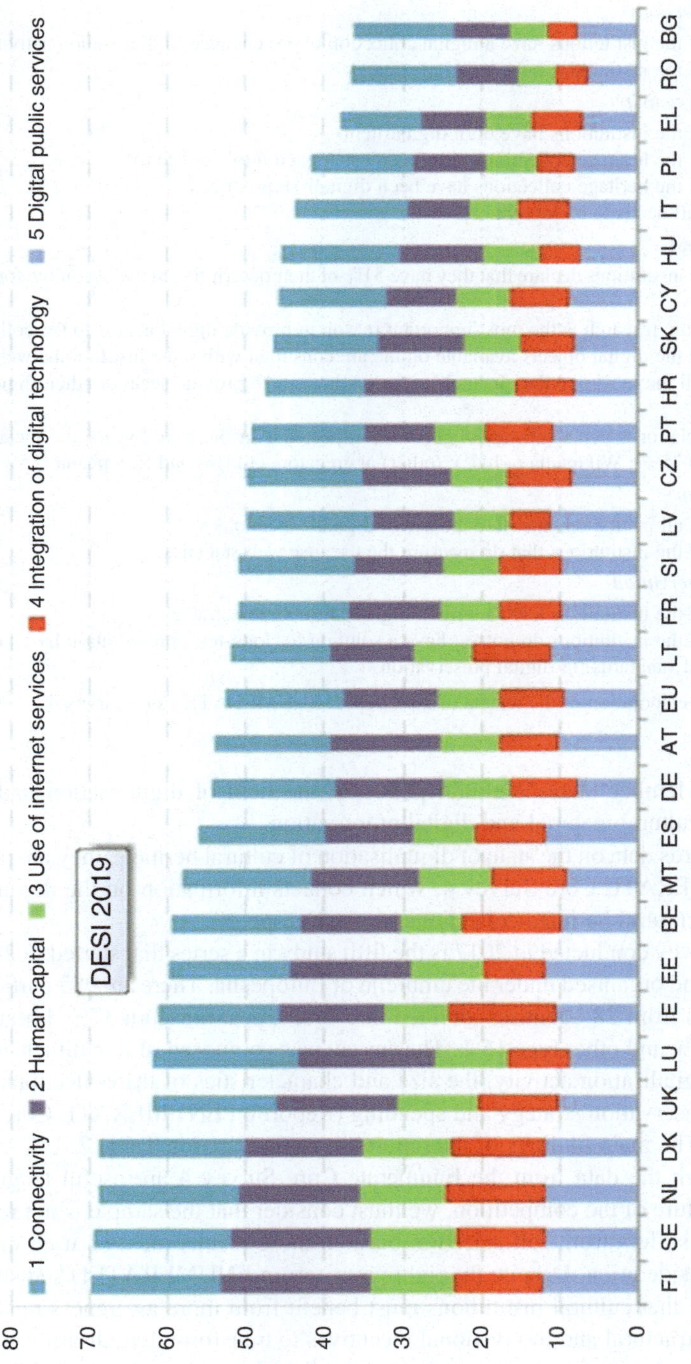

Fig. 3 Digital Economy and Society Index (DESI) ranking 2019. Source: DESI 2019, European Commission

Table 2 Five dimensions of the survey

Digital collections
– 82% of the institutions have a digital collection or are engaged in digitisation activities – 42% of the institutions have a digital strategy
Digitisation activity
– 59% of the institutions have born-digital items – 58% of the heritage collections have been catalogued into a collection – 22% of the heritage collections have been digitally reproduced – 54% still needs to be reproduced
Digital access
– Overall institutions declare that they have 51% of their descriptive metadata online for general use – Academic research is the most important reason to provide digital access to the collection – 51% of the digital objects available online are consulted within the institutional website, which is still the most popular channel for the institutions to provide access to their digital collections; – Channels for which a substantial growth is expected over the next 2 years are Social media platforms (+25%); Wikipedia (+14%); (other) aggregators (+11%) and Europeana (+5%);
Participation
– 49% of all institutions measure the use of digital collections – 90% of the institutions that do measure the use, use web statistics
Digital preservation
– 27% of the institutions have a written digital preservation strategy – 45% of the institutions do not yet have a solution for long-term preservation based on international standards for digital preservation

Source: Our elaboration of the results of the "Report ENUMERATE Core Survey 4"

integrate Member States' cultural policy in the field of digitalisation and online access to cultural material and digital preservation.

As regards data on the state of digitalisation of cultural heritage, they are provided by ENUMERATE Core Survey 4, which collects information on the digitalisation efforts of cultural heritage collections across Europe.

The survey conducted in 2017 is the fifth study in a series that started in 2008 and is the second organised under the umbrella of Europeana. There are 983 participating institutions from 28 countries, divided into four types: museum 37%, library 27%, archive 21% and other type 15%. The survey was composed of five dimensions: the state of digitalisation activity, the size and characteristics of the collections, digital access, conservation strategy and spending (Report on ENUMERATE Core Survey 4, 2017). The main findings of the report are summarised in Table 2.

Although the data from the Enumerate Core Survey 4 are useful to give us a clearer picture of the competition, we must consider that the sample is not representative and suffers from self-selection bias, so these results must be used under the proper consideration. Despite the right premises on ENUMERATED statistics, it is undeniable that cultural institutions must benefit from more awareness-raising policies and structural and motivational incentives to transform digitalisation strategies in the cultural sector from exceptions to established practices (Rizzo 2016).

4.2 Digitalisation of Culture Heritage

It is well known that cultural heritage generates a positive impact from a social and economic point of view; for this reason, many studies have been carried out on the topic. Regarding social impact, the first study was conducted by Matarasso (1997). According to his theories, the arts and culture offers six benefits: *personal development, social cohesion, community empowerment and self-determination, local image and identity, imagination and vision, health and well-being*. Subsequently, Belfiore (2006) highlighted some critical points of Matarasso's work, such as the evaluation method and research field, and proposed another study in 2002 in which she recognised the positive effects of art on social inclusion. From an economic impact view, Throsby (1999, 2003) claims that the existing values of cultural goods cannot be monetised. Although he identifies a positive correlation between economic and cultural value, he states that it is characterised by imperfection and the difficulty of measuring.

Instead, Trimarchi (2004) distinguishes between direct and indirect benefits of culture. The direct benefits are related to issues such as employment or technology innovation, while indirect benefits involve the immaterial sphere, which includes aesthetics, symbolism and spirituality.

Other studies analyse the general impact of digitalisation on cultural policy (Flew and Swift 2013), with particular interest in museums and libraries (Navarrete 2013a, b; Paolini et al. 2013; Salaun 2013). Nevertheless, this flourishing literature, with a few exceptions (Borowiecki and Navarrete 2015), has rarely been used to investigate the implications of digitalisation for archives that retain data and preserve cultural heritage. Our contribution tries to integrate this issue since, in recent years, several cultural institutions in European countries, from libraries to archives and museums, have worked on digitising many contents of their cultural heritage. Their efforts have gone in the direction of creating digital reproductions and copies of cultural artefacts, such as documents, paintings and audiovisual materials. According to Marttila and Botero (2017), for institutions, digitalisation initiatives are useful thanks to their contribution to the long-term preservation of their collections, but also in another aspect, such as expanding public access and the use of their heritage. This contribution is in line with the one proposed by Marttila because our case study presents these dual tasks. In particular, it promotes the inclusion of as many users as possible. Moreover, given the current debate on the importance of introducing systematic innovation to the cultural field, the digitalisation process seems to be useful in encouraging public participation and civic engagement in the cultural field and multi-actor inclusion, not only of end-users, but also of public and private institutions and the third sector.

For example, crowdsourcing practices have great potential when used in the cultural field for public participation.

Bonacchi (2019) defines crowdsourcing in the cultural field as the creation, enhancement, analysis and interpretation of cultural heritage objects (including data) and places by relatively large groups of people, through the completion of individual, small tasks on the Internet.

Crowdsourcing is also useful for fostering participatory heritage creation, enhancing and interpretation museums and galleries. Moreover, it is able to engage and sustain participation among people who have neither a general nor a strong interest in the cultural field, but who are very much interested in carrying out certain types of activities. While it is now apparent that cultural heritage is the central element of identity and memory, much less clear is the recognition of its importance in the production of value processes.

Alternatively, according to Seitsonen (2017), participation in the crowdsourcing projects is useful for creating cultural and social capital.

Instead, Campagna et al. (2020) has shown the effect of cultural participation on civic engagement and how participation in arts and cultural activities is highly correlated with participation in civic life. It is particularly worth noting that at higher levels of civic participation, cultural participation has a positive and strong effect on the civic engagement of people who have a low level of education.

Although there are many contributions on the digitalisation of cultural heritage, most of them focus mainly on three points of view: technical methodologies for content digitalisation, intellectual property and copyright issues and, finally, the storytelling in the cultural field developed through the digital perspective. Concerning the first aspect, Pavlidis et al. (2007) illustrate the main methods for three-dimensional digitalisation applicable to cultural heritage, and Yilmaz et al. (2007) highlight the importance of close-up digital photogrammetry in cultural heritage documentation. However, Bakirman et al. (2020) have reported on the implementation of ultra-light UAV systems for cultural heritage documentation.

Concerning the second perspective, intellectual property and digitalisation of cultural heritage, Borissova (2018) reinforces the question of the presence or absence of intellectual property on materials, orphan works and out-of-print works and what their position is in the public domain.

The digitalisation process involves the formalisation of intellectual property rights models, which should avoid the assignment of conflicting rights and permissions during the formalisation of the model (Bellini 2015).

Menard (2016) argues that the legislative results are divided between modest protections for right holders and clear gains for open access rights, which are evidences of the growing complexity of intellectual property policy in response to mass digitalisation.

Finally, storytelling can revolutionise the way the public interacts with cultural heritage and is now considered an important tool for attracting and satisfying the public at museums and other cultural heritage sites (Vrettakis et al. 2019). Paolini and Blas (2014) reported that, "Most people today agree that digital stories are the best way to engage users through different devices: from tablets to desktops, from smartphones to mobile phones, for audio only. Different narrative styles can be used, and different situations can be imagined, including immersive storytelling or augmented reality storytelling".

Compared to this literature on the digitalisation of cultural heritage, this chapter takes a completely new and innovative position; the encounter between the digitalisation of cultural heritage and sustainable development is an original and novel feature. To the best of our knowledge, our work is the first contribution that highlights the role of the digitalisation process as a driver of sustainable development in the cultural field. Our case study promotes and realises both the cultural, economic and social sustainability of digitalisation applied to cultural heritage.

4.3 Sustainable Development and Cultural Heritage

While the notions of economic or environmental sustainability have long been known, the path of cultural sustainability has only been traced more recently.

The process that defines culture as the fourth pillar of sustainable development started in 2002, when Agenda 21 for Culture was proposed, which is the first document supporting governments for cultural development and the link with sustainable development.

A new category has emerged, reminding us that both the cultural heritage of each community and its cultural and creative industries and cultural practices are strategic frameworks for current and future generations.

The idea of sustainable cultural development derives from the economic significance that can be attributed to culture. Therefore, it can be considered as a capital, whose stock is composed of the cultural value contained in tangible and intangible assets. Whatever way it is formed, the stock of cultural capital available to a community or nation is an available resource that must be managed, and this management function can be oriented towards sustainability criteria (Throsby 2017).

The theoretical basis of this concept derives from the theory of cultural capital, which can be linked to that of natural capital from which the concept of sustainable development started (Tisdell 2003). According to Throsby (1999), the concept of cultural capital refers to *"an asset that contributes to cultural value"*. It is composed of tangible and intangible assets, showing us that the sustainable development strategy cannot be separated from the cultural aspect because it must understand and respect cultural specificities, identities, diversity and involve different stakeholders.

The importance of the cultural dimension linked to sustainable development stems from the fact that cultural factors influence several variables, such as human relations and economic development. Although numerous studies have demonstrated the positive economic benefits of cultural heritage preservation (Gražulevičiūtė 2006), cultural heritage is organised to maintain and enhance cultural values.

Moreover, for Rypkema (2005), the role of cultural heritage in sustainable development is to learn how preserving cultural heritage provides environmental sustainability, cultural sustainability and economic sustainability.

From this point of view, digitalisation can contribute in many ways; digitised cultural material "can be used to improve the visitor experience, develop educational content, create documents, tourism applications and games" and gives online accessibility a chance to revolutionise traditional models.

From the point of view of environmental sustainability, the role of the digitalisation of cultural heritage is to preserve it. It is well known that climate change, urban development, natural disasters and armed conflicts continue to damage us and deprive us of the cultural heritage that has been left to us. It could be useful to think about how the digitalisation of cultural heritage not only allows us to share and find new ways to use cultural heritage, but also in some cases, the only way to keep it with us.

Finally, thanks to digitalisation, it is possible to make cultural heritage more accessible to all, reducing problems related to the inequality of access to information. Cultural institutions themselves are committed to adopting this new way of communicating with their audiences. Digitalisation can bring benefits such as the democratisation of culture and knowledge, but to achieve full cultural sustainability with particular regard to digital content, organisational priorities and cultural management strategies are needed. Digital content needs specific tools, for example, unique platforms for access to cultural heritage, such as Euopeana. At the same time, it is essential to monitor barriers to the economic sustainability of these services, policies and technological solutions. Key issues are content reliability, investments in business models, public and private partnerships and sponsorships, the use of open software and the development of Digital Rights Management systems for controlled access to content in certain contexts (Amari 2012).

5 Case Study: The Experience of Europeana

Our analysis starts from the study of the cultural sector that looks at digitalisation not only as a promotional tool but also and above all as a dynamic and relational platform for the development of a strategy of proximity to active cultural actors (not only employees and consumers but also public/private decision-makers). This relational approach, based on the interaction and cooperation between the digital portal and its large community, strongly characterises the Europeana case study.

Europeana is a platform that is part of the Digital Agenda for Europe and is a major aggregator of digital resources (Purday 2009). It is the largest European cultural heritage portal, was launched in November 2008 and is co-funded by the European Commission, as most publicly funded cultural and artistic institutions operate as non-profit organisations. Bakhshi and Throsby (2012) propose a generalised model of the behaviour of a non-profit cultural institution composed of five objectives, two of which are: access objectives (making their results available to many and a wide range of consumers) and educational services (often through specific programs to involve and educate young people or to more generally improve their consumers' learning experience) are part of Europeana's wider mission.

Europeana provides universal access to cultural heritage (museums, libraries, archives and cultural centres) through the digitalisation of cultural content and the main cultural heritage providers across Europe to adapt their practices to meet international standards for indexing and data retention. Data providers grant Europeana the right to publish image previews, but third parties may not reuse such previews.

Digitalisation of cultural content and publication in Europeana involves a very complex and time-consuming process, as all information should be collected under due-diligence conditions.

Concordia et al. (2010) have tried to make it clear that Europeana is the central agenda of our effort to make API-based data and rich functionality available. Application Programming Interfaces (APIs) allow developers to build new tools to ensure access to data for the creation of new products and services in the light of innovation (Dietrich and Pekel 2012). Application Programming Interfaces are web services that allow us to find visualisation content relevant to Europeana in an online environment and external environments such as applications, websites and games. Today, Europeana provides four free APIs after simple registration access:

- Rest API is the most widely used API that allows users to filter records for a variety of data repositories such as data, creator, etc.;
- The Europeana-Linked Open data service allows users to explore, access and download metadata;
- The Europeana OAI-PMH service supports users in collecting all Europeana metadata;
- The Europeana Annotations API allows users to generate, update and retrieve object annotations.

Another aspect that Europeana wants to offer is a rich semantic contextualisation for the representation of objects so that it is not a traditional digital library interface. To enable this functionality, object representations in Europeana must be systematically linked to the open data or semantic contextualisation resources contained in the Europeana data space.

The easiest way to give access to the data is to open a digital portal. In fact, the Europeana project uses the same method as many institutions to make cultural data available, but at the European level. On the homepage, there is a search box that allows you to access all the contents of Europeana. Digital collections may have multiple uses compared to the portal. They can be connected, for example, with other collections, geographical points, or be accessible via mobile phones with a GPS tracker. Currently, the Europeana platform provides access to over 50 million objects from European libraries, museums, archives, galleries and audiovisual collections. More than 3500 cultural institutions contribute by sharing their cultural content with Europeana. These range from international museums to regional archives or other smaller realities in each EU member country. Some of the most important institutions include the Rijksmuseum in Amsterdam, the Louvre in Paris and international libraries such as the British Library in London. The digital objects and collections that users can find within Europeana are not stored on a central computer but remain available to the cultural institutions that have hosted them on their networks. Europeana collects contextual information—or metadata—about objects, including a small image. Users search for this contextual information. Once they find what they are looking for, if they want access to the full content of the object, they can click on the original site containing the content. It is important to note that Europeana accepts metadata on digital objects; it does not make decisions about digitalisation. The decision on which objects are digitised is the responsibility of the organisation holding the material. For this reason, the process of digitalisation by cultural institutions appears to be pre-eminent and useful.

6 Europeana Against Social Media

Social networks are a significant resource available to the community. According to the specific rules of each platform, users can share information, opinions, knowledge and interests, but in general, they can make use of them as they prefer, without particular limitations. In fact, user modelling techniques based on social network activities are gaining more and more attention from the research community (Carmagnola et al. 2007). Specifically, the purpose for which Europeana uses social networks is to increase access to digitised cultural heritage and make it easily sharable. On this basis, the second phase of this work aims to understand better how Europeana communicates with the social media principles it sponsors on its official website: Facebook, Pinterest and Twitter.[5] The platform uses different tools to communicate its activities, projects, news and so on, but the social pages have the great advantage of being more direct and easily usable, allowing the maximum dissemination of information, objectives, mission and the vision of the institution.

From the analysis of these channels, carried out through a non-participant observation,[6] the following data presented in Table 3 have emerged.

If you look at Europeana's analysis on social media, you can see that, currently, social media represent a small part of its traffic, but statistics show that end-users coming to Europeana through social media are more engaged and more likely to share and use content. This tendency is in line with Europeana's main mission which is "to transform the world with culture and to build and make Europe's rich heritage easier for people to use, whether for work, learning or just for fun".

[5]Although Eureopena has other profiles on other social networks, in this work, only those that it advertises on the official website are considered. The data reported are related to the observations made in March 2019.

[6]It refers to the fact that the observation took place without direct intervention on social media; the collection of information was carried out through the study of posts, comments, photo and video galleries. No direct interaction with users and no intervention have been carried out on social channels.

Table 3 Official social media used by Europeana

Facebook
The Europeana Facebook page includes links for the official platform and other general information such as their telephone number and address. The page offers the same features as any other page; thereby, only the page's manager can post and share all types of content on its wall. The users can comment and also post photo, videos and links in the comment section. Posts vary from the "likes" to short comments or emoticons, and even short personal experiences. Currently, the page has more than 109,061 likes, 112,101 total followers, 282 reviews, a score of 4.7 out of 5.47, advertised events and more than 2000 photos and 77 videos.
Pinterest
The Europeana profile includes links for the official platform and a brief explanation of the content. The platform allows us to create thematic virtual blackboards on numerous topics. Through Pinterest, you can also explore all the favourite categories and pin (that is, a virtual pin) all images that will be then saved and categorised under a virtual theme board. The Europeana Pinterest page has 9884 pin and 15,032 followers. On the board, the pins are proposed in temporal order and are generally chosen on the basis of a specific theme or historical period.
Twitter
The Europeana Twitter account includes links for the official platform and other general information. It is interesting to note that, on Twitter, Europeana does not have a specific location; in fact, the location given in the account is "Europe". Other general information refers to the financing source of the platform. Currently, the account has more than 35,000 followers, and more than 17,000 tweets. Moreover, in the account, there are about 3467 photos of cultural images and public events related to cultural policies. Instead, the tweets refer to strategic and cultural information and events, but the account also provides content exchange.

Source: Our elaboration

7 Impact Assessment

The third and final phase of this research concludes with a proposal for an analysis to define Europeana's possible social and economic impacts. Impact assessment is a very interesting but complex area of research in the cultural field, especially when considering culture in a digital context. The questions to consider are, for example: Has the training of students improved thanks to the publication of digital resources? Has the well-being of the community benefitted? Has the economy benefitted directly or indirectly? Do people feel more connected to their national and European identity? First of all, it is necessary to identify the Europeana users in which they can be grouped (Fig. 4).

Secondly, it is necessary to identify a theoretical framework, which is useful for drawing up our evaluations and interpretations of Europeana's impact. According to Trimarchi (2004), the cultural sector generates a multiplicity of benefits (and therefore involves many costs), so measuring them seems like a very difficult task. The problem is building a hierarchy between the different benefits, which are divided into direct and indirect benefits. Direct benefits are linked to issues such as employment or technological innovation, while indirect benefits also involve the immaterial sphere, such as aesthetics, symbolism and spirituality. Based on the information, institutional documents and publications about Europeana available on the web, we

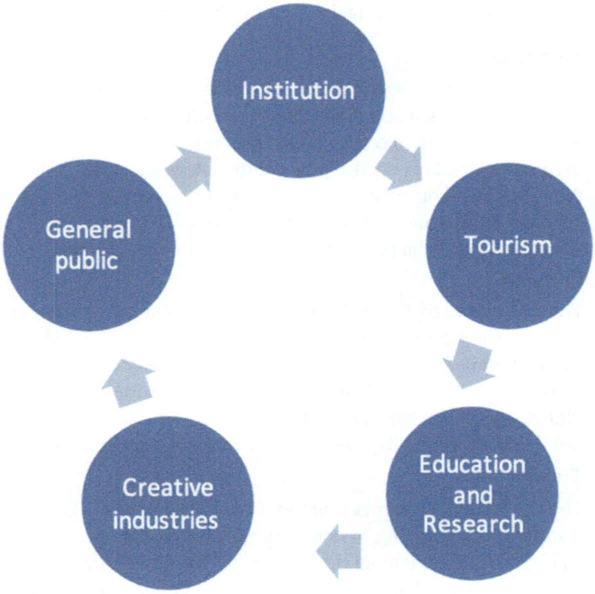

Fig. 4 Europeana users. Source: Our processing

are able to analyse those that, according to our reasoning, could be considered the impacts of the platform.

From a direct point of view, Europeana has an impact on employment. Many people are involved in the management and implementation of the platform. The system is very articulate and requires the involvement of several figures. They are not only engaged in the practical tasks of the technical management of the platform but also in an articulated system of the contents' organisation. In addition, professional figures are involved in the maintenance of relations with the data supplier partners.

Moreover, the aspect of technological innovation is also fully reflected in the Europeana project since, through it, a large part of Europe's cultural heritage has been digitised and made available online. Europeana has accelerated the process of technological innovation in the cultural sector because it has been able to collect contributions that are already digitised, making the fruition easier. At the same time, it has built a sustainable best practice network of stakeholders to aggregate, enrich and share a critical mass of material.

As far as the sphere of intangible benefits are concerned, they seem more difficult to quantify, however aspects such as aesthetics and symbolism are certainly intrinsic and connected to the idea of culture. As a result, Europeana has only contributed to further spreading these two aspects. In fact, all the collections accessible through Europeana have a great value because they aim to enclose the European culture, releasing beauty through the circulation of the culture itself.

Matarasso (1997) tried to shed light on the social impact of culture through his famous study "Use or Ornament?", in which he divided the social impact of participation in the arts into six different parameters. His indicators have been designed with reference to non-virtual participation in the arts; nevertheless, they can be used to understand whether Europeana has produced some form of social impact. Our deductions go in the direction of assimilating virtual participation in the arts to the real one, always considering that the two types of fruition are not substitutable but complementary, as also evidenced by Evrard and Krebs (2017).

The first and second parameters are personal development and social cohesion. Certainly, Europeana contributes to increasing user spheres such as education, skills and intergenerational understanding, as the collections of Europeana cover different historical periods. Those who use the platform certainly benefit personally by acquiring information and data that can then be used in different fields, from their private life to their work life.

We understand social cohesion as behaviours and ties between individuals or communities, aimed at attenuating inequalities related to social, economic and cultural situations. It is implemented through the platform precisely because anyone with an internet connection can use Europeana's services, without restrictions.

Social cohesion is also facilitated by the connections developed on the social channels that Europeana uses to exchange content and opinions between user groups. On the hand, it is not possible to assess more intimate aspects, such as the trust and management of social change, or fear of crime.

The third parameter, community empowerment and self-determination, are understood by Matarasso as the strengthening of organisational capacities, consultation and involvement in democratic processes and support for community-led initiatives. Above all, users such as Institutions and Creative Industries see the platform as a useful tool that is able to strengthen their organisational skills, thanks to the use of contents from the platform, which will then be used for other purposes. Europeana also offers support to all communities that want to make their cultural heritage visible and knowable; this activity must necessarily be supported by democratic processes.

The fourth parameter, local image and identity, is greatly impacted by Europeana, which, through its collections, reinforces the sense of belonging to a community that sees its art preserved and enhanced. Europeana strengthens even more aspects like these because it collects, catalogues and classifies content in an orderly way, making everything easier to explore and attributable to a particular community.

The fifth parameter, Fantasy and vision, concerns creativity, professional practice and symbols, involving all users who are inspired by Europeana and its collection. The platform becomes a multiplier of creativity used in both public and private environments.

Finally, the last parameter is represented by health and well-being. It is well known that cultural consumption increases well-being and health, so digital consumption made through Europeana can be assessed in the same way as the traditional way of consumption. Art and culture are useful for fighting stress and anxiety and, as a result, benefit our psychophysical health (Grossi and Ravagnan 2013).

Table 4 Non-monetary effects

Usage value: Consumer or producer surplus derives from using digital cultural heritage
Option value: Option to make use of material and offered service
Existence value: Contributions derive from the existence of Europeana in terms of its contribution to a person's pride or identity
Social value: Positive effect deriving from consuming arts, culture and heritage on education, quality of life and health

Source: Our elaboration of "The value of Europeana"

In addition, it is possible to make a monetary (economic) and non-monetary valuation of the value of Europeana.[7] From the first point of view, the most significant impact that the cultural platform produces is cost savings for institutions and tourism. These results refer to indirect effects we assume are related to Europeana, e.g. content that can be used to create tools for tourism purposes. Then the economic value, in general, can be associated with tourism expenditure (a better visibility of European cultural heritage that can act as a tourist attraction for international tourism) and cost-savings for institutions. As regards the last point, the analysis refers to smaller institutions, which have difficulties in developing adequate digital service platforms on their own and helps institutions to develop digital services that enable them to integrate their digital collections. Finally, non-monetary effects, such as usage value, option value, existence value and social value, are summarised in the following Table 4.

8 Interview Results

Consistent with the purpose of our chapter, we conducted some interviews with privileged observers of the phenomenon studied. The interviewee is an expert in cultural economy and cultural heritage management and an expert in organisation and business management and digitalisation processes in the public sector.[8]

The first part of the interview focused on the importance of digitalisation in the cultural sector and its connection with the concept of sustainable development.

The second part of the interview focused on the role that the Europeana platform plays in the process of the digitalisation of cultural heritage.

The questions from each section are presented in tabular form with the relative answers (Table 5).

[7]This assessment is largely covered by a cost-benefit analysis carried out by a research institute of the Faculty of Economics of the University of Amsterdam see, commissioned by Europeana to assess the socio-economic value of its action. see http://www.seo.nl/en/home/

[8]In order to simplify the reading of the interview results, from this point on, we refer to them as cultural heritage experts and digitalisation experts.

Table 5 Interview

First part of the interview: Focus on digitalisation	
First question: What is, in your opinion and based on your experience, the meaning of the sustainable development of cultural heritage?	
Cultural heritage expert: *The issue of sustainable development has long been considered one of the most crucial controversies in the field of conservation and dissemination of cultural heritage. Returning to Brundtland's simple statement, the main concern has been dimensional, and consequently, the relevant recommendation has focused on the need to keep people away from cultural heritage. Indeed, many monuments and sites are still brutally overcrowded and congested, due to the prevailing attraction of their iconic perception. This has ended up favouring a sort of mechanical enjoyment of sites such as the Colosseum or Petra, of works of art such as the Mona Lisa, of entire cities such as Florence or Venice, recording a visible reduction in the time taken and transforming the visit into a basic assessment of presence. Over the years what we will pass on to future generations (and also to ourselves in the future) will be objects rather than interpretations, critical elaborations, technical knowledge, historical evaluations, in a word the connection of every work of art, monument or site to its spirit of the time. If, on the one hand, it is necessary to keep the physical form of works of art, monuments and sites solid and inviolate, on the other hand, it seems fundamental to enrich their material life with a growing stream of interpretations aimed at transmitting values and generating critical thinking, something that a stone—even if noble and ancient—cannot achieve without a cognitive endowment.*	**Digitalisation expert**: *In reality, sustainability is one of the issues that cuts across any economic sector, so even the cultural sector cannot ignore it. However, sustainability means, above all, paying attention to the impact that the use of goods and services has on the environment and the community. In this way, it is necessary to follow, through the lens of sustainability, all stages of the process of the production, supply and enjoyment of cultural heritage, creating additional value added to the resulting offer.*
Second question: How can we interpret the digitalisation of the cultural sector?	
Cultural heritage expert: *Digitalisation is essential for every level of management action related to cultural heritage to be proven effective: from monitoring to conservation, from cataloguing to comparison, from exhibition to dissemination, digital tools are useful in supporting the cultural offer with an easily interchangeable and cross-referenced data archive, so that our knowledge—from occasional visitors to scholars and academic technicians—is collected, fertilised, improved and exchanged.*	**Digitalisation expert**: *Digitalisation is one way we can enable greater enjoyment of cultural heritage, both in terms of the quantity of the public potentially interested and in terms of quality, as it can operate differently because of the diversity of profiling and therefore of the different user segments.*

(continued)

Table 5 (continued)

Third question: Do you agree that digitalisation can be interpreted as a powerful driver of the sustainable development of culture?	
Cultural heritage expert: *It welcomes and spreads knowledge, and at the same time, offers many elaborations among the collected data, which imply a deeper and wider analysis by scholars, the society's fruition, and consequently, a coherent and motivated will to support it.*	**Digitalisation expert**: *Digitalisation represents an excellent opportunity for the Italian Cultural Heritage to channel new business and work opportunities, working towards greater decongestion of the main Italian cultural sites. In this way, the tourist offer could also be improved and adapted to new audiences.*
Fourth question: What are the most important reasons for providing users with digital access to the collections	
Cultural heritage expert: *Users can benefit from the breadth, multidimensionality and cross-fertilisation of cultural heritage, enjoying a more stable and reliable basis on which to build their research paths, theoretical elaborations and empirical analyses.*	**Digitalisation expert**: *Culture is an excellent engine for the growth of society, but only if you can understand its value and importance concerning the identity of the communities it serves. Expanding the enjoyment of cultural heritage to those who express an interest in it could therefore be an engine of greater development and growth for the population served. Naturally, all this requires the launch of new entrepreneurial activities capable of making this heritage effectively usable and accessible at a distance and in a different way.*
Second Part of the interview: focusing on the role the Europeana platform plays in the process of the digitalisation of cultural heritage	
Fifth question: What is the role of Europeana in the process of collecting the digitised cultural heritage?	
Cultural heritage expert: *The digitalisation of cultural heritage, like any other object, requires a central midpoint for the correct and consistent collection and processing of information. It has exactly the same function as the Alexandria library, which has admitted to the existence and connections with other libraries (e.g. private libraries) but has been recognised as the general convergence centre, what we still need for information to avoid contradictions and prejudices.*	**Digitalisation expert**: *Europeana provides an excellent search engine for many cultural objects of European origin, of which a broader and more in-depth view is possible thanks to the use of appropriate, precise and specific filters.*
Sixth question: What could be/will be the contribution of Europeana for the sustainable development of cultural heritage?	
Cultural heritage expert: *In the logic of the prevailing role of cognitive sustainability (answer 1), a digital repository of knowledge is crucial. No single institution can provide cultural heritage with such a powerful, extensive and evolutionary endowment of technical, critical and cultural knowledge.*	**Digitalisation expert**: *Europeana could represent an excellent opportunity for discussion and study that could achieve a real development of the system of digitalisation and enhancement of the continental cultural heritage. The hope is the awareness that it can also be a stimulus and an enlargement to neighbouring countries. In this way, it could be a real pillar in the process of growth of an increasingly enlarged and definitively global society.*

From the interviews carried out, some strengths can be highlighted and, in our opinion, could make Europeana a best practice to be replicated. Both respondents highlighted that the digitalisation of cultural heritage is an important tool for disseminating culture, and a practical example of this is the Europeana platform. Digitalisation could be interpreted as an engine for the sustainable development of culture. What we have deduced from the interviews is the confirmation of what is proposed in the theoretical framework and with the chapter's research question. Therefore, the methodological process used in this research could be used to build guidelines that permit others to replicate the analysis for other platforms, thanks also to the use of the variables proposed in the following paragraph named "impact assessment".

9 Conclusion

Europeana has created a service platform for digital cultural heritage and content from libraries, archives, galleries and museums that generate spill-over effects on the economic and social aspects, but also on the process of knowledge creation and the value chain. These aspects assume great relevance from the point of view of sustainable development. Based on our findings, we have drawn several conclusions on the role that the digitalisation process, analysed through Europeana, can play in the sustainable development of cultural heritage. As far as the impact aspect is concerned, we deem that the digital platform studied, guarantees the widest possible use of its contents and, at the same time, re-use in different areas, thus ensuring an acceptable and economic use of resources. These processes act as multipliers of creativity in which the innovation resulting from digitalisation promotes shared knowledge. Europeana's experience with the provision of cultural services, as already underlined above, improves social inclusion and cohesion through its capacity to develop intercultural dialogue.

Therefore, the platform generates returns in the form of social benefits, because inclusive institutions provide access to different groups as audiences and address issues of representation and participation in the institutions. It works thanks to three aspects of social inclusion, which are: access, representation and participation. If the first two are discounted, for example, the third is achievable thanks to the provision of participation tools through shared initiatives on social networks. Europeana also reinforces the aspect of intercultural dialogue because it is understood as a two-way process that actively involves both natives and internationals. Its role of cultural service delivery makes all participants equal within a social space of multi-ethnic and multi-centric interaction.

Therefore, is possible to affirm that the benefits of Europeana are manifold. It provides a shared infrastructure that saves costs for use and cooperation and gives the possibility to visit virtual and offline exhibitions, either directly through itself or through the institutions that own the content.

For these and other reasons, Europeana can be considered an added value to the EU's priorities for the cultural policies of its member States because they are able to create a network of experts and institutions for cultural heritage, increasing the reuse of European cultural heritage even beyond national borders.

Finally, Europeana appears a useful tool to encourage tourism both at the European and international level. Cultural tourism represents an ancient phenomenon in continuous evolution, and with its own originality, it has the potential to influence an individual's motivational state. Europeana has positive implications in the experience of cultural tourism, especially when visiting a point of cultural interest, as this has contributed to planning a more successful trip, even contributing to adding value to the trip and the learning experience associated with the acquisition of cultural knowledge. The platform could improve the attractiveness of lesser-known territories, which could become new tourist destinations.

Europeana has a significant number of user profiles; this distinction means that it must offer quality services specifically aimed at the users. The users range from institution to the wider community. In fact, the institution categories include galleries, libraries, archives and museums, but also educational institutions. The community is represented by the general public, people inside and outside Europe who are interested in art, culture and heritage, tourists and researchers. Finally, creative industries are a distinct category that is composed of institutions and people, ranging from publishers who make books on arts, culture, heritage or travelling. The sustainability of the fruition by several users also passes through specific tools such as social media, where Europeana can exchange cultural contents even in a virtual community. This expansion makes the public that generally interested in art, culture and cultural heritage broader and more diverse in age, social background, interests and education.

Nevertheless, this particular type of use must be supported by strategic and political plans based on the development of connectivity through the implementation of solid and adequate infrastructures such as broadband.

Europeana has played a significant role in the development of the digitised cultural heritage in Europe. Its role has been to facilitate collaboration between cultural institutions across Europe, through the coordination of actions to standardise digital infrastructures, to reduce the possible operations of adaptations to the standards of non-European countries.

This activity has been made possible, for example, by providing access to cultural objects from all over Europe, and by providing access to the cultural object description format (metadata).

At the same time, for the Institutions involved in the process of content sharing with Europeana, the benefits of the digitalisation process derive from an improvement in the value of their material, giving them a different position in terms of visibility in society. Moreover, without Europeana, the fragmentation of the databases used by the different institutions leads to development costs and under-investment in the valorisation and maintenance of digital resources; its presence balances this under-investment leading to sustainable use of resources.

In the near future, in which the physical and digital world will continue to merge, Europeana will face major challenges such as the limited presence of digitised material and problems related to the copyright of digitised content. These challenges certainly deserve to be properly investigated. Our final observations should be considered preliminary; further analysis is needed to obtain further insights into the process of the digitalisation of cultural heritage in the light of sustainable development. This chapter, in fact, should only be considered a basis for an in-depth analysis since it suffers from certain constraints.

Due to budget and time constraints, the Europeana case study framework is purely descriptive, so the analysis of its use and impact should only be considered indicatively.

The limited number of interviews and the choice of respondents are due to the difficulties of obtaining answers from people directly involved in Europeana. Moreover, the lack of data limited the scope of the literature review and comparative analysis.

It was also impossible to access statistical and internal Europeana data, such as the analysis of access points to the platform, which would have been very useful in understanding the geographical locations of users. No data on the number of employees and/or working environment in Europeana were accessible to us. Finally, since our analysis was only proposed for Europeana, we recommended repeating it with similar platforms, to discover the limits and advantages of a process that can have a great impact on the sustainable development of cultural heritage.

Acknowledgements The authors are grateful to Professor Michele Trimarchi and Professor Rocco Reina for their contribution to this research work as privileged observers. The usual caveat applies.

References

Amari, M. (2012). *Manifesto per la sostenibilità culturale: e se, un giorno, un ministro dell'economia venisse incriminato per violazione dei diritti culturali?* (Vol. 139). Milano: FrancoAngeli.

Bakhshi, H., & Throsby, D. (2012). New technologies in cultural institutions: Theory, evidence and policy implications. *International Journal of Cultural Policy, 18*(2), 205–222.

Bakirman, T., Bayram, B., Akpinar, B., Karabulut, M. F., Bayrak, O. C., Yigitoglu, A., & Seker, D. Z. (2020). Implementation of ultra-light UAV systems for cultural heritage documentation. *Journal of Cultural Heritage, 44*(7), 174–184.

Belfiore, E. (2006). *The social impacts of the arts–myth or reality? Culture vultures: Is UK arts policy damaging the arts* (pp. 20–37). London: Policy Exchange.

Bellini, O. E. (2015). *Three new paradigms for student housing: Cost, time and quality*. In "Abitareilfuturo" 3rd edition of "Inhabiting the future" the International Study Days-International Conference (pp. 972–985). Clean.

Bengtsson, R., Sverige, & Regeringskansliet. (2014). *Rapport: Strategiskatrenderiglobaltperspektiv – 2025: enheltannanvärld? Kanslietförstrategiskanalys (SA)*. Stockholm: Regeringskansliet.

Bonacchi, C., Bevan, A., Keinan-Schoonbaert, A., Pett, D., & Wexler, J. (2019). Participation in heritage crowdsourcing. *Museum Management and Curatorship, 34*(2), 166–182.

Borissova, V. (2018). La digitalizzazione del patrimonio culturale e le questioni relative alla proprietà intellettuale. *Giornale dei Beni Culturali, 34*, 145–150.

Borowiecki, K. J., & Navarrete T., (2015) *Digitization of heritage collections as indicator of innovation.* Discussion papers on business and economics (No. 14). University of Southern Denmark.

Borowiecki, K. J., & Navarrete, T. (2017). Digitization of heritage collections as indicator of innovation. *Economics of Innovation and New Technology, 26*(3), 227–246.

Brennen, J. S., & Kreiss, D. (2016). Digitization. In *The international encyclopedia of communication theory and philosophy* (pp. 1–11). Hoboken, NJ: Wiley.

Campagna, D., Caperna, G., & Montalto, V. (2020). Does culture make a better citizen? Exploring the relationship between cultural and civic participation in Italy. *Social Indicators Research*, 1–30.

Carmagnola, F., Dinner, F., Cortassa, O., Gena, C., & Torre, I. (2007, July). Towards a tag-based user model: How can the user model take advantage of tags? In *International conference on user modeling* (pp. 445–449). Berlin: Springer.

Collin, J., Hiekkanen, K., Korhonen, J. J., Halén, M., Itälä, T., & Helenius, M. (2015). *IT leadership in transition: the impact of digitisation on Finnish organisations*. Research report. University of Aalto, Department of Informatics.

Concordia, C., Gradmann, S., & Siebinga, S. (2010). Not just another portal, not just another digital library: A portrait of Europeana as an application program interface. *IFLA Journal, 36*(1), 61–69.

Corbetta, P. (1999). *Methodology and techniques of social research* (pp. 405–435). Il Mulino, Editore.

Daniel, T. C., Muhar, A., Arnberger, A., Aznar, O., & Boyd, J. (2012). Cultural services to the ecosystem services agenda. *Acts of the National Academy of Sciences, 109*(23), 8812–8819.

Dietrich, D., & Pekel, J. (2012). *Open data in cultural heritage institutions*. European Public Sector Information Platform Topic Report No.4.

Eisenhardt, K. M. (1989). Building theories from case studies. *Academy of Management Review, 14* (4), 532–550.

ENUMERATE the Core Survey 4 Report. (2017). *Make the case for investments in our digital activities*. Europeana-pro.

Evrard, Y., & Krebs, A. (2017). The authenticity of the museum experience in the digital age: The case of the Louvre. *Journal of Cultural Economics, 42*, 1–11.

Flew, T., & Swift, A. (2013). Cultural policy. In R. Towse & C. Handke (Eds.), *Handbook on the digital creative economy* (pp. 155–161). Cheltenham: Edward Elgar.

Gaskell, G. (2000). Individual and group interviews. In *Qualitative research with text, images and sound* (pp. 38–56). London: Sage.

Gražulevičiūtė, I. (2006). Cultural heritage in the context of sustainable development. *Environmental Research, Engineering And Management, 37*(3), 74–79.

Grossi, E., & Rvagnan, A. (2013). *Cultura e Salute*. Milano: Springer Editore. Isbn:978-88-470-2780-0.

Levy, P. (2002). Electronic mediation series. In *Cyberculture* (Vol. 4). Minneapolis: University of Minnesota.

Loulanski, T. (2006). The revision of the concept of cultural heritage: The subject of a functional approach. *International Cultural Heritage Magazine, 13*(2), 207–233.

Macri, E., & Trimarchi, M. (2015). The emerging non-profit for a sustainable economy. *RevueEuropéennedudroit social, 5/2015*. Special issue. issn:1843-679X.

Marchegiani, L., & Rossi, G. (2016). The technology applied to the cultural heritage sector has not (yet) surpassed our humanity. In *Innovation and organizational change* (pp. 53–62). Cham: Springer.

Marcum, D. B. (2007). Digitization for access and preservation strategies of the Library of Congress. *First Monday, 12*(7).

Marttila, S., & Botero, A. (2017). Infrastructure for common cultural heritage. *Computer Supported Cooperative Work (CSCW), 26*(1–2), 97–133.

Matarasso, F. (1997). *Use or ornament. The social impact of participation in the arts* (Vol. 4(2)). Stroud: Comedia.

Menard, G. (2016). Copyright, digital sharing and liberal order: Sociolegal constructions of intellectual property in the age of mass digitization. *Information, Communication and Society, 19*(8), 1061–1076.

Moher, D., Liberati, A., Tetzlaff, J., & Altman, D. G. (2009). Favorite voices for systematic reviews and meta-analysis: The PRISMA declaration. *Annals of Internal Medicine, 151*(4), 264–269.

Navarrete, T. (2013a). Digital cultural heritage. In I. Rizzo & A. Mignosa (Eds.), *Handbook on the economics of cultural heritage* (pp. 251–271). Cheltenham: Edward Elgar.

Navarrete, T. (2013b). Museums. In R. Towse & C. Handke (Eds.), *Handbook on the digital creative economy* (pp. 330–343). Cheltenham: Edward Elgar.

Paolini, P., & Di Blas, N. (2014). Story for the cultural heritage. In *Tecnologie innovativi nella mappatura urbana* (pp. 33–45). Cham: Springer.

Paolini, P., Mitroff Silvers, D., & Proctor, N. (2013). Technologies for cultural heritage. In I. Rizzo & A. Mignosa (Eds.), *Handbook on the economics of cultural heritage* (pp. 272–289). Cheltenham: Edward Elga.

Pavlidis, G., Koutsoudis, A., Arnaoutoglou, F., Tsioukas, V., & Chamzas, C. (2007). Methods for 3D digitization of cultural heritage. *Journal of cultural heritage, 8*(1), 93–98.

Preuss, U. (2016). Sustainable digitisation of cultural heritage—Report on initiatives and projects in Brandenburg, Germany. *Sustainability, 8*(9), 891.

Purday, J. (2009). Think about culture: Europeana. EU from concept to construction. *BibliothekForschung und Praxis, 33*(2), 170–180.

Report on Sustainable Development in the European Union Monitoring report on progress towards SDG in an EU context. (2019).

Rizzo, I. (2016). Technological perspectives for cultural heritage. In *The artful economist* (pp. 197–214). Cham: Springer.

Rypkema, D. (2005). *Cultural heritage and sustainable economic and social development.* European Cultural Heritage Forum, Cultural Heritage Counts for Europe, Brussels. http://www.europanostra.org/downloads/speeches/donova-rypkema_keynote_address_07dec_05.pdf

Salaun, J.-M. (2013). The immeasurable economy of libraries. In G. Rizzo Foam (2011) (Ed.), *The value of art for business*. Cambridge: Cambridge University Press.

Seitsonen, O. (2017). Crowdsourcing cultural heritage: public participation and conflict legacy in Finland. *Journal of Community Archaeology and Heritage, 4*(2), 115–130.

Sinkovics, R. R., Richter, N. F., Ringle, C. M., & Schlaegel, C. (2016). A critical look at the use of SEM in international corporate research. *International Marketing Review, 33*(3), 376–404.

Throsby, D. (1999). Cultural capital. *Journal of Cultural Economics, 23*(1–2), 3–12.

Throsby, D. (2003). Determining the value of cultural goods: How much (or how little) does contingent valuation tell us? *Journal of Cultural Economics, 27*(3–4), 275–285.

Throsby, D. (2017). Culturally sustainable development: theoretical concept or practical policy tool? *International Journal of Cultural Policy, 23*(2), 133–147.

Tisdell, C. (2003). *Ecological and environmental economics: Selected themes and policy responses*. Cheltenham: Edward Elgar.

Trimarchi, M. (2004). Regulation, integration and sustainability in the cultural sector. *International Journal of Heritage Studies, 10*(5), 401–415.

Vecco, M. (2010). Una definizione di patrimonio culturale: Dal tangibile all'intangibile. *Giornale dei Beni Culturali, 11*(3), 321–324.

Vrettakis, E., Kourtis, V., Katifori, A., Karvounis, M., Lougiakis, C., & Ioannidis, Y. (2019). Narralive-Create and experience mobile digital narration in cultural heritage. *Digital Applications in Archaeology and Cultural Heritage, 15*, e00114.

WCED, S. W. S. (1987). World commission on environment and development. *Our Common Future, 17*, 1–91.

Yilmaz, H. M., Yakar, M., Gulec, S. A., & Dulgerler, O. N. (2007). Importance of close-up digital photogrammetry in heritage documentation. *Journal of Cultural Heritage, 8*(4), 428–433.
Yin, R. (1994). *Case study research: Design and methods*. Thousand Oaks: Sage.
Yin, R. (2002). *Applications of case study research* (2nd ed., pp. 1–92). London: Sage.

Emanuela Macrì is a post-doc Research Fellow at the Department of Law, Economics and Sociology of the "Magna Graecia" University of Catanzaro (Italy) where she also teaches political economy and cultural economics. She holds a PhD in Health Economics and Management at the "Magna Graecia" University. She has been awarded with the title of Doctor Europaeus for her PhD thesis on "Health and the Arts: the value of no-profit action for patients' welfare in Niguarda Experience". She has been visiting fellow at Institute of Economics (IdEP) of the University of Svizzera Italiana. Her research interests lie in the fields of cultural economics and management with a focus on digitalisation of cultural heritage and the impact of culture on health and education. Her publications appeared in academic journals and presented her research at several international conferences. She is a member of Research Center "Economics and Management of Services" of the "Magna Graecia" University.

Concetta Lucia Cristofaro is a post-doc Research Fellow at the Magna Graecia University (Catanzaro) in Italy, where, she received her PhD in Health Economics and Management. Currently, she is adjunct Professor of in Business Organisation, Conflict Management and Human Resource Management, Public and Private Organisation at the Magna Graecia University. Her research is focused on the fields of: intellectual capital, small and medium business, ITC, digitalisation, family business, performance, gender diversity and knowledge management. Her publications appeared in academic journals and presented her research at several international conferences. She is a member of Research Center "Economics and Management of Services" of the "Magna Graecia" University.

The Contribution of Crowdfunding Regulation to Cultural Entrepreneurship in a Supportive Ecosystem

Elisabetta Lazzaro and Douglas Noonan

Abstract The rapid emergence of donation and reward crowdfunding for cultural and creative projects and ventures (CCCF) points to many benefits and barriers. Appreciating these benefits and barriers with a view of regulating crowdfunding's operations and growth is central to promoting sustainable development and entrepreneurial processes. We first present and discuss the major benefits (lower transaction costs, broader geographic distribution of financing, better market research data, and, hence, higher odds of later commercializing the product and attracting more traditional venture capital, democratization, and cocreation and peer-to-peer innovation) and barriers (information asymmetries, moral hazard, favoring superstars, and discrimination). We explain CCCF's main reasons for its untapped potential. A key role is played by national regulatory frameworks differently binding or supporting a cultural entrepreneurial ecosystem. We then outline the major areas over which regulatory action ought to be expected, including taxation and matching subsidies, consumer and investor protection, and information provision. As in most of these areas, public policy specific to CCCF is relatively underdeveloped, we discuss critical design features and identify how efficiency and equity might be affected by policy. We finally recommend key policy priorities toward a harmonized, balanced, and supportive regulation and suggest directions for future research in this emerging arena.

Keywords Cultural industries · Creative industries · Reward crowdfunding · Donation crowdfunding · Policy design

---

E. Lazzaro (✉)
Business School for the Creative Industries, University for the Creative Arts, Epsom, Surrey, UK
e-mail: elisabetta.lazzaro@uca.ac.uk

D. Noonan
O'Neill School of Public and Environmental Affairs, Indiana University – Purdue University Indianapolis (IUPUI), Indianapolis, IN, USA
e-mail: noonand@iupui.edu

P. Demartini et al. (eds.), *Cultural Initiatives for Sustainable Development*, Contributions to Management Science,
https://doi.org/10.1007/978-3-030-65687-4_18

1 Introduction

The cultural and creative sector includes a variety of sectors relying on cultural and creative inputs, such as architecture, archives and libraries, artistic crafts, audio-visual (including film, television, software and video games, and multimedia and recorded music), cultural heritage, design, creativity-driven high-end industries and fashion, festivals, live music, performing arts, books and publishing (newspapers and magazines), radio and visual arts, and advertising (Flew 2011). For these sectors, both for-profit and not-for-profit oriented, in the face to the global shrinking of public and private funding, in the last few years, crowdfunding has emerged as a valid alternative or complementary mode of funding. Crowdfunding in the cultural and creative sector (CCCF) is mainly oriented toward the financing of projects, rather than structural finance. CCCF usually takes the form of donation or reward crowdfunding, where only in the latter case, the backer expects a reward, either symbolic or in kind. Developed not only initially and especially in the USA but also in the UK and continental Europe, CCCF represents a market in significant expansion and evolution.

CCCF is becoming increasingly important to finance cultural and creative projects and ventures, untapping entrepreneurial potential (Lazzaro 2017). As a digital organizing infrastructure part of a competitively increasing digital entrepreneurship, CCCF contributes to unbind the cultural entrepreneurial process and outcomes and to convert a predefined entrepreneurial agency into a more distributed and social agency, where entrepreneurial resources, ideas, and opportunities are garnered from and output is participated and cocreated by an increasingly broader, more diversified in goals and capabilities and a more dynamic collective (group of actors) (Nambisan 2017). In particular, CCCF favors community involvement (Josefy et al. 2016), market research and audience development, the sharing of information, and promotion, besides the development of professional skills (European Commission 2017). Yet, it is not free of barriers and risks. The rapid emergence of donation and reward crowdfunding, especially for cultural and creative projects and ventures, points to many benefits and barriers.

Appreciating benefits and barriers of CCCF with a view of regulating crowdfunding's operations and growth is central to promoting sustainable development and entrepreneurial processes. Ultimately, crowdfunding's ability to achieve its potentials and overcome its barriers depends directly on how it is regulated. Through a literature-review approach, we first present and discuss major benefits and barriers of CCCF in the light of the current scholarly debate.

If we look at the rapidly evolving marketplace of CCCF, it looks fragmented in terms of platforms' business models, geographical concentration, relevance of CCI sectors covered, available information, implied skills to operate and access this market, and, correspondingly, regulation. In such a, so far, diverse soft-policy environment, we advance propositions for a creative and supporting intervention by regulators and policymakers, aimed not just at reducing CCCF barriers but also at fostering its benefits, and in particular those supporting the activities of cultural entrepreneurs and ventures. In this sense, the major contribution of this paper is to offer recommendations for policy design grounded on the economic analysis of existing regulations.

In the next two sections, we organize and examine barriers and benefits of CCCF in the light of the existing literature, respectively. In Sect. 4, also by means of examples, we critically discuss major features of regulation that may limit barriers and support benefits of creators, consumers, and platforms. In Sect. 5, we point to areas of lacking regulation, which may represent possible opportunities for regulatory intervention and, more in general, policy design.

2 Benefits of CCCF

Major benefits of CCCF lower transaction costs (Belleflamme et al. 2015), broader geographic distribution of financing (Agrawal et al. 2015; Sorenson et al. 2016), better market research data, and, hence, higher odds of later commercializing the product (Butticè and Noonan 2020) and attracting more traditional venture capital (Xu 2018), cocreation (Thürridl and Kamleitner 2016) and peer-to-peer innovation (Bruton et al. 2015), democratization of funding (Mollick and Robb 2016), and increased synergies with complementary financing.

2.1 *Transaction Costs*

At its core, crowdfunding's key distinctive innovation owes to its dramatically lower transaction costs for fundraising associated with its online platform and associated technologies. By lowering the transaction costs, the creator can now market her project to a much larger and possibly better targeted audience ('the crowd'), ultimately lowering delivery costs (Belleflamme et al. 2015). This can be especially important in the creative and cultural sector as lower transaction costs enable artists to expand their audience *and* possibly bypass traditional (high-cost) middlemen, studios, or other gatekeepers to more directly engage their market.

The growth in crowdfunding as a preferred fundraising technology comes as some creators are better able to reach their market. The lower transaction costs can affect the intensive margins by lowering fundraising costs among existing contributors, which helps the viability of previously marginal projects (Galuszka and Bystrov 2014). Perhaps most dramatically, crowdfunding's reduced transaction costs affect the extensive margin, enabling some artists and creators the opportunity to market their projects where previously they lacked a sufficient audience or identifiable market. Cultural entrepreneurs reliant on in-person venues, such as performing artists and other exhibitors, may benefit most from the wider reach that online crowdfunding platforms enable.

Crowdfunding can reduce the relative minimum efficient scale (MES) of startups raising capital in financial markets to enable new entry. As crowdfunding lowers the transaction costs of reaching the crowd of investors or of consumers, these crowds enable new producers to succeed in new markets, especially in small markets. The relationship between crowdfunding platforms and minimum efficient scale points to the technology's potential to spur entrepreneurship in the creative and cultural sector, especially in smaller markets. Removing trade barriers and expanding markets can also complement crowdfunding's technological advances in reducing the minimum efficient scale (Lazzaro and Noonan 2020). This reduction in minimum efficient scale can have profound implications for freelance artists and other small-scale creative workers who previously depended on working through larger-scale organizations to enter the market.

2.2 *Geographic Distribution of Financing*

As lower transaction costs have expanded the size of the crowd, more interested contributors can be identified and previously unviable ('niche') projects can receive a "green light." Broader geographic distribution of financing for entrepreneurial projects often leads the list of crowdfunding's benefits (Sorenson et al. 2016). How crowdfunding platforms fit with a "world is flat" perspective on the global economy continues to receive scholarly attention (e.g., Agrawal et al. 2015; Fleming and Sorenson 2016; Mendes-Da-Silva et al. 2016; Giudici et al. 2018), with a preliminary conclusion that new venture financing is far less geographically concentrated under crowdfunding (Yu et al. 2017). Yet, location continues to matter a great deal, especially at the extensive margin, as crowdfunding has enabled more geographic localization (local specialization?) of projects as ventures in smaller markets are now viable (Davidson and Poor 2018; Rykkja et al. 2020). This geographic 'flattening' effect reaches not just where projects are located but also where the creators and the backers need to locate—as increasingly the producers and consumers need not collocate.

2.3 *Market Research*

Crowdfunding's lower transaction costs include improved 'market research' data from crowdfunding information in these online fundraising markets. Project creators can better convey the qualities of their ideas, products, services, etc. in a rich, multimedia environment. Crowdfunding research routinely shows that projects with better information describing their venture are more likely to succeed in their fundraising goals (Mollick 2014; Courtney et al. 2017).

Availability of market information results also increased for prospective bakers, as crowdfunding ecosystems take cues from other online sales platforms to create and maintain information about creator reputation, possibly reducing information asymmetries (Butticè and Noonan 2020).

2.4 *Cocreation and Peer-to-Peer Innovation*

In some cases, CCCF also fosters more interaction between the creator and prospective backers, allowing for even richer exchanges of information (Thürridl and Kamleitner 2016) and peer-to-peer networks to harness communities of entrepreneurs and investors (Bruton et al. 2015). Spillovers from CCCF to other more traditional forms of financing new ventures (Martínez-Climent et al. 2020) and IP protection (Erickson 2018) remain important areas for future research.

Bringing creators more directly in contact with their audience or crowd may offer benefits other than just improved market research. It might improve the experience for the backers, give stakeholders more influence over production, and even identify more 'wisdom of the masses' (Mollick and Nanda 2015) and backers' engagement (Marchegiani 2018).

2.5 *Complementary Financing*

Allowing creators to collect better market research data through a crowdfunding campaign can yield great value to creators, including enhancing their odds of later receiving more traditional venture capital (Butticè and Noonan 2016; Xu 2018). In essence, testing prototypes in a crowdfunding marketplace with real stakes (money) involved marks a substantial improvement in market research over running focus groups and expert opinion. In general, projects that receive a higher amount of pledged money through crowdfunding attract more subsequent interest from investors (particularly in the world of technology), although "this positive effect emerges (and becomes more intense) only when it is complemented by the presence of patents granted for the new product idea, which proves technological viability and exclusive protection, or when the entrepreneur has built a large network of social relationships and thus can benefit from a large pool of strategic resources" (Roma et al. 2017, p. 1607).

CCCF is increasingly incorporated into public-private match-funding (European Commission 2017) or civic crowdfunding schemes, fostering citizens' engagement in public local governance (Brent and Lorah 2019) and, more in general, an ecosystem perspective for value cocreation (Quero and Ventura 2019).

2.6 *Democratization of Funding*

Indirect benefits of crowdfunding also warrant attention, as these often grab the limelight. Advocates often point to crowdfunding's ability to "democratize funding" (Mollick and Robb 2016) and "levelling the playing field" (De Buysere et al. 2012; Bruton et al. 2015). In a sense, these refer to the promise of crowdfunding to expand the margins of viable projects and allow new entrants. Insofar as there are severe barriers to entry for funding—for venture capital, for government funding, for charitable donations, etc.—then crowdfunding may indeed offer a platform that bypasses those monopolized funding markets. In this sense, crowdfunding closely resembles other platform-based disruptive markets, such as ridesharing and AirBnB, whose benefits partly derive from overcoming barriers to entry. It even helps reduce barriers' entry into more conventional financing. A correlation exists between "Kickstarter projects in a region" and "increased angel investing activity, even after instrumenting with projects that should not be of interest to investors" (Yu et al. 2017).

Diversifying the pool of funders, and thus diversifying the projects that ultimately receive funding, may hold some of the largest benefits from a broader, social perspective. Fears of "mob rule" and "tyranny of the majority" aside, crowdfunding as a mechanism to mitigate elitism in new cultural and creative ventures may yet hold benefits for society. Each of these benefits, however, directly depend on barriers that crowdfunding faces (Galuszka and Brzozowska 2016). After all, if certain groups are disenfranchised from participating in crowdfunding, these indirect benefits are greatly limited.

3 Barriers

Despite its rapid rise, CCCF faces significant barriers and poses some serious risks. Major barriers to CCCF include information asymmetries (Ganatra 2015) and their consequences of adverse selections and, especially, moral hazard (Strausz 2017), favoring superstars (Agrawal et al. 2014), and discrimination (Younkin and Kuppuswamy 2017; Johnson et al. 2018).

3.1 *Information Asymmetries*

The most immediate issue, mentioned above, concerns information asymmetry and the lack of transparency. Allowing creators to share more information can reduce uncertainty in the market, but does little to address information asymmetries. Although information asymmetry in fundraising is not exclusive to crowdfunding (Kaun 2001; Safner 2015), merely shifting the marketplace to an online forum hardly resolves the matter, although platforms' reputation may compensate for the problem of signaling quality (Davies and Giovannetti 2018). The creative and cultural sector faces additional (information asymmetry) challenges given the risky nature of creative processes and the resulting output often being 'experience' or 'credence' goods whose quality is hard to verify prior to funding or consuming it. While this was true prior to the rise of CCCF, a shift to online settings can exacerbate these informational challenges.

Adverse selection, whereby good projects would fail because observably identical bad projects are the only projects that sustain on the crowdfunding platform, remains a concern in theory at least. Although the growth of crowdfunding platforms suggests that either this concern is overstated or that there is a great deal of 'lemons' (Akerlof 1970) waiting for the crowdfunding opportunity, the fundamental challenge of better enabling transparency remains. Again, this is true for all fundraising campaigns for new ventures, while crowdfunding may act as a legitimization signal for arts organizations (Alexiou et al. 2020).

Perhaps a bigger concern for crowdfunding is the moral hazard problem (Strausz 2017) or the possibility that creators fail to adequately deliver to their backers. Once funded, in at least most crowdfunding platforms, projects may face suboptimal pressure to deliver on their promises (Agrawal et al. 2014). Although reputational effects may be severe, one-off creators who simply scam the market before disappearing into anonymity pose a threat. Subtler, and potentially more threatening, are the biases that crowdfunding might support.

From a principal-agent perspective, adverse selection and moral hazard issues can also affect creators with respect to consumers, such as the commercial power exercised by consumers over which products will come to the market and consumers' creative power over products they support (Chaney 2019).

3.2 *Superstars*

Although it might promise to 'democratize' funding, its theoretical promise might be replaced by a reality of discrimination (Rhue and Clark 2018; Younkin and Kuppuswamy 2017) and other biases (Greenberg and Mollick 2017; Johnson et al. 2018). The rise of crowdfunding over the past decade has been characterized by several key features. Agrawal et al. (2014) find that crowdfunding is not geographically constrained and tends to support 'market-for-superstar'-style distributions with fundraising highly skewed to a few creators. According to Jon Leland, vice president of insights at Kickstarter, disproportionate success of large-scale CCCF ventures remains a concern (Leland 2020). Small-scale creators in CCCF still most crave publicity, lower fees, and director support according to a recent survey of entrepreneurs on Kickstarter (Leland 2020).

3.3 *Funding Substitute*

Crowdfunding capital may substitute for other traditional sources of financing, although this research continues to evolve (see, e.g., Roma et al. 2017; Yu et al. 2017). In addition, Agrawal et al. (2014) also observe a prominent role of friends and family in early stages of fundraising. The availability of nontraditional capital sources can be critical for entrepreneurial success during recessionary times. CCCF entrepreneurs themselves report that direct financial support and other resources are most critical to accelerating the creative process (Leland 2020). As they pursue what is often a part-time venture for them, they look for capital from crowdfunding as a complement rather than a substitute for other resources.

4 Existing Regulation

As, in general, CCCF does not imply, contrary to investment CF, financial investments and returns, the US and the EU (at both Union and country levels) legislations are directly less concerned about it. Overall, this results in a rather liberal regulation in the US, although in the EU, characterized by a lack of harmonization, the picture is more heterogeneous at country level. In particular, CCCF markets are diversely developed in EU member states, in terms of history, size, and types of regulation in place (ECN 2017; Lazzaro and Noonan 2019, 2020). Noticeably, more CCCF liberal regulations, such as Sweden, do not necessarily correspond to more developed CCCF markets, such as the UK, whose supportive policy includes a progressive regulation and tax reliefs (CrowdfundingHub 2016). Overall, national regulatory frameworks differ in whether they treat specifically about crowdfunding or not (Patti and Polyák 2017) and the degree through which they facilitate or, on the contrary, restrain crowdfunding (European Commission 2017).

Major reasons for a still lightly and indirectly[1] regulated (in the USA and some European countries) and scarcely harmonized (in the EU) CCCF market may be reconducted to the typical nonspeculative nature of CCCF, its relatively small turnover, and its relatively less severe asymmetric information, as opposed to equity and lending CF (Lazzaro and Noonan 2020).[2]

A relative lack of regulation implies not only opportunities but also challenges for those engaging in CCCF. Lacking a regulatory system in place at the outset, the crowdfunding approach enjoyed some rapid growth without (legal) barriers and associated compliance fixed costs (Calcagno and Sobel 2014), as it is the case of the United States. But unregulated growth can pose some risks, especially if participants in the crowdfunding marketplace are victims of fraud, predation, or bias or if

[1]Noticeably, the indirect intervention of US federal regulators on CCCF, leaving it to the state level, could be assimilated, to some extent, to the limited harmonization of the EU.

[2]For a detailed comparison of US-EU CCCF regulation, please see Lazzaro and Noonan (2020).

the inherent risk on the platform deters optimal investments. The absence of clear standards and stable regulation in these markets can dissuade creators and backers from participating, and it can lead to distortions in the marketplace as some projects, or even some platforms, are favored through market design, taxation, or technological limitations.

4.1 Harmonization

The lack of harmonization between national systems stands as a major barrier. This may be especially true in Europe, where variation in national regulatory frameworks limits growth of what is essentially a global marketplace. Compounding this lack of harmonization in regulatory systems is the important difference in fiscal regimes, financial incentives, payment systems, intellectual property rights, data protection, professional standards, languages, and more. Crossing borders or switching among crowdfunding platforms can require significant adjustment costs. Market fragmentation in crowdfunding, especially in a post-Brexit world, remains a serious barrier to its continuous expansion.

CCCF is not harmonized at the EU level, and the EU also lacks an authority in charge of that. Crowdfunding regulation is rather the object of a variety of recent and developing national regulations, although these regulations tend to generally and indirectly concern CCCF, rather than being specific about CCCF. EU harmonization intervenes at a more indirect and general level on fewer specific—though EU considered prominent issues also concerning CCCF—such as data protection and consumer protection (read backers in CCCF). EU policies are mainly aimed at indirectly improving the CCCF regulatory framework, besides its awareness and hence its adoption by users. The EU approach toward CCCF regulation remains rather soft, by means of workshops and public consultations, communications, a stakeholders' forum, reviews, reports, and commissioned studies (European Commission 2017).

Various issues and aspects connected with CCCF fall in more general national measures, and they vary from country to country, as they are barely impacted by a limited corresponding EU harmonization. Yet, across a number of EU member states, there are various exceptions to missing or reduced specific regulation of CCCF at national level.

The lack of regulation harmonization at the EU level favors a fragmentation of the crowdfunding market, leaving the sizes of national markets relatively small and limiting crowdfunding crossing borders. Furthermore, a lack of a clear and comprehensive regulation favors uncertainty and hinders consumers' confidence and, hence, the upscaling of the market. Therefore, the majority of EU countries, especially the most experienced ones, are in favor of a regulation of crowdfunding and its EU harmonization, at the same time avoiding too heavy, nontransparent, and onerous regulation. On the other hand, a minority of relatively less experienced countries (e.g., Malta, Slovakia, etc.) are in favor of a more liberal framework to facilitate the market expansion.

4.2 Taxation

The fiscal regulatory framework governing donation- and reward-based cultural and creative crowdfunding ventures within the United States is still emerging. Although more limited than the European legal and regulatory framework(s) for maintenance of crowdfunding, the existing laws and regulatory climate in the United States provide opportunities and challenges for regulators and those engaging in CCCF.

Taxation and tax relief of CCCF remain unclear and complex in the United States, with the former especially for crowdfunding ventures (Metrejean and McKay 2015) and the latter especially for artists (Gruba et al. 2015). Given the significant competitive advantage for online marketplaces derived from their avoiding sales tax (Goolsbee 2000), treating reward-based crowdfunding pledges as essentially taxable presales, as it is the case of the US, could hinder the attractiveness of these platforms. Furthermore, a taxation discriminating on the type of reward being offered (Gruba et al. 2015) and interstate shipping of rewards could increase transaction costs (related to fiscal information) and limit the choice about reward options and, hence, of CCCF attractiveness.

4.3 Intellectual Property Rights

An adequate intellectual property rights (IPR) enforcement controls for the risk of copyright infringement and other IP protection associated with sharing early-stage ideas (Roberts and Nowotarski 2013).

4.4 Fraud

While in the EU, fraud falls within the general regulation harmonization of consumer protection, in the United States, it captures considerable public attention in reward-based crowdfunding. This includes acts of fraud (intentional or otherwise) and failing to deliver the rewards or perks promised by the crowdfunding project. According to one Mollick (2015), only around 9% of projects fail to deliver promised rewards, with the majority of those failures being from smaller projects and no noticeable links between failure rates and demographics such as education, gender, and family status. Crowdfunding platforms generally refuse to refund money directly to backers when projects fail, and limited options exist for these to ensure reception of deliverables or a refund of their donation. Legal actions are rare, probably due to the high legal costs relative to the small donation amounts typically made to artists and vendors (Moores 2015). Regulatory frameworks for reward-based crowdfunding in the US do not oversee the platforms themselves, but they do include general regulatory mechanisms for fraud (intentional or otherwise), such as consumer protections.

4.5 Entry Barriers

In the EU, as a consequence of missing harmonization, overall crowdfunding regulation is mainly competence of member states, a number of which have developed specific legislations in the last couple of years. This results in a rather EU fragmented market (Patti and Polyák 2017). Given the different regulatory systems, crossborder activities are limited (European Commission 2017) and so the geographical scope of CCCF (ECN 2017).

Despite the absence of crowdfunding harmonization in the EU, there are regulation similarities and/or crossborder activities between neighboring countries and in common language areas (e.g., Austria and Germany; Belgium, France, and the Netherlands; the Czech Republic and Slovakia; Slovenia and Austria). Larger national crowdfunding markets are also the most interested by international (in/outbound) activity.

5 Areas for Regulatory Intervention

In comparing the US and EU settings for regulation of CCCF, stark differences are immediately evident. The US context is characterized by one substantial harmonization in a largely unregulated, single market for CCCF across the country. This condition owes primarily to the absence of substantial state-level regulations—a situation that is always likely given the constitutional limits on US state governments regulating interstate commerce, which CCCF typically is—and a reluctance on the part of the federal government to specifically regulate this sort of activity. Aside from recent national rules that enable equity-based crowdfunding in the US for the first time, the US government has overlooked crowdfunding, especially those that most appeal to the cultural and creative sector. The EU approach contrasts sharply with this situation. Because there has been no strong effort at harmonization at the EU level, EU member states have been free adopting different approaches to regulating CCCF. The result is that the member states take on a primary role in regulating CCCF, unlike the US, and a fragmented patchwork of continually evolving state-level rules.

5.1 Self-regulation

The crowdfunding environment for the creative and cultural sector is perhaps most simply described as a self-regulated marketplace. This "wild west" sort of environment has been core to development. CCCF has experienced rapid growth, booming over a little more than a decade from virtually nothing to hundreds of millions of dollars being pledged every year. This rise in prominence, perhaps even more notable given the scarcity of private funding in the arts and cultural industries, has been accompanied by very little regulatory attention. CCCF has emerged in a vacuum of regulation, rather than because of it. Regulators have been slow to react in the US and in many EU member states, arguably because this activity is still too (economically) minor to warrant their attention. This contrasts with equity-based crowdfunding, an area that more quickly attracted regulators' attention as it directly relates to the finance sector's regulatory apparatus for investments. Furthermore, it is unclear how much harm is being caused by CCCF activity in the absence of regulatory attention. Early indications (e.g., Mollick 2015) suggest that fraud, for example, is not rampant. The lack of government regulation, combined with how experimental much of the cultural and creative sector is by its nature, has helped create this environment noted for its creativity, openness, and ability to reach diverse "crowds" of the public. Reward-based crowdfunding has demonstrated considerable innovativeness, especially compared to the more regulated equity-based crowdfunding (Gutiérrez and Sáez 2018). This strength of reward-based CCCF stems from the predominance of nonpenalty contracts that enable CCCF markets to serve an important role in identifying and supporting emerging talent. Critical to this function are the smaller stakes and ability to use funds raised in CCCF to leverage 'downstream' benefits, as seen in successful CCCF creators' ability to succeed in their next campaign (Butticè et al. 2017; Li and Martin 2019) and to subsequently raise funds in other ways (Mollick and Kuppuswamy 2014; Roma et al. 2017). Gutiérrez and Sáez (2018) thus argue that these reward-based platforms should not be regulated as investment-based crowdfunding already is, because doing so risks losing its innovativeness and ability to support the discovery of new talent and creative ideas. This role of CCCF as a talent discovery tool hinges on it *not* operating like the regulated equity and debt financing systems do. Of course, some regulation, especially as CCCF begins to mature, could help CCCF continue to grow (Ganatra 2015), especially if they bolster the platforms' ability to identify and support new talent and ideas.

The absence of formal regulation does not mean that CCCF activity is truly unregulated. It is regulated; it is just predominantly self-regulation. In the US, given the limited scope of federal and state laws targeting CCCF for regulation, crowdfunding platforms typically engage in considerable self-regulation. This includes some design features of the platforms themselves as well as some routine policing and enforcement activity by the platform operators. The design of crowdfunding platforms generally involves considerable information sharing and public information capacity regarding key features of creators and their campaigns.

Creators must share information about themselves and their projects, often in great detail, which influences their current and future success. In addition, the success of projects (e.g., amount of funds raised) is also public knowledge. Combined with online discussion boards and other records of feedback from the public, these CCCF platforms make information available that supports the development of reputations, both positive and negative, which affects future outcomes. This creates a powerful incentive for creators to behave responsibly, even beyond the conclusion of a fundraising campaign. For examples, Butticè and Noonan (2020) demonstrate a strong influence of very active backers on the likelihood that creators actually deliver on their product, and Li and Martin (2019) show how entrepreneurs who ultimately fail to deliver will suffer penalties on future campaigns. When the platform design lacks sufficient information, other third-party providers can attempt to fill this gap. A variety of websites (e.g., Kicktraq and Krowdster) have emerged adjacent to the Kickstarter platform to provide creators and potential backers with more information about campaigns. This approach to regulation via reputational effects is highly informal and relies on decentralized, voluntary practices to enforce.

This sort of self-regulation "by design of the platform" complements the more active efforts that crowdfunding platforms can engage in themselves. Some of the larger platforms require terms-of-use that stipulate prohibited offerings and business practices as well as how entrepreneurs should report the funding for tax purposes. Each platform sets their own terms-of-use and determines their own enforcement strategies, reflecting the primarily decentralized and self-regulatory approach in the US. Yet, the potential for—and past examples of—entrepreneurs using these lax standards to defraud investors via attractive projects they have no intent to actually produce remains a concern and a potential barrier to attracting larger "crowds" to the platforms. As an example, Kickstarter has suspended projects on its site for noncompliance with these terms-of-use. This often involves violation of the rule that working prototypes are needed if a physical product is to be delivered as a reward. Enforcement of this rule is inconsistent, however, and it does nothing to restrict rewards without physical products. Of course, platforms like Kickstarter could increase their enforcement efforts (Ganatra 2015), but their incentives to do so are limited by moral hazard (as the platform's revenue depends on funds raised rather than success deliveries), and rapid growth in activity from all over the world[3] tests their capacity. The risk, and history, of scams on these platforms is sufficient that online public databases exist to record these instances and allow potential backers to conduct due diligence.

[3]Kickstarter now draws over 40% of its projects from outside of the US.

5.2 *Harmonization*

A critical issue in CCCF regulation remains harmonization of regulatory systems. The variety in the designs and operations of different crowdfunding platforms can pose a serious barrier to entrepreneurs in launching their fundraising campaign. The many different systems to learn imposes a startup cost and a learning cost to switch among the platforms. Separate and different regulatory frameworks for those different platforms compound these costs and generate additional uncertainty for entrepreneurial creators. Perhaps even more important is the fragmentation of the markets served by these disjoint crowdfunding platforms. The potential of crowdfunding—tapping into the finances and wisdom of the masses—is hampered by fragmenting markets and accessing smaller "crowds."

As regulations limit who can participate and raise barriers to operating fundraising campaigns in additional platforms, the smaller crowds generate less financing. Harmonizing across jurisdictions and markets can help crowdfunding attain its economies of scale. It can also assist with regulatory bodies' enforcement (Ganatra 2015). The lack or harmonization can especially impact small and mid-size projects that need large crowds to achieve enough scale to overcome fixed costs. Thus, the study of EU cultural crowdfunding by the European Commission (2017) culminates in recommendations for the EU to improve transboundary crowdfunding and upscale platforms, essentially calling for greater regulatory harmonization as platforms can reach larger crowds. Some of the fragmentation challenges in the EU extend to systems beyond the scope of crowdfunding (e.g., languages, payment systems, and tax regimes), issues also not present in the US context. Regardless, as the European Commission (2017) study finds, greater transparency and harmonization in procedures, taxation, payment systems, and how organizations and projects employ crowdfunding can promote greater trust and sustainability of crowdfunding in the EU. Along these lines, the call for more widespread and shared understandings of crowdfunding platform functions (e.g., charitable fundraising tools? storefronts for presales?) can reduce frictions and misguided expectations in CCCF transactions (Ganatra 2015).

5.3 *Intellectual Property Rights*

Another challenge for CCCF relates to intellectual property rights and protections for innovations and new creations brought to such a public forum. Crowdfunding offers no exemption or relaxation of the broadly applicable laws to protect intellection property (IP) in both the US and the EU. Thus, while these laws still apply in crowdfunding contexts, no special regulation or additional support or protection exists in either region. The potential insufficiency of self-regulation combined with no special regulatory oversight remains. After all, a common CCCF campaign involves a creator proposing a novel idea along with making freely available to the public a sample, prototype, or details of its design. This public disclosure can constitute prior art and thus curtail its patentability.

Whether the fundraising campaign is successful or not, the (potentially) valuable idea is floated out to anyone to see, and copy, for their own private benefit. Many of the novel ideas contained in these campaigns lack prior formal legal protection or involve fundraisers who lack resources to defend their legal rights. Entrepreneurs seeking funding in CCCF have very limited recourse in the event they are sued for patent infringement. The risk of IP theft or fraud can deter creators and funders from using crowdfunding platforms. Low barriers to entry and public information disclosure, however, are core strengths of CCCF. A balance between encouraging innovation and protecting IP needs to be struck (Miao 2015). More research is needed to identify regulatory approaches that can enhance efficiency. A decentralized and self-regulatory system like in the US relies on users and third-party providers to monitor and report possible instances of IP theft and fraud. The efficacy of these voluntary and largely reputation-based mechanisms is not established, especially for IP protection. Recommended IP protection improvements can improve the balance by protecting innovators on crowdfunding platforms, while protecting their experimental status and precluding permanent injunctions against (patent infringing) crowdfunding projects that have already been delivered (Miao 2015). Enacting these reforms would mark a strong movement away from the status quo where crowdfunding receives no special treatment under patent law.

5.4 Recommendations

Several recommendations emerge from the comparative analysis of the development of CCCF and regulatory frameworks in the EU and the US. These recommendations accompany a recognition of centrally important features of this emerging form of fundraising. The largely unregulated setting, reliant predominantly on self-regulation and general commercial law protections, has supported remarkable growth in use and impact of crowdfunding. Yet, what has worked previously may not suffice if CCCF is to mature and grow into a more robust, sustainable tool. The abilities to tap into (and serve) bigger crowds, identify innovative talent and ideas, and reward those entrepreneurs have proven essential attributes of the crowdfunding systems—attributes that policy reforms should protect and enhance. To that end, the following recommendations arise:

1. *Harmonization*. Fragmentation of crowds and markets across different legal jurisdictions poses a major barrier to the development of CCCF. Removing these barriers and harmonizing CCCF regulatory regimes stand to boost the continuous growth of CCCF and its positive impacts on the sector. This recommendation applies mainly to the EU and its divergent regulatory approaches among member states. Harmonizing this patchwork can lower the cost of using the tool while expanding market sizes and enabling economies of scale. Even in the US, harmonizing state rules can improve protections for funders, reduce costs to the legal system, avoid a situation where many states begin developing their own, divergent regulatory regimes, and further fragment already niche markets.
2. *Strike a balance*. Whatever new regulations develop for CCCF, the interests of entrepreneurship, fairness, and openness should be balanced to support more creativity. Too easily can be costly, limiting regulation be imposed to undermine the core attributes of crowdfunding that enabled its remarkable emergence. For example, protecting intellectual property of innovators by requiring patents might put off a great many potential creators, just as offering no IP rights to entrepreneurs disclosing their ideas can deter some from using the platforms. Balance must be struck. For CCCF, where the unstructured and open nature of crowdfunding, in general, holds enormous appeal, the starting point of self-regulation seems appropriate. Regulations above that level should also take into account context and recognize that entrepreneurs are responsive to a country's relative procedural (legal) formalism (Stephen et al. 2009). A regulatory shift to increase transparency and disclosure requirements—and monitor against fraudulent information—offers one way to both tighten regulation while also enhancing the value of crowdfunding. For a platform that inherently depends on public information, mitigating information asymmetries stands to benefit creators and their funders alike. This emphasis on transparency and disclosure is reflected in the JOBS Act approach to regulating investment crowdfunding in the US, and such an approach ought not be overly burdensome to reward-based CCCF contexts (Ganatra 2015). Even today, launching a campaign on a crowdfunding platform is serious endeavor that involves one-time costs, learning, and skills different from what many individuals and organizations in the creative or cultural sector possess. Regulations that raise costs to enter crowdfunding platforms even more will dissuade some projects from launching. Some general regulatory approaches can be counterproductive here (Gutiérrez and Sáez 2018). According to CCCF entrepreneurs, the main factors in accelerating creative processes on Kickstarter are funding, time, market demand, and physical resources (Leland 2020). Regulations to attract more funding and demand would be welcome, but any regulation risks restricting the creative processes in CCCF. Again, striking a balance means tailoring the regulations to promote information without excessively burdening or limiting new ventures. This balance involves sacrificing some openness for accessibility, and vice versa, and the result will inevitably contain risk for creators and funders alike. The vibrant crowdfunding ecosystem is not risk-free, just as innovation and entrepreneurship are not. Regulations should

strike a balance in how they implicitly define the risk tolerance of the marketplace—one that accepts some risk in an attempt to support even more creativity.

3. *Expand the use of crowdfunding*. Public officials should also explore ways to expand the use of crowdfunding as a tool to support innovation. That innovation includes civic innovation or using the crowd to support creative solutions to public problems. Crowdfunding's two most important features—identifying promising new ideas and finding and building public support for projects—surely can serve the goals of public policymakers who seek the same. CCCF already has a strong track record in identifying innovative ideas, as a sort of 'market research and development' platform (e.g., Roma et al. 2017; Thies et al. 2019), which relies on the wisdom of the crowd (e.g., Mollick and Nanda 2015) as backed by their own real financial commitments. These attributes hold great promise for government projects or ventures. Some government agencies have already taken advantage of crowdfunding to replace or supplement tax revenues with voluntary contributions (Brent and Lorah 2019). Yet, more could be done. This includes take-up by more agencies for a greater variety of purposes. Civic crowdfunding holds considerable promise for more participatory planning and raising more resources. For instance, matching private contributions with public funds can both amplify the fundraising capacity of the venture and potentially be used to support policy goals (Brent and Lorah 2019). Crowdfunding mechanisms can also generate more engagement and activity from backers, beyond their financial investment (Butticè and Noonan 2020). Letting communities' preferences more directly influence funding of public projects stands to improve governance, despite the potential obvious problems with allowing voting with pocketbooks. Tailoring matching schemes to weight by income, using numbers of backers rather than funds spent, and other techniques exist to address this concern. Beyond learning more about citizen preferences, civic crowdfunding can "crowd in" extra private support to complement public expenditures. Government agencies should explore using CCCF to generate more funding for public projects, better aligned with public preferences. Efforts to integrate with and grow other support for CCCF entrepreneurs can bring the financial and physical resources and small communities of support important to developing this area (Leland 2020). Effective use of crowdfunding in new domains such as arts and culture projects can make positive impacts on support, innovativeness, and engagement.

6 Discussion and Conclusions

Over the past decade, crowdfunding has emerged to play an important role in financing for the cultural and creative sector. This alternative approach to fundraising has grown increasingly mainstream—as a Pew Research Center highlights that, as of a few years ago, almost a quarter of all Americans have contributed to a crowdfunding project (Smith 2016). This popularity extends beyond the funders, as crowdfunding platforms have generated billions of dollars worth of funding for hundreds of thousands of projects. Most important for CCCF are reward- or donation-based designs, rather than investment- or equity-based designs. Yet, crowdfunding provides more than just financial transactions. Crowdfunding offers advantages in terms of catalyzing more information provision, better identification and promotion of innovators and creative projects, and greater engagement with and codesigning opportunities with audiences and stakeholders. Crowdfunding also holds promise in democratizing or 'leveling the playing field' (Breznitz and Noonan 2020), rather than concentrating funding allocation decisions to a few elites or technocrats. Our review of the literature has critically assessed the benefits and limitations of CCCF and how various institutional elements permit or enable crowdfunding to produce those benefits. Challenges like tax rules, IPR protection, and fraud remain for crowdfunding platforms and regulators.

CCCF's emergence brings with it not only a considerable variety of benefits but also some limitations and risks. A brief catalog of the benefits offered by CCCF includes:

- Lower transaction costs in reaching supporters and receiving funding, implying
  - lower overhead or administrative costs,
  - much wider outreach—a bigger crowd,
  - feasibility of financing more "niche" projects—a longer, fatter tail of creative diversity, and
  - more geographic diversity in where creative ventures are supported
- More engagement with a broader set of stakeholders, offering more opportunities for cocreation, feedback to creators, and network building among creators and funders
- More information about new ventures made public, including
  - evidence about the extent of public support or demand,
  - past projects' ideas and successes, allowing for a new mechanism for gaining reputations, and
  - the resulting positive spillovers to successful creators' subsequent fundraising endeavors and ventures
- Opportunities for greater diversification of funding sources, allowing for reduced risk for individual creators and organizations
- Democratized funding allocations

Many of the benefits, however, can be more "potential" than "actual" without effective platform design and regulation, including self-regulation. The list of limitations to the potential of CCCF includes:

- Limited self-enforcement due to moral hazards faced by platform operators and by creators themselves, who may not deliver on promises
- Costly access to platforms, especially with designs and regulations varying across platforms and jurisdictions

- Costly access to technologies, which can skew the benefits from CCCF to certain types of individuals, groups, projects, and regions
- Costly (social) networking with potential for bias
- Information asymmetries, where would-be funders lack critical information about the quality of proposed projects, including
  - difficulties in monitoring, detecting, and enforcing rules about fraud and
  - potential for adverse selection or a "market for lemons" to dominate crowdfunding markets as high-quality ventures cannot differentiate themselves sufficiently from low-quality projects
- Theft of or inadequate protections for IPR.

Many of these limitations, however, are not unique to crowdfunding. Some (e.g., IPR protections) may be worse than more traditional fundraising approach, while others (e.g., information asymmetries) may be mitigated by relatively information-rich crowdfunding settings. Much of this naturally depends on the "crowd" attracted to the platforms and how much due diligence they exercise.

Well-designed regulation stands to help CCCF mitigate some of these challenges. Regulations can address bias, fraud, and costs within and across platforms. These regulations can take the form of self-regulation, which is predominant in the US context or more formal government regulations. Protecting IPR, discouraging fraud and compensating victims, and guarding against various forms of bias are primary areas for regulatory attention, especially because regulators are often in no better (informational) position than the crowd to address information asymmetries *ex ante*. The differing standards and regulations, however, can bring their own costs and barriers—some of which undermine the very benefits of CCCF. Much of the rapid rise in CCCF has been attributed to the lack of costly formal regulation. Moreover, rules that limit the size of the crowd (e.g., limiting transboundary transactions) also pose one of the greatest barriers for crowdfunding. In comparing the US and EU settings, the performance of CCCF also highlights the importance of harmonization in regulatory approaches—as the EU system of fragmented and evolving rules can limit CCCF's applications among European states. The centralized EU approach is mostly limited to passive "consultations" and informational efforts, leaving its stricter regulations for equity and invest-based crowdfunding. The largely federal—extremely limited and very harmonized—regulatory approach to reward-based crowdfunding in the US has fostered its robust growth. This contrasts with the greater variety of regulatory regimes and platform models across the EU, leading to fragmented "crowds" and greater costs of moving between systems.

Going forward, the expansion of CCCF can be supported by several reforms. One area beyond the control of self-regulation is IPR protection, where laws specific to crowdfunding platforms might be needed to encourage more platform use. Reducing the risk that creators take in publicly disclosing their innovative ideas might help attract even more creators to these platforms. In addition, expansion of the crowdfunding approach to new domains or finding new applications for crowdfunding models represents new frontiers. Public officials, for example, can make more use of crowdfunding for civic innovation, informing the allocation of public funds, and catalyzing additional financial support via matching programs. Most of all, it appears that additional harmonization in regulation can lower the cost of using platforms and expand the size of the markets or crowds. Expanding the market reach for a given platform, rather than fragmenting markets, leverages the scale economies and stands to serve more diverse, niche sorts of creative ventures.

This critical review of the literature has revealed support for several recommendations. First, harmonization—especially in the EU—can catalyze more use and greater benefits from crowdfunding. Supportive, rather than oppressive, regulations should be prioritized to support transparency, while striking a balance in tolerating some risk to maintain CCCF's core strength in attracting, and financing, innovation, and creative talent. Self-regulation, and a softer hand by government authorities, can support some essential features of the CCCF marketplace: its accessibility and its ability to find and support creative talent. Governments should explore and even experiment with leveraging the features of crowdfunding to better pursue policy goals. This likely means more public-private partnerships, introducing matching programs and using information (about both innovative ideas and public preferences) in CCCF platforms to guide policy.

Finally, in this paper, we have highlighted and suggested particular areas for future research, including: spillovers from CCCF to more traditional forms of fundraising, besides complementary and substitutes' effects; potential biases in CCCF platform dynamics and implications of multiple overlapping platforms, alternative platform models or designs, and implications of alternative regulatory approaches on innovation, IP protection, and financing generated; and additional evidence about how CCCF catalyzes individual creativity of a given creator or project versus how it impacts the creativity or novelty of the aggregate distribution of projects.

References

Agrawal, A., Catalini, C., & Goldfarb, A. (2014). Some simple economics of crowdfunding. *Innovation Policy and the Economy, 14*, 63–97.

Agrawal, A., Catalini, C., & Goldfarb, A. (2015). Crowdfunding: Geography, social networks, and the timing of investment decisions. *Journal of Economics and Management Strategy, 24*(2), 253–274.

Akerlof, G. A. (1970). The market for "lemons": Quality uncertainty and the market mechanism. *Quarterly Journal of Economics, 84*(3), 488–500.

Alexiou, K., Wiggins, J., & Preece, S. B. (2020). Crowdfunding acts as a funding substitute and a legitimating signal for nonprofit performing arts organizations. *Nonprofit and Voluntary Sector Quarterly, 49*(4), 827–848.

Belleflamme, P., Omrani, N., & Peitz, M. (2015). The economics of crowdfunding platforms. *Information Economics and Policy, 33*, 11–28.

Brent, D. A., & Lorah, K. (2019). The economic geography of civic crowdfunding. *Cities, 90*, 122–130.

Breznitz, S., & Noonan, D. (2020). Crowdfunding in a not-so-flat world. *Journal of Economic Geography, 20*(4), 1069–1092.

Bruton, G., Khavul, S., Siegel, D., & Wright, M. (2015). New financial alternatives in seeding entrepreneurship: Microfinance, crowdfunding, and peer-to-peer innovations. *Entrepreneurship Theory and Practice, 39*(1), 9–26.

Butticè, V., & Noonan, D. (2016, October 17). *Do not bite more than you can chew: The side effects of crowdfunding on entrepreneurial performances*. Bergamo: Riunione scientifica nazionale AIIG.

Butticè, V., & Noonan, D. (2020). Active backers, product commercialisation and product quality after a crowdfunding campaign: A comparison between first-time and repeated entrepreneurs. *International Small Business Journal, 38*(2), 111–134.

Butticè, V., Colombo, M. G., & Wright, M. (2017). Serial crowdfunding, social capital, and project success. *Entrepreneurship Theory and Practice, 41*(2), 183–207.

Calcagno, P. T., & Sobel, R. S. (2014). Regulatory costs on entrepreneurship and establishment employment size. *Small Business Economics, 42*, 541–559.

Chaney, D. (2019). A principal–agent perspective on consumer co-production: Crowdfunding and the redefinition of consumer power. *Technological Forecasting and Social Change, 141*, 74–84.

Courtney, C., Dutta, S., & Li, Y. (2017). Resolving information asymmetry: Signaling, endorsement, and crowdfunding success. *Entrepreneurship Theory and Practice, 41*(2), 265–290.

CrowdfundingHub. (2016). *Current state of crowdfunding in Europe*. Amsterdam.

Davidson, R., & Poor, N. (2018). Location, location, location: How digital platforms reinforce the importance of spatial proximity. *Information, Communication and Society, 25*(1), 1–15.

Davies, W. E., & Giovannetti, E. (2018). Signalling experience & reciprocity to temper asymmetric information in crowdfunding evidence from 10,000 projects. *Technological Forecasting and Social Change, 133*, 118–131.

De Buysere, K., Gajda, O., Kleverlaan, R., Marom, D., & Klaes, M. (2012). *A framework for European crowdfunding*. https://www.fundraisingschool.it/wp-content/uploads/2013/02/European-Crowdfunding-FrameworkOct-2012.pdf

Erickson, K. (2018). Can creative firms thrive without copyright? Value generation and capture from private-collective innovation. *Business Horizons, 61*(5), 699–709.

European Commission. (2017). *Crowdfunding. Reshaping the crowd's engagement in culture*. Brussels: European Commission.

European Crowdfunding Network. (2017). *Review of crowdfunding regulation*. Brussels.

Fleming, L., & Sorenson, O. (2016). Financing by and for the masses: An introduction to the special issue on crowdfunding. *California Management Review, 58*(2), 5–19.

Flew, T. (2011). *The creative industries: Culture and policy*. London: Sage.

Galuszka, P., & Brzozowska, B. (2016). Crowdfunding and the democratization of the music market. *Media, Culture and Society, 39*(6), 833–849.

Galuszka, P., & Bystrov, V. (2014). Crowdfunding: A case study of a new model of financing music production. *Journal of Internet Commerce, 13*(3–4), 233–252.

Ganatra, J. H. (2015). When a Kickstarter stops: Exploring failures and regulatory frameworks for the rewards-based crowdfunding industry. *Rutgers University Law Review, 68*, 1425–1472.

Giudici, G., Guerini, M., & Rossi-Lamastra, C. (2018). Reward-based crowdfunding of entrepreneurial projects: The effect of local altruism and localized social capital on proponents' success. *Small Business Economics, 50*(2), 307–324.

Goolsbee, A. (2000). In a world without borders: The impact of taxes on Internet commerce. *The Quarterly Journal of Economics, 115*(2), 561–576.

Greenberg, J., & Mollick, E. (2017). Activist choice homophily and the crowdfunding of female founders. *Administrative Science Quarterly, 62*(2), 341–374.

Gruba, J., McAfee, L., & Ashby, B. (2015). Crowdfunding contributions and state sales and use taxes. *The Tax Adviser*. Retrieved August 29, 2018, from https://www.thetaxadviser.com/issues/2015/jun/tax-clinic-08.html

Gutiérrez, M., & Sáez, I. (2018). The promise of reward crowdfunding. *Corporate Governance: An International Review, 26*, 355–373.
Johnson, M. A., Stevenson, R. M., & Letwin, C. R. (2018). A woman's place is in the... startup! Crowdfunder judgments, implicit bias, and the stereotype content model. *Journal of Business Venturing, 33*(6), 813–831.
Josefy, M., Dean, T. J., Albert, L. S., & Fitza, M. A. (2016). The role of community in crowdfunding success: Evidence on cultural attributes in funding campaigns to "save the local theater". *Entrepreneurship Theory and Practice, 41*(2), 161–182.
Kaun, J. (2001). The phantom profits of the opera: Nonprofit ownership in the arts as a make-buy decision. *Journal of Law, Economics and Organization, 17*(2), 507–520.
Lazzaro, E. (2017). Cultural and creative entrepreneurs. In B. Mickov & J. E. Doyle (Eds.), *Culture, innovation and the economy* (pp. 33–37). London: Routledge.
Lazzaro, E., & Noonan, D. (2019). US vs. EU regulatory frameworks of crowdfunding for the arts and culture. In G. Schiuma, P. Demartini, & M.-Y. Yan (Eds.), *Knowledge ecosystems and growth—Proceedings of the 14th international forum on knowledge asset dynamics* (pp. 2059–2071).
Lazzaro, E., & Noonan, D. (2020). Cultural and creative crowdfunding: A comparative analysis of regulatory frameworks in Europe and the USA. *International Journal of Cultural Policy*, 1–17. https://doi.org/10.1080/10286632.2020.1776270
Leland, J. (2020, June 5). *COVID-19 impact on creative work: Implications for creators, their work, and supporting organizations*. https://www.youtube.com/watch?v=_5AQNZGWe2o
Li, E., & Martin, J. S. (2019). Capital formation and financial intermediation: The role of entrepreneur reputation formation. *Journal of Corporate Finance, 59*, 185–201.
Marchegiani, L. (2018). From Mecenatism to crowdfunding: Engagement and identification in cultural-creative projects. *Journal of Heritage Tourism, 13*(2), 143–151.
Martínez-Climent, C., Mastrangelo, L., & Ribeiro-Soriano, D. (2020). The knowledge spillover effect of crowdfunding. *Knowledge Management Research and Practice*. https://doi.org/10.1080/14778238.2020.1768168.
Mendes-Da-Silva, W., Rossoni, L., Conte, B. S., Gattaz, C. C., & Francisco, E. R. (2016). The impacts of fundraising periods and geographic distance on financing music production via crowdfunding in Brazil. *Journal of Cultural Economics, 40*(1), 75–99.
Metrejean, C., & McKay, B. (2015). Crowdfunding and Income Taxes. *Journal of Accountancy*. https://www.journalofaccountancy.com/issues/2015/oct/crowdfunding-and-income-taxes.html
Miao, L. (2015). *Patent issues of crowdfunding—Exploring patent strategies of crowdfunding projects and the need for patent reform*. Master's thesis. Munich Intellectual Property Law Center.
Mollick, E. R. (2014). The dynamics of crowdfunding: An exploratory study. *Journal of Business Venturing, 29*(1), 1–16.
Mollick, E. R. (2015). *Delivery rates on Kickstarter*. Available at SSRN: https://doi.org/10.2139/ssrn.2699251
Mollick, E. R., & Kuppuswamy, V. (2014). *After the campaign: Outcomes of crowdfunding*. University of North Carolina Kenan-Flagler Research Paper 2376997. https://doi.org/10.2139/ssrn.2376997
Mollick, E. R., & Nanda, R. (2015). Wisdom or madness? Comparing crowds with expert evaluation in funding the arts. *Management Science, 62*(6), 1533–1553.
Mollick, E. R., & Robb, A. (2016). Democratizing innovation and capital access: The role of crowdfunding. *California Management Review, 58*(2), 72–87.
Moores, C. (2015). Kickstart my lawsuit: Fraud and justice in rewards-based crowdfunding. *University of California-Davis Law Review, 49*, 383–424.
Nambisan, S. (2017). Digital entrepreneurship: Toward a digital technology perspective of entrepreneurship. *Entrepreneurship Theory and Practice, 41*, 1029–1055.

Patti, D., & Polyák, L. (2017). *Funding the cooperative city: Community finance and the economy of civic spaces*. Vienna: Cooperative City Books.
Quero, M. J., & Ventura, R. (2019). Value proposition as a framework for value cocreation in crowdfunding ecosystems. *Marketing Theory, 19*(1), 47–63.
Rhue, L., & Clark, J. (2018). *The consequences of authenticity: Quantifying racial signals and their effects on crowdfunding success*. Available at SSRN: https://doi.org/10.2139/ssrn.2837042
Roberts, G., & Nowotarski, M. (2013). The IP issues of crowdfunding. *Managing Intellectual Property, 229*, 36–38.
Roma, P., Petruzzelli, A., & Perrone, G. (2017). From the crowd to the market: The role of reward-based crowdfunding performance in attracting professional investors. *Research Policy, 46*(9), 1606–1628.
Rykkja, A., Munim, Z. H., & Bonet, L. (2020). Varieties of cultural crowdfunding: The relationship between cultural production types and platform. *Baltic Journal of Management, 15*(2), 262–280.
Safner, R. (2015). *Do Patronize Me: The comparative political economy of arts patronage, copyright, and crowdfunding*. Available at SSRN: https://doi.org/10.2139/ssrn.2499621
Smith, A. (2016, May 19). *Shared, collaborative and on demand: The new digital economy*. http://www.pewinternet.org/2016/05/19/the-new-digital-economy/
Sorenson, O., Assenova, V., Li, G. C., Boada, J., & Fleming, L. (2016). Expand innovation finance via crowdfunding. *Science, 354*(6319), 1526–1528.
Stephen, F., Urbano, D., & van Hemmen, S. (2009). The responsiveness of entrepreneurs to working time regulations. *Small Business Economics, 32*, 259–276.
Strausz, R. (2017). A theory of crowdfunding: A mechanism design approach with demand uncertainty and moral hazard. *American Economic Review, 107*(6), 1430–1476.
Thies, F., Huber, A., Bock, C., Benlian, A., & Kraus, S. (2019). Following the crowd—Does crowdfunding affect venture capitalists' selection of entrepreneurial ventures? *Journal of Small Business Management, 57*(4), 1378–1398.
Thürridl, C., & Kamleitner, B. (2016). What goes around comes around? Rewards as strategic assets in crowdfunding. *California Management Review, 58*(2), 88–110.
Xu, T. (2018). *Learning from the crowd: The feedback value of crowdfunding*. Available at SSRN: https://doi.org/10.2139/ssrn.2637699
Younkin, P., & Kuppuswamy, V. (2017). The colorblind crowd? Founder race and performance in crowdfunding. *Management Science, 64*(7), 2973–3468.
Yu, S., Johnson, S., Lai, C., Cricelli, A., & Fleming, L. (2017). Crowdfunding and regional entrepreneurial investment: An application of the CrowdBerkeley database. *Research Policy, 46*(10), 1723–1737.

Elisabetta Lazzaro, PhD, is a Professor of Creative and Cultural Industries Management at the Business School for the Creative Industries, University for the Creative Arts, UK, and Executive Board Member of the Association for Cultural Economics International (ACEI). She has a track record of about 70 publications in international peer-reviewed journals, books, and policy reports, with a focus on art markets, crossovers and socioeconomic impact of culture, entrepreneurship, innovative business models and sustainable financing of arts professionals and organizations, digitization and technology transfer in the arts, cultural heritage and creativity, and international-related policy aspects. She has formerly been a professor of economics and business of the arts, culture, and creative industries at the HKU University of the Arts Utrecht (Netherlands), Université Libre de Bruxelles (Belgium), and Southern Methodist University in Dallas (USA).

Douglas Noonan PhD, is a Professor of Public and Environmental Affairs at the O'Neill School at Indiana University-Purdue University Indianapolis. He is a founding faculty director of the Center for Cultural Affairs at Indiana University, and he is the codirector of the Arts, Entrepreneurship, and

Innovation Lab in partnership with the US National Endowment for the Arts. He serves as the co-editor-in-chief of the Journal of Cultural Economics. His current research and teaching include a variety of policy and economics issues related to cultural policy, the urban and built environment, and support for innovation and economic development. He is the elected chair of the Urban-Serving University section of the Network of Schools of Public Policy, Affairs, and Administration.

The manufacturer's authorised representative in the EU is Springer Nature Customer Service Centre GmbH, Europaplatz 3, 69115 Heidelberg, Germany. If you have any concerns regarding our products, please contact ProductSafety@springernature.com

Printed and bound by CPI Group (UK) Ltd, Croydon, CR0 4YY
22/09/2026
02221048-0006